ADVANCES IN MULTI-PHOTON PROCESSES AND SPECTROSCOPY

ADVANCES IN MULTI-PHOTON PROCESSES AND SPECTROSCOPY

Volume 10

Edited by

S H Lin
Institute of Atomic and Molecular Sciences, Taiwan & Arizona State University, USA

A A Villaeys
Université Louis Pasteur, France

Y Fujimura
Tohoku University, JAPAN

World Scientific
Singapore • New Jersey • London • Hong Kong

Published by

World Scientific Publishing Co. Pte. Ltd.
P O Box 128, Farrer Road, Singapore 912805
USA office: Suite 1B, 1060 Main Street, River Edge, NJ 07661
UK office: 57 Shelton Street, Covent Garden, London WC2H 9HE

British Library Cataloguing-in-Publication Data
A catalogue record for this book is available from the British Library

ADVANCES IN MULTI-PHOTON PROCESSES AND SPECTROSCOPY — Vol. 10

ISBN 981-02-2716-7

This book is printed on acid-free paper.

Printed in Singapore by Uto-Print

PREFACE

In view of the rapid growth in both experimental and theoretical studies of multiphoton process and multiphoton spectroscopy of atoms, ions and molecules in chemistry, physics, biology, materials science, etc., it is desirable to publish an Advanced Series that contains review papers readable not only by active researchers in these areas, but also by those who are not experts in the field but who intend to enter the field. The present series attempts to serve this purpose. Each review article is written in a self-contained manner by the experts in the area so that the readers can grasp the knowledge in the area without too much preparation.

The topics covered in this volume are "Energetic and Molecular Dynamics in Solution Studied by the Third Order Nonlinear Optical Spectroscopies; Transient Lens and Transient Grating", "Electronic Interactions & Interchromophore Energy Transfer" and "Analysis of Optical Spectra of Photosynthetic Reaction Centers Absorption, Hole-Burning and Resonance Raman Spectra". The editors wish to thank the authors for their important contributions. It is hoped that the collection of topics in this volume will be useful not only to active researchers but also to other scientists in biology, chemistry, materials science and physics.

S.H. Lin

A.A. Villaeys

Y. Fujimura

Contents

PART ONE

Energetic and Molecular Dynamics in Solution Studied by the Third Order Nonlinear Optical Spectroscopies; Transient Lens and Transient Grating

Energetic and molecular dynamics in solution studied by the third order nonlinear optical spectroscopies; Transient lens and transient grating

Masahide Terazima

Department of Chemistry, Graduate School of Science,
Kyoto University, Kyoto, 606, Japan

Abstract;

In this article, chemical applications of the third order nonlinear optical spectroscopies are described; Mainly, studies of intramolecular energy dynamics after photoexcitation to electronic excited states and of molecular dynamics in solution using the transient grating (TG) and transient lens (TrL) spectroscopies, which have been developed by our group recently are reviewed. The TrL method can detect a variety of nonlinear optical phenomena with a rather simple experimental setup. Several methods to discriminate various $\chi^{(3)}$ processes in the signal are demonstrated. Using the TG technique, the dynamics of the excited states of organic molecules are detected through the thermal effect after the nonradiative transitions, and the molecular dynamics of intermediate (chemically reactive) species appearing during chemical reactions in a liquid phase are studied. Unique photophysical and photochemical properties revealed by these nonlinear optical spectroscopies are discussed.

Contents

1. Introduction

1-1. Overview

In the field of chemistry, we are very much interested in dynamics of molecules with hoping that we can ultimately understand chemical reactions completely; the 'dynamics' include dynamical processes of molecules in space as well as energy transfer processes among various intramolecular states or intermolecular systems.(Fig.1)

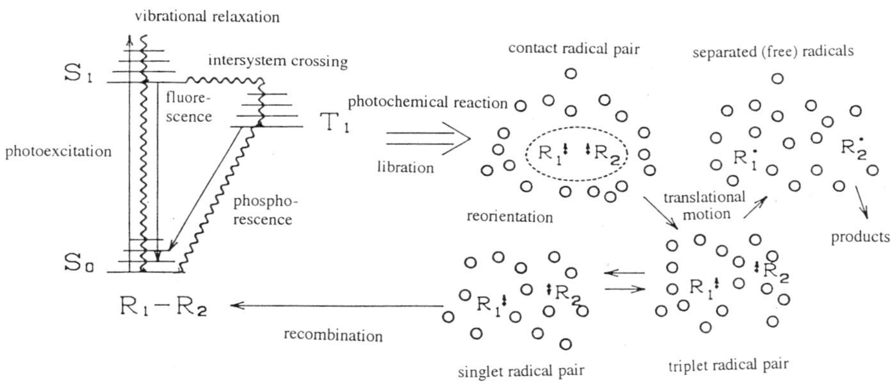

Fig.1 Dynamics in chemistry. After the photoexcitation, various photophysical processes take place, such as the vibrational relaxation, internal conversion, intersystem crossing, fluorescence, and phosphorescence. If a photochemical reaction occurs, radicals in the pair should be closely located each other (contact radical pair) just after the creation. After the reorientational relaxation of the radicals, they separate each other by the random movement of solvent molecules (solvent separated radical pair). The spin multiplicity will change during the movement and in the singlet spin state, they can recombine. The un-recombined radicals further separate to lead separated (free) radicals and eventually it is converted to some products. Therefore, in the photochemical reaction, many events on energetics and molecular dynamics are involved.

In order to understand such dynamics, it has been well known and indeed demonstrated that the light (electromagnetic radiation)-matter interaction provides rich and valuable information. Even before invention of lasers, many spectroscopic methods such as luminescence, absorption, light scattering techniques have been applied for that purpose. These techniques have been dramatically improved and even new spectroscopies have been developed quickly by applying the laser technology owing to novel characters of the lasers; highly coherent, intense, monochromatic, and ultrashort pulsed light. Nonlinear optics; the studies of how light propagates through and interacts with matter, belong to such newly developed spectroscopies and they enable us to study the processes that could not be reached by the linear spectroscopies.

Here I review investigations on energy and molecular dynamics in solution by using the transient grating (TG) and transient lens (TrL) methods. The TrL method, which was developed by our group recently is one of the cross phase modulation techniques to observe a spatial profile of a probe beam propagating in a medium with another intense pump beam. We have applied this method to the studies of various $\chi^{(3)}$ processes as well as intramolecular energy transfer processes in solution. The TG method, one of the well known four-wave mixing spectroscopies, has been well developed over the years and many reviews have been written on this technique from different points of views. I concentrate on two topics of the TG applications; ultrasensitive and ultrafast detection of heat releasing processes after photoexcitation of molecules in a liquid phase and detection of molecular motions in space. Both of these topics are very important for understanding photochemical reactions.

The organization of this article is as follows; later of this section outlines a basic introduction of the nonlinear spectroscopies and summarizes various methods for studying the phenomena, although they are limited within what they will be needed in a later part of this review. Section 2 starts with a brief introduction on the TrL method followed by applications of this technique and the results are discussed in each subsection. In section 3, the principle and applications of the TG method are described. In particular, I would like to concentrate on recent results on detection of heat energy and on investigations of molecular (stable and transient species) dynamics in solutions.

1-2. Nonlinear optics

All electromagnetic phenomena which depend on electric, **E**(r,t), and magnetic field, **B**(r,t), are governed by the Maxwell's equations. For nonmagnetic and nonconducting dielectric materials, the equations are

$$\mathrm{rot}\mathbf{E}=-\partial\mathbf{B}/\partial t \qquad \mathrm{rot}\mathbf{H}=-\partial\mathbf{D}/\partial t \tag{1-1}$$

$$\mathrm{div}\mathbf{D}=0 \qquad \mathrm{div}\mathbf{B}=0$$

Electric induction field, **D**, and magnetic induction field, **B**, can be written as

$$\mathbf{D}=\varepsilon_0\mathbf{E}+\mathbf{P}=\varepsilon\mathbf{E}, \qquad \mathbf{B}=\mu_0\mathbf{H} \tag{1-2}$$

from the material equations by using the dielectric constant in vacuum, ε_0, the permeability of vacuum, μ_0, and the electric dipole polarization P. Generally, P, which describes the response of the medium to the field is a nonlinear function of the electric field. When the electric field is not too large, the polarization can be expanded as a power series of E [1],

$$(1/\varepsilon_0)P_j(\omega)=\chi^{(1)}{}_{jk}(\omega)E_k(\omega)+\frac{1}{2\pi}\int_{-\infty}^{\infty} d\omega_1 d\omega_2 \chi^{(2)}{}_{jkl}(\omega;\omega_1,\omega_2)E_k(\omega_1)E_l(\omega_2)+$$

$$\frac{1}{(2\pi)^2}\int_{-\infty}^{\infty} d\omega_1 d\omega_2 d\omega_3 \chi^{(3)}{}_{jklm}(\omega;\omega_1,\omega_2,\omega_3)E_k(\omega_1)E_l(\omega_2)E_m(\omega_3)+\ldots$$

$$= \varepsilon_0\chi^{(1)}E+P_{NL} = \varepsilon_0\chi E \tag{1-3}$$

where χ is the susceptibility and P_{NL} is the nonlinear term of the polarization. Frequently, this quantity is expressed by the refractive index ($\tilde{n}$) which is given by

$$\tilde{n}^2 = 1+\chi$$

Since χ is generally a complex quantity,

$$\chi=\chi'+i\,\chi'' \tag{1-4}$$

so the refrative index is

$$\tilde{n} = n + i\kappa . \tag{1-5}$$

The imaginary part of χ leads to an attenuation of the electric field when χ'' is positive and is related to the absorption process. Both quantities χ' and χ'' are related each other through the Kramers-Kronig relation.

The $\chi^{(n)}$ terms correspond to tensor of order n+1. Many of these components are related to each other depending on the symmetry of the medium. For example, in an isotropic medium, $\chi^{(2)}$ is identically zero and it is $\chi^{(3)}$ that contains information of the light-matter interaction of the lowest order. $\chi^{(3)}$ is a fourth rank tensor with up to 81 nonvanishing elements but in isotropic solution, which I am going to deal with, only three $\chi^{(3)}$ components are independent.

$$\chi^{(3)}{}_{xxxx}=\chi^{(3)}{}_{yyyy}=\chi^{(3)}{}_{zzzz}$$
$$\chi^{(3)}{}_{yyzz}=\chi^{(3)}{}_{zzyy}=\chi^{(3)}{}_{zzxx}=\chi^{(3)}{}_{xxzz}=\chi^{(3)}{}_{xxyy} = \chi^{(3)}{}_{yyxx}$$
$$\chi^{(3)}{}_{yzzy}=\chi^{(3)}{}_{zyyz}=\chi^{(3)}{}_{zxxz}=\chi^{(3)}{}_{xzzx}=\chi^{(3)}{}_{xyyx} = \chi^{(3)}{}_{yxxy}$$
$$\chi^{(3)}{}_{yzyz}=\chi^{(3)}{}_{zyzy}=\chi^{(3)}{}_{zxzx}=\chi^{(3)}{}_{xzxz}=\chi^{(3)}{}_{xyxy} = \chi^{(3)}{}_{yxyx}$$
$$\chi^{(3)}{}_{xxxx}=\chi^{(3)}{}_{xxyy}+\chi^{(3)}{}_{xyxy}+\chi^{(3)}{}_{xyyx}$$

An equation which describes the propagation of the electric field can be obtained as

$$\nabla^2 E - (1/c^2)\partial^2 E/\partial t^2 - \nabla(\nabla E) = (1/\varepsilon_0 c^2)\partial^2 P/\partial t^2 \tag{1-6}$$

Induced polarization in the medium, P_i does not instantaneously respond to the light field and is given by

$$P^{(3)}{}_i(t) = \int_{-\infty}^{\infty}\int_{-\infty}^{\infty}\int_{-\infty}^{\infty} dt_3\, dt_2\, dt_1\, \chi^{(3)}{}_{ijkl}(t-t_3, t-t_2, t-t_1) E_j(t_3) E_k(t_2) E_l(t_1) \tag{1-7}$$

The time dependence of $\chi^{(3)}{}_{ijkl}$ is related to the energy and molecular dynamics of the sample, which we are interested in. Due to the presence of the nonlinear polarization, the refractive index is now a function of the electric field of light

$$n = n_0 + n_2 \langle E2 \rangle \tag{1-8}$$

where n_0 is the weak-field refractive index and n_2 is the second order (nonlinear) index of refraction and $2n_0n_2=3\chi^{(3)}(\omega)$.

In many cases, the light-matter interaction is induced by an intense light field (pump) and monitored by a weaker light (probe). Assuming negligible absorption of light and applying the slowly varying approximation, we can express the amplitudes of the pump, A_1, and the probe, A_2, fields as

$$\partial A_1/\partial z-(i/2k_1)\nabla_T^2 A_1=(ik_1n_2/n_0)(|A_1|^2+2|A_2|^2)A_1\cong ik_1n_2/n_0|A_1|^2A_1 \qquad (1\text{-}9a)$$

$$\partial A_2/\partial z-(i/2k_2)\nabla_T^2 A_2=(ik_2n_2/n_0)(|A_2|^2+2|A_1|^2)A_2\cong 2ik_2n_2/n_0|A_1|^2A_2 \qquad (1\text{-}9b)$$

where we assume that the pump pulse is much more intense than the probe pulse, ∇_T is the transverse component of the Laplacian, $k_i=n(\omega_i,|A_i|^2)\omega_i/c$ and n_0 is the linear refractive index.

1-3. Third order susceptibilities in chemistry

Since the four electric fields are involved in $P^{(3)}$ (Fig.2), the nonlinear process of $P^{(3)}$ is called the four wave mixing. For measuring the third order nonlinear susceptibility, $\chi^{(3)}$; i.e., the amplitude, the sign, wavelength dependence, and temporal profile of $\chi^{(3)}$, a variety of four-wave mixing methods have been developed so far.

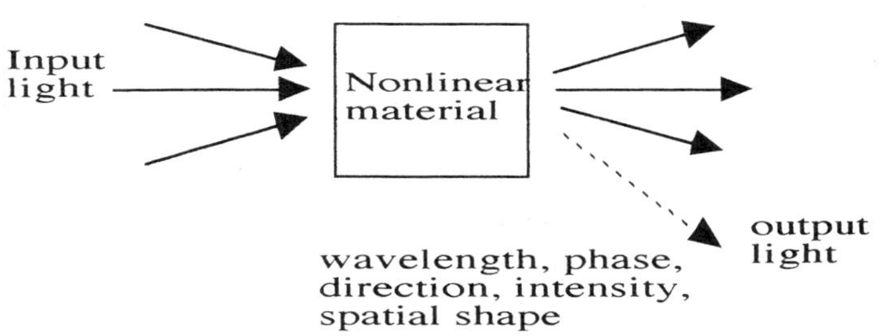

Fig.2 The third order nonlinear polarization creates the fourth light field from input three fields. The created light could be different from the original one in the wavelength, phase, direction of the propagation, intensity, and spatial beam shape.

Although the principles of some techniques are common and overlapped, they are named differently depending on target properties and experimental methods such as

(a) Third harmonic generation
(b) Electric field induced second harmonic generation
(c) Coherent anti-Stokes Raman scattering (CARS) and coherent Stokes Raman scattering (CSRS)
(d) Raman induced Kerr effect spectroscopy (RIKES)
(e) Optical Kerr effect (OKE) (Optically induced birefringence)
(f) d.c. Kerr effect
(g) Stimulated light (Raman, Brillouin, Rayleigh) scattering
(h) Transient grating (for example, degenerative four wave mixing)
(i) Photon echo
(j) Z scan (Self-(de)focusing)
(k) Transient lens (Induced-(de)focusing)

In an isotropic molecular liquid, at least 7 distinct mechanisms are expected to contribute to $\chi^{(3)}$.

(1) The electronic contribution, which arises through optical field induced distortion of the electronic charge distribution of the molecule. In particular, it is known that delocalized π electrons give rise to large optical nonlinearities.[2]

(2) Molecular reorientation, which occurs by the reorientation of anisotropic polarizable molecules through the electric field-induced dipole interaction. This and above mechanisms are generally called the nuclear and electronic responses of the optical Kerr effect (OKE), respectively. In a special case, this orientational effect will appear without directly interacting to the electric field. One notable example in liquid phase is the mesophase of liquid crystalline samples. When molecules are oriented, any light induced perturbations which disturb the orientation will change the macroscopic polarizability indirectly and contribute to $\chi^{(3)}$. This is a cooperative effect of many molecules.

(3) Interaction induced effect. When more than two atoms or molecules interact with each other, the polarizability of the system is no longer described by a simple sum of those of the isolated atoms or molecules, but the interaction induces new

polarizability. It could contribute to the anisotropic part as extensively studied by the depolarized light scattering [3] as well as the isotropic part.

(4) Thermal effect, which arises due to the heating effect of the medium after the radiationless transition of photoexcited molecules. The thermal effect contributes to $\chi^{(3)}$ through the density change and temperature change (section 2-5). If there is a temporal modulation in that effect, it induces a pressure wave, sound, and hence, the acoustic effect of the thermal origin can be included in this category. Since the released energy by the radiationless transition in fluids will eventually flow into the translational freedom, this effect is generally observed after any types of photoexcitations (resonant condition) and is closely related to the energy dynamics in the system.

(5) Creations of photoexcited molecules or intermediate species (or reaction products) after photochemical reactions. When a molecule is excited to the excited electronic (or vibrational) states, it should have a different polarizability from that of the ground state and the macroscopic polarizability of the system should be altered from the original state. By the same reasons, if the photoexcited molecule reacts to lead intermediate species (or stable final products), the different chemical species should create different polarizabilities and they contribute to $\chi^{(3)}$.

(6) Electrostriction. This effect arises because the optical field intensity gradient induces a force on the molecules to minimize the free energy of the system in the presence of the field. As the consequence, the molecules tend to move into a region of the high field intensity. The spatial movement changes the density of the molecules and induces a pressure wave. Since the polarizability is a function of the density, it contributes to $\chi^{(3)}$.

(7) Volume change. When a molecule is excited to the excited state and the molecular geometry is changed by a photochemical reaction, for example, a photo-isomerization, the molecular volume should be different from that of the original state. Even when only the electronic character such as the dipole moment is changed, the reorientation of the solvated molecules induces the partial molar volume change in the medium. Then $\chi^{(3)}$ will be induced as the consequence of the density change. More pronounced effect is expected when several molecules are combined together or a molecule is decomposed to several parts by

a photochemical reaction. Similar to the thermal effect, this volume change induces the acoustic wave.

Generally some of these mechanisms contribute to the nonlinear optical signal simultaneously, and the fact that more than one photophysical processes contribute to the signals makes these nonlinear spectroscopies more complicated than linear spectroscopies. It is, however, the very fact that makes these techniques very unique and valuable in studying the photophysics or molecular dynamics. Fortunately, since the various mechanisms presented above contribute to $\chi^{(3)}$ differently in magnitude and in time scale and also they depend on the various characters (wavelength, polarization and so on) of the light, each contribution can be separated out experimentally from the observed signal. For

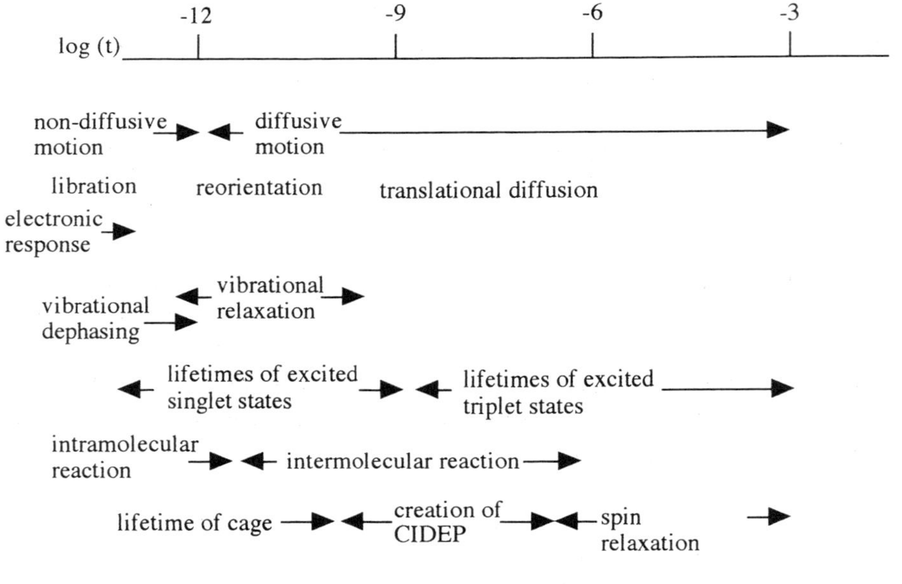

Fig. 3 Time scales of various photophysical and photochemical processes which contribute to $\chi^{(3)}$

example, $\chi^{(3)}$ processes relax in various time scales and have distinct temporal signatures. Figure 3 shows some physical processes we are interested in here. By specifying the time scale of the observation, the pump and probe wavelengths (real excitation or non-resonant), direction of the propagation, and the polarizations, we can selectively study various phenomena.

2 Transient lens spectroscopy

2-1. Transient lens effect

When the spatial shape of the pump beam has a Gaussian form, the light field at the central part interacts stronger with matter than at the edge of the beam and, as a result, a probe beam which is spatially overlapped with the pump beam is perturbed spatially inhomogeneously. The spatially inhomogeneous interaction induces the distortion of the phase of the probe beam and then of the beam spatial profile, too. The intensity dependent perturbation can be treated as the

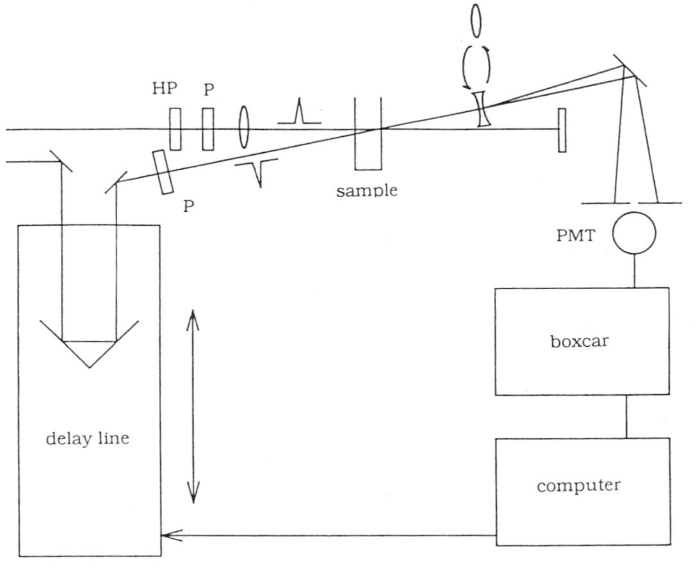

Fig.4 One of experimental setups for detection of the TrL effect in picosecond time scale. HP; half-wave plate, P; Polarizer, PMT; photomultiplier.

intensity dependent refractive index (eq.(1-8)) or absorption change of the medium. This effect can be detected by the same light (the self phase modulation) or by another weaker probe beam (the cross phase modulation).

In particular, the cross phase modulation is important for the studies of molecular and/or energy dynamics. One of the sensitive methods to detect such cross phase modulation is so-called 'thermal lens (TL)' method,[4] which detects the probe beam expansion (or focusing) induced by the intensity dependent refractive index in a far field with a pinhole (Fig.4). Ever since the discovery of the TL effect (next section), this method has been applied to the detection of heat released processes by nonradiative transitions but never applied to investigations of the dynamics of energies and those of molecules faster than submicrosecond time scale until recently, because of the reasons discussed in the next section. However as expected from a fact that self-focusing and defocusing phenomena have been observed for a long time with a short pulsed laser system [1], it is reasonable to think that the origin of the lens is not limited to this mechanism. Recently we have applied this technique to detect the cross phase modulation for studies of more general light-matter interactions; i.e., the lens is not limited to the thermal origin. At the same time, the time range in which this method is applicable has been expanded dramatically. Indeed we have successfully detected the lens signal from femtosecond to millisecond range. Now, we should call this lens method transient lens (TrL). In the following, I briefly summarize the principle of the detection of the TrL signal.

The transient lens (TrL) effect can be expressed by the coupled equations, (1-9). If n_2 is negative, the first and the second equations of (1-9), respectively, describe the self-focusing for the pump beam and the induced-focusing for the probe beam. If n_2 is positive, the self-defocusing and induced-defocusing should be observed. Although the signal intensity and the temporal dependence of the TrL signal can be predicted by the coupled equations, I try to show the TrL effect by using a simple optical ray theory for the help of intuitive understanding.[5] The complex refractive index will vary by the light-matter interaction through the third order of optical nonlinearity.

$$\delta n(r,t) = \frac{2\pi}{n_0} \int_{-\infty}^{t} dt' \chi^{(3)}(t') E^2_{pump}(r)$$

where E_{pump} is the light field of the pump beam, n_0 is the unperturbed refractive index and $\delta n(t)$ represents the variation of the refractive index due to several $\chi^{(3)}$ contributions which is presented in the previous section. We assume that the shape of the pump beam is a Gaussian type and the pump beam induces a Gaussian type refractive index distribution which should be time dependent.

$$n(r,t)=n_0 - \delta n(t)\exp(-r2/w2) \qquad (2\text{-}1)$$

where w is the beam radius of the excitation beam, r is the distance from the center of the excitation beam. If the refractive index that varies due to multiphoton processes of the excitation light or some higher order kinetics are involved in the temporal dependence, the spatial profile, n(r,t) will be different from eq.(2-1). In that case, a more complex treatment is necessary, but it will be straightforward.

For monitoring the TrL signal, we use a collimated probe beam (Fig.4). After the probe beam passes through the TrL region and travels a distance d' in the free space, the beam is expanded by a concave (focal length $f_{cc}<0$) or convex ($f_{cv}>0$) lens (cf. Fig.10). After traveling another distance, d, for allowing the expansion, the central part of the probe beam is detected through a pinhole. The TrL signal is a change in the probe beam size at a far distance and it is detected as the change of the light density at the center of the beam using a pinhole. When the spatial size of the probe beam is wider than that of the pump beam, an interference pattern between the perturbed (diffracted) and unperturbed probe beam appears as an interference ring at the edge of the beam.[6] Here we consider a weak pump beam condition and the detection only at the central part of the probe beam so that we can neglect the interference effect due to the aberrational nature of the TrL.

According to the paraaxial ray optics, the propagation of the probe beam can be expressed in terms of the ABCD matrix.[7] In the case of $\delta n(t)<0$, the equation is given by

$$\begin{pmatrix} r_{out} \\ r'_{out} \end{pmatrix} = \begin{pmatrix} 1 & d \\ 0 & 1 \end{pmatrix} \begin{pmatrix} 1 & 0 \\ -1/f & 1 \end{pmatrix} \begin{pmatrix} 1 & d' \\ 0 & 1 \end{pmatrix} \times$$

$$\begin{pmatrix} \cosh(a(t)^{1/2}L) & a(t)^{-1/2}\sinh(a(t)^{1/2}L) \\ -a(t)^{1/2}\sinh(a(t)^{1/2}L) & \cosh(a(t)^{1/2}L) \end{pmatrix} \begin{pmatrix} r_{in} \\ r'_{in} \end{pmatrix} \qquad (2\text{-}2)$$

where r and r' are the distance from the central axis and the slope of the ray, respectively, subscripts 'in' and 'out' mean the input and output of the rays respectively and $a(t)=2\delta n(t)/w^2$. The thickness of the sample is L, which is assumed to be thin. The input beam is collimated before entering the sample, $r'_{in}=0$. When n(t)>0, the fourth matrix of the right hand side of eq.(2-2) becomes

$$\begin{pmatrix} \cos(a(t)^{1/2}L) & a(t)^{-1/2}\sin(a(t)^{1/2}L) \\ -a(t)^{1/2}\sin(a(t)^{1/2}L) & \cos(a(t)^{1/2}L) \end{pmatrix} \tag{2-3}$$

A photodetector measures the probe light intensity through a pinhole, which is inversely proportional to $\pi(r_{out})^2$.

$$I(n(t)) \propto \{\pi(r_{out})^2\}^{-1}$$

Therefore, if the beam expansion or the beam focusing occurs, the resultant light density change can be detected as a variation of the detector output. The TrL signal, S(t) is defined by

$$S(t)=\{I(n(t))-I(0)\}/I(0) \tag{2-4}$$

where I(0) is the light intensity when there is no TrL effect (Fig.5). Assuming that a(t) L is far less than unity and $L+d' \ll f \ll d$, S(t) is approximated by

$$S(t)=4f\delta n(t)L/w^2 \tag{2-5}$$

Therefore the TrL signal is proportional to f as well as $\delta n(t)L/w^2$. The sign of the TrL signal is determined by the signs of f and $\delta n(t)$. The temporal profile is determined by that of $\delta n(t)$. This TrL method provides a sensitive and quantitative measurement of $\chi^{(3)}$ with a simple setup.

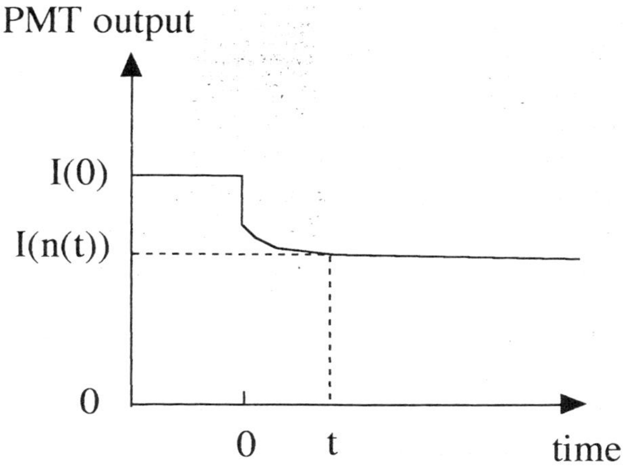

Fig. 5 Illustration of the TrL signal. The system is irradiated by the pump light at t=0 and the intensity 0 is the dark level of the probe light.

2-2. Historical background of the 'thermal lens' method

The TL effect was first observed at Bell Laboratories in 1964 as the expansion of a laser beam spot size induced by optical absorption of a sample which is placed inside the laser cavity.[8] This phenomenon was correctly interpreted in terms of the creation of the 'thermal lens' produced by the heating effect of the Gaussian laser beam.[9] Soon after that, numerous researches have been done on theoretical treatments [10] and studies of thermal properties,[11] reaction kinetics,[12] energy transfer processes,[13] excited state dynamics,[14] measurements of quantum yields of luminescence[15] and, in many cases, detections of weak absorptions of one photon [16,17] and multiphoton processes.[18] In these studies, investigations on the dynamics have been limited in rather long time regions. The reason of the limited studies for the dynamics is that the lens signal is believed to come only from the density variation. Even if the medium is heated instantaneously by the nonradiative transition, it takes a characteristic time before the medium expands. Roughly the time for the expansion of the medium is given by τ_{ac}, (τ_{ac}=w/v; w; beam radius, and v; sound velocity in the medium) which is around 100 ns under a normal experimental setup. Due to this prejudice, it is widely believed that the time resolution of this method cannot be faster than the acoustic transit time from the irradiated region. This is actually true as long as the signal origin is the density change due to the heat (this component is now called the density lens as will be discussed later). The dynamic range of the thermal lens method has been limited from the submicrosecond to millisecond (limit of the longer time scale comes from the thermal diffusion time across the irradiated region).

Nevertheless, this method has provided unique information on the dynamics of the photoexcited state so far. [12-14] For example, kinetics of a photochemical reactions were monitored calorimetrically early in the development of the thermal lens technique.[12-a] Deactivation of the singlet oxygen formed by the quenching process of the excited triplet states of organic molecules as well as the quantum yield of the creation were detected and studied quantitatively by the TL method.[12-b,d] Here let us consider simplified excited state dynamics among the excited singlet (S_1) state, the excited triplet (T_1) state, and the ground (S_0) state schematically shown in scheme 1 and examine the temporal profiles of the TL signal. (cf. Fig.1).

Scheme 1

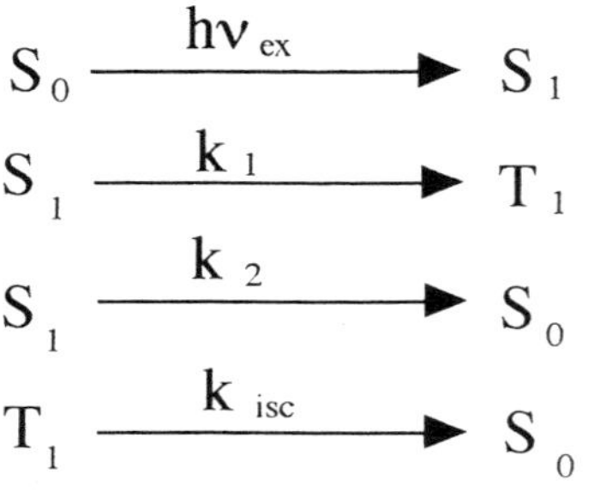

If any luminescence from the S_1 or T_1 states, and any transient absorption cannot be detected, the dynamics of the excited states are totally unclear by conventional linear spectroscopies. However, since the thermal energy always participates in the relaxation processes more or less, the TL method is generally applicable to study the excited state dynamics. As long as the rate constant of each step is faster than that of the thermal diffusion time (around tens of milli-second order), the thermal energy due to the radiationless transition is accumulated in the irradiated region. If the lens signal is completely composed of the thermal effect, the signal rises with two time constants, k_1+k_2 and k_{isc}. Generally k_1+k_2 is much faster than τ_{ac} and the rise due to the decay of the S_1 state is determined by the instrumental response. From the energy conservation law, the relative intensity of this simplified system is given by

$$U_{slow}/U_{tot} = \phi_{isc} E_T/(h\nu_{ex} - h\nu_f\phi_f) \tag{2-6}$$

where U_{slow} and U_{tot} are, respectively, the slower rising component, (which is due to the intersystem crossing from the T_1 to S_0 states) and the total signal intensity, $h\nu_{ex}$ photon energy of the excitation light, ϕ_{isc} the quantum yield of the triplet formation, ϕ_f the quantum yield of luminescence, $h\nu_f$ average photon energy of the luminescence, E_T energy of the triplet state. Therefore by measuring the temporal profile of the 'TL' signal, various photophysical properties in eq.(2-6) can be determined as well as the lifetime of the triplet state (k_{isc}).

A typical example of the TL signal is shown in Fig.6 after the photo-excitation of phthalazine in benzene under degassed condition with a nanosecond pulsed laser light (10 ns width) and a cw probe beam (He-Ne laser).[19-a]

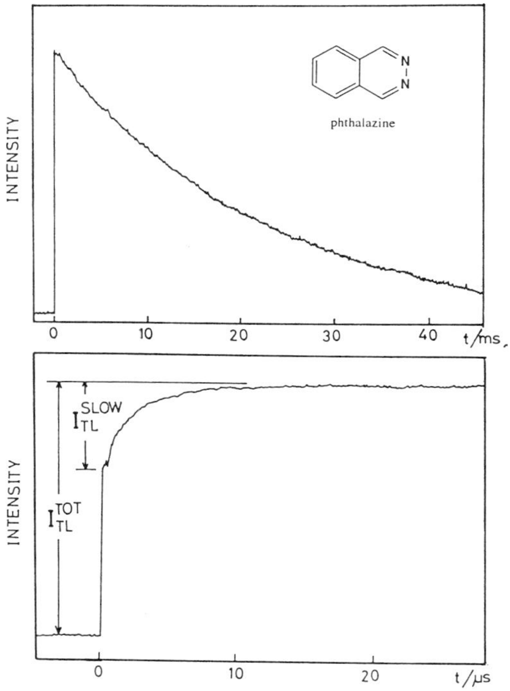

Fig.6 Typical 'TL' signal after the photoexcitation of an organic molecule (phthalazine in benzene) in a longer time scale (a) and the rise part of the signal (b).

Phthalazine (2,3-diazanaphthalene) has attracted many spectroscopists because of the unique photophysical characters. It shows rather strong phosphorescence with a few 100 ms lifetime in polar solvents but it is not luminescent at all in nonpolar solvents.[20] Even though many scientists have been working on the mysterious luminescent nature experimentally or theoretically, the mechanism and the dynamics of the excited states had remained unclear for a long time. We studied the photophysical properties of this molecule by the conventional TL method.[19-a]

In a millisecond time scale, the signal decays by the time constant of the thermal diffusion. (From the decay rate, the thermodynamical properties of the medium can be studied.) In the initial part of the signal, there are two types of the rises; the fast and the slower rise components. The faster rise time is determined by τ_{ac} and an instrumental response rise. Since the slower rise time is sensitive to the oxygen concentration in the solution, and phthalazine is known to be photochemically stable, the slow rise component is assigned to the heating effect by the radiationless transition from the T_1 state. The observed intensity ratio of U_{slow}/U_{tot} provides the absolute value of the quantum yield of the triplet formation, which is usually very difficult to be measured by other means and it was found to be ca. 0.5 regardless of the polarity of the solvents. The lifetime of the T_1 state in nonpolar solvents under the degassed condition is ca. 3 μs, which is more than 5 orders of magnitude shorter than those in polar solvents. Based on these observations, we revealed that the non-phosphorescent character in nonpolar solvents is not due to the inefficient S_1 to T_1 intersystem crossing but due to the exceptionally very fast nonradiative transition from the T_1 state. Time-resolved EPR studies [19-b] show that the triplet character of phthalazine is very sensitive to the matrix. It continuously changes from the $^3\pi\pi^*$ state to $^3n\pi^*$ state by decreasing the polarity of the solvents. The very efficient nonradiative transition was found to occur from the $^3n\pi^*$ state and to be common in the members of the ortho-diazaaromatics. The mechanism of the fast radiationless transition from the excited states has been elucidated based on these information and theoretical studies on an analogous molecule; pyridazine (1,2-diazabenzene) as shown in Fig.7.[21]

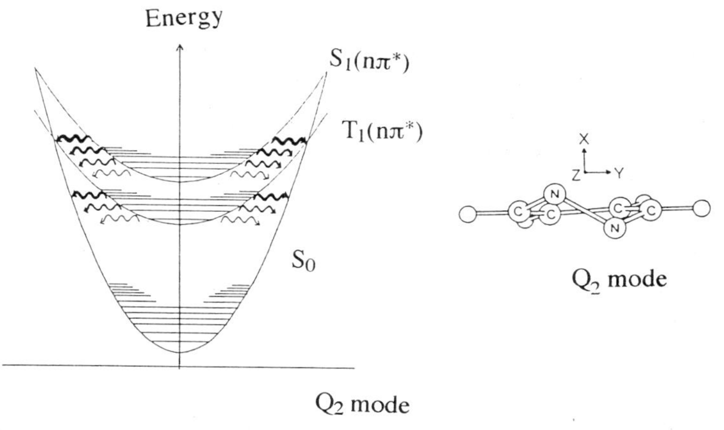

Fig.7 Potential surfaces of the low lying excited states of pyridazine along the Q_2 mode. The different curvatures between the excited states and the ground states are the origins of the very efficient nonradiative transitions. The efficiencies of the transitions from the higher vibrational states are higher than those from the lower states. The thicknesses of the arrows represent the rate of the transitions.

2-3. Separation of the real and imaginary parts of the refractive index

After creations of the excited states or photochemical intermediate species, the complex refractive index of the system should be altered. The nonlinear optical response probed by any four-wave mixing experiments has both real and imaginary components (eq.(1-5)) and the separation of these parts are important for studying the molecular dynamics. Reflecting the importance, several separation methods have been developed in each spectroscopy.

In the TrL spectroscopy, when the probe wavelength is close to any absorption bands, the imaginary part of the refractive index (the transient absorption and/or bleach of the ground state absorption) will change the probe light intensity under the TrL configuration. This imaginary part of the refractive index should be separated from the probe beam distortion which is due to the real part of the refractive index.

The separation can be performed by monitoring the beam profile directly, since absorptive component (hereafter we use 'TA' for describing this component including the transient bleaching effect) changes the probe beam intensity but the real part of the refractive index just changes the light density keeping the total intensity constant. However, monitoring the spatial profile requires a beam profiler, which is still not convenient to be used for recording the profiles at various times after the pumping. For the widely used setup, a pinhole and a photodetector system (Fig.4), sometimes, the TA contribution has been checked by monitoring the probe light intensity without the pinhole (or with a widely opened iris). However, the probe beam does not usually overlap with the excitation beam perfectly, and as a result, the monitored TA intensity by this method would not necessarily come from the same region where the TL signal comes. Hence, it is difficult to evaluate the TA component quantitatively by this conventional method; in other words, The very small area in the probe beam is monitored in the TrL experiment so that the quantitative measurement of the pure TA component only from that area is almost impossible. Furthermore, since the time development of the refractive index due to the photoexcited state should be the same as that of the absorption, the time profile of the signal does not help for the separation. To achieve the separation, we developed two methods; (a) coincident focal point method and (b) the concave-convex lens (CCL) method.

(a) Coincident focal point method [22]

When both of the excitation and the probe beams are focused before entering to the sample, the TrL effect due to the refractive index lens is sensitive to the location of the probe beam waist with respect to the excitation waist.[23] When a concave type TrL is induced by the pump beam inside the sample which is usually placed at the focal point of the pump beam and when the focal point of the probe beam is located between the focusing lens and the sample (opposite side to the detection system), the probe beam is expanded by the TrL effect (Fig.8). Therefore the light intensity through the pinhole at the far distance should decrease. On the other hand, if the focal point of the probe is located at the detection system side, the probe beam is focused by the concave type TrL effect, and hence, the light density should increase. In a special case that the focal point of the probe beam coincides with that of the excitation beam, the lens signal disappears and only the TA signal is detected. The light intensity change due to TA can be monitored at this point, and then the signal intensity can be subtracted from the TrL signal intensity measured at the different focal point. A result of the first demonstration of this method is shown in Fig.8.[22]

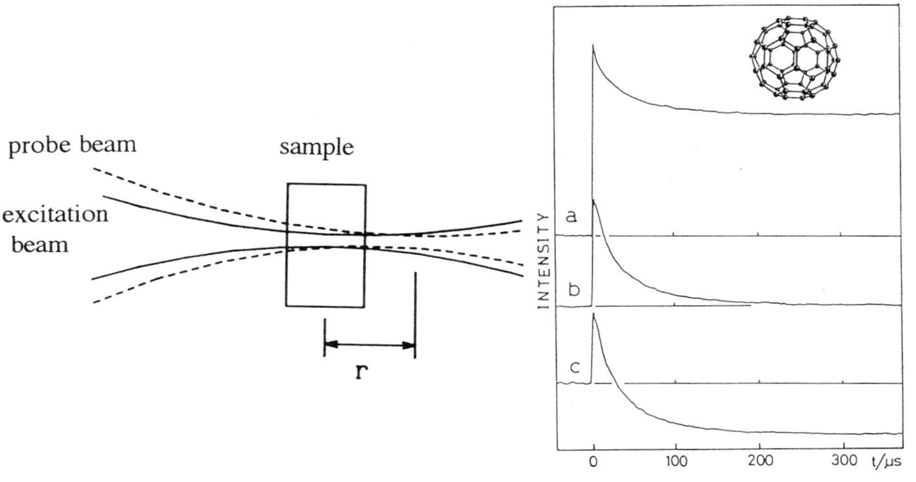

Fig.8 Distance r between the focal points of the pump and probe beams and the TrL signal after photoexcitation of C_{60} in benzene (a) r=5 mm, (b) r=0 mm, (c) r=-3 mm. The signal at r=0 is due to the pure TA effect.

A method based on a similar idea is recently used in the z scan experiment. [24] This z scan method has been developed to measure a magnitude of $\chi^{(3)}$ sensitively by scanning the sample along the z direction (propagation direction of the light) of a focused pump beam (Fig.9).

If the pump beam creates a convex lens in the sample ($n_2>0$), the pump beam is expanded before the focal point (the sample is placed closer to the light source) and is focused after the focal point. This beam expansion or focusing is detected at a far distance from the sample with a pinhole and a photodetector (Fig.9). The difference between the transmission maximum and minimum allows to evaluate the magnitude of $Re(\chi^{(3)})$. Principally, this z scan is based on the self (de)focusing effect and it will give $\chi^{(3)}_{xxxx}(-\omega;\omega,-\omega,\omega)$ component. When the sample is at the

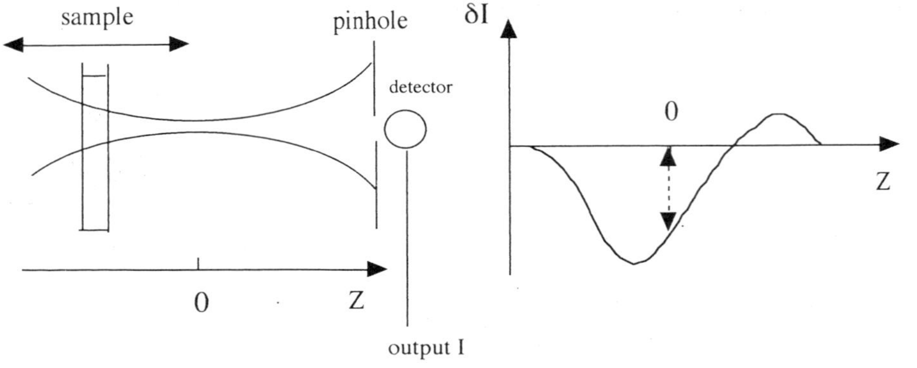

Fig.9 (left) A Z scan experimental configuration and (right) a typical signal. The intensity at z=0 (shown by an arrow) is due to the TA effect.

focal point of the beam, (which is equivalent to the coincidence of the focal points of the pump and probe beams in the TrL case), the self-focusing or self-defocusing should not be observed. The variation of light intensity at this position should be due to the absorption effect $\mathrm{Im}(\chi^{(3)})$. Therefore, the imaginary and real parts of the refractive index can be separated in this z scan experiment. Our TrL method is one type of the z scan method, but it has more abilities than the z scan method; such as the capability of measuring the temporal profile, polarization dependence, wavelength dependence (two colors experiment) of the $\chi^{(3)}$ processes.

(b) The CCL method

When the probe beam is collimated before the TrL region, the TrL signal is described by eq.(2-5) in the previous section. It should be noted that the sign of the signal is determined by not only the sign of $\delta n(t)$ but also that of f; (focal length of the external lens). If the probe beam is expanded by the concave lens, the beam intensity through the pinhole should be decreased by negative $\delta n(t)$. On the other hand, the probe beam light density is increased by the negative $\delta n(t)$, when the probe beam is expanded by the convex lens. The TA contribution always decreases (increases for transient bleaching) the probe light intensity. Using eqs.(2-4) and (2-5), the TrL signal intensity is given by

$$S(t)=\{I(n(t))-I(0)\}/I(0)=\{I_0(n(t))(1-U_{TA})-I(0)\}/I(0)= 4f\delta n(t)/w2 - I_0(n(t))U_{TA}/I(0) \quad (2\text{-}7)$$

where U_{TA} is the signal intensity due to TA, $I_0(n(t))$ is the TrL signal without the TA effect. If the TrL signal is sufficiently small, the observed signal is expressed by

$$U(t)=4f\delta n(t)/w2-U_{TA}(t) \quad (2\text{-}8)$$

with $f=f_{cv}>0$ for the convex lens case and $f=f_{cc}<0$ for the concave lens. Then U_{TA} and the lens signal ($U_{TrL}=4f\delta n(t)/w2$) can be obtained from the TrL signals measured by the concave lens (U_{cc}) and by the convex lens (U_{cv})

$$U_{TA}(t)=\{U_{cv}(t)+U_{cc}(t)\}/2 \quad (2\text{-}9\ a)$$

$$U_{TrL}(t)=\{U_{cv}(t)-U_{cc}(t)\}/2 \quad (2\text{-}9\ b)$$

Then, the contributions of the real and imaginary parts of the refractive index can be separated.[5]

The applicability of this method has been demonstrated by measuring the time profiles of the TrL signals from several samples whose photophysical properties have been well characterized, for example, the decay of the singlet oxygen in solution or the quantum yields of triplet formation of some organic molecules.[5] After that demonstration, this method is applied to obtain a pure lens signal of C_{60} in benzene probed at 633 nm. Since C_{60} has a rather strong T-T absorption band around 700 nm, the TrL signal probed by the HeNe laser cannot be free from the TA effect (cf.Fig.8). The probe light intensity measured with the external concave lens is decreased by the TrL effect as shown in Fig.10(c), whereas it is increased with the convex lens (Fig.10(d)). If the signal is composed of only the lens effect, the temporal profile of the signal should be identical except the sign. However the profiles (c) and (d) are very different each other and the discrepancy should be due to the TA contribution. Using eqs.(2-9), the rise part

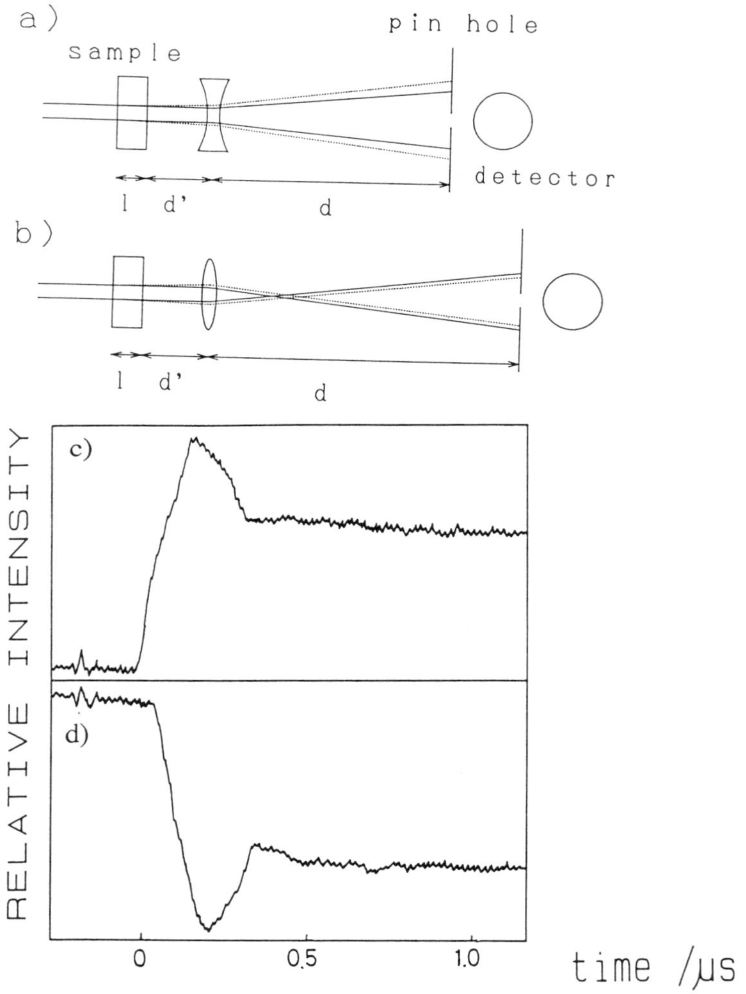

Fig.10 Schematic showing of the CCL method. When a concave lens is used for the expansion of the probe beam (a) ($U_{cc}(t)$), the dispersive TrL effect further expands the beam size at a far distance. On the other hand, (b) when a convex lens is used ($U_{cv}(t)$), the dispersive TrL lens acts to focus the probe beam at a far distance. (c and d) An example of the CCL method for C_{60} in benzene probed at 633 nm.

of the TrL signal from the real part of the refractive index is obtained and analyzed as seen in section 2-5-1. This method provides a useful and convenient way for subtracting TA from the observed TrL signal

2-4. Population lens

After subtracting the TA effect from the observed TrL signal, only the TrL signal due to the purely refractive index lens can be obtained. However, there are still a lot of sources which could contribute to the creation of the lens signal. When the photoexcited states are present during the measurement, one of commonly observed lens signals besides the thermal effect is so called the population lens (PL). After the pump light excites a molecule to the excited states, the refractive index could vary due to the different polarizabilities between the ground and excited states. Contribution of this effect in the lens signal was first reported for the excited triplet state of C_{60} probed by the He-Ne laser in the microsecond time scale.[22] This state dependent lens signal is named PL. (We limit the name of PL within the refractive index change due to the electronic origin, which is concomitant with the absorption change. When the excited states are created or photochemical reaction takes place, the molecular volume will change and then it induces the 'population dependent' refractive index. However in this case, we call it the volume lens for more clear sense.) Since the dispersion range of the refractive index is much wider than the absorption spectrum range, this component is useful for monitoring the dynamics of the excited state with less limitation for the probe wavelength. As I will show later, the excited state can be detected at the tail of the absorption band or even at a far wavelength from the absorption band. However, at the same time, it will disturb the TrL measurement if one wants to examine other properties in the lens signal, for example, the thermal effect. In such a case, it is very important to eliminate or subtract the PL contribution from the observed lens signal. Generally, however, it is not an easy task. The following five methods will work for that purpose.

(a) probe wavelength dependence

The probe wavelength dependence of the refractive index change due to the thermal effect is usually negligible, whereas that due to PL is not. The relationship between the absorption spectrum and the dispersion of the concomitant refractive index change is given by the Kramers-Kronig relation. Using a simple harmonic oscillator model for the line shape of n and k, the complex index of refraction can be written as

$$n^2-1 = (Ne^2/3\varepsilon_0 m)\sum_i \frac{f_i}{\omega_i^2-\omega^2-i\gamma_i\omega} \tag{2-10}$$

where N is the density, e; charge of the electron, m; mass of the electron, f_i; oscillator strength of the i-th band, and ω_i and γ_i are the frequency at the absorption maximum and the band width of the i-th band, respectively. Separating eq.(2-10) into real and imaginary components yields

$$n = \frac{N_1}{2n_0}\sum_i \frac{4\pi e^2 f_i(\omega_i^2-\omega^2)}{m[(\omega_i^2-\omega^2)^2+\gamma_i^2\omega^2\}} \tag{2-11}$$

where N_1 is the density of the excited state. Therefore if we know the absorption spectrum from the ground state as well as those from the excited states, the relative contribution of PL should be able to be calculated.

Let us consider the simplified system described in section 2-2 (Scheme 1) to examine the effect of PL. Since the lifetimes of the excited singlet states are shorter than the time resolution of the normal TL experiment, it is not necessary to consider the PL contribution of the excited singlet states. If PL due to the T_1 state (intensity; U_{PL}) contributes to the signal, the effect will change the signal intensity of the slow component, U_{slow}. The PL contribution should not exist after all of the excited states relaxed. Hence, the total signal intensity, U_{tot} should not be altered by PL. The intensity ratio of the slow rising component to the total intensity (eq.(2-6)) is written as

$$U_{slow}/U_{tot} = (U_{slow}^{TL}+U_{PL})/U_{tot} \tag{2-12}$$

Since U_{PL} could be positive or negative depending on the dispersion, the ratio of U_{slow}/U_{tot} could be larger or smaller than what is expected only from the thermal contribution (U_{slow}^{TL}/U_{tot}) (Fig.11). In this equation, U_{PL} depends on the probe wavelength, while U_{slow}^{TL} dose not. Therefore the temporal profile should depend on the wavelength (Fig.11).

For example, we have measured U_{slow}/U_{tot} of anthracene in benzene excited with an excimer laser (308 nm, 10 ns pulse width) probed at various wavelengths using multi-lines of an Ar ion laser and the value is plotted in Fig.12 with a reported T-T absorption spectrum.[5] The value decreases with increasing the probe wavelength. The probe wavelength dependence calculated from eq.(2-11) agrees fairly well with the experimental value. This method was applied to several systems and it was found that in many cases this calculation method works fine.[5]

It should be noted, however, that this calculation is not always correct. Sometimes, very strong absorption bands in far UV region, which we cannot measure by ordinary ways, contribute to PL and give a 'background' contribution to PL.

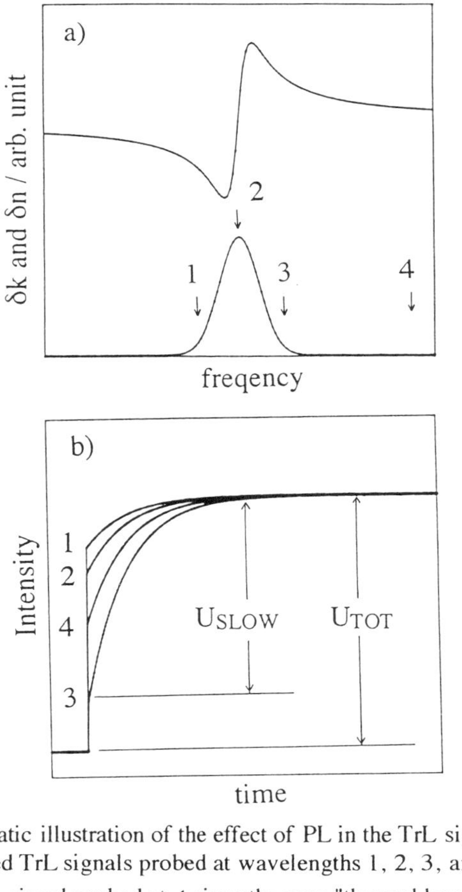

Fig.11 Schematic illustration of the effect of PL in the TrL signal from the thermal effect. Expected TrL signals probed at wavelengths 1, 2, 3, and 4 in (a) are shown in (b). The TrL signal probed at 4 gives the pure "thermal lens" signal without PL.

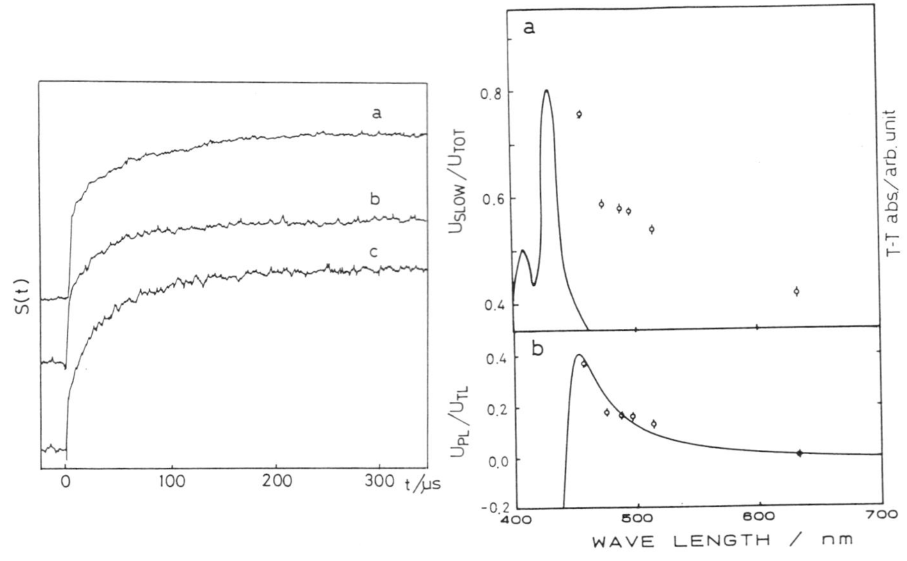

Fig.12 (left) Probe light wavelength dependence of the temporal profile of the TrL signal after excitation of anthracene in benzene under a nitrogen bubbled condition probed at (a) 633 nm, (b) 497 nm, and (c) 458 nm. (right) (a) Observed probe light wavelength dependence of U_{PL}/U_{TL} (circles) and the T-T absorption spectrum, (b) the calculated U_{PL}/U_{TL}(line)

(b) Variation of thermodynamical parameters

If not all of the absorption spectra of the ground and excited states are known, the PL contribution cannot be estimated from the calculation described above and we should somehow 'measure' it experimentally. For that purpose, we might use a different dependence of the TL and PL intensities on the thermal properties of the medium.

The TL signal intensity depends on the thermal properties of the medium as

$$U_{TL} = AQ(dn/dT)/\rho C_p = A'SC_{ex} \qquad (2\text{-}13)$$

where A and A' are proportional constants which depend on the experimental setup, S is defined by $(dn/dT)/\rho C_p$ and C_{ex} is the concentration of the excited state which should be proportional to the thermal energy Q from the excited state. Therefore the following relation can be obtained from eqs.(2-12) and (2-13).

$$U_{slow}/U_{tot} = U^{TL}_{slow}/U_{tot} + A''U_{PL}/Sn \qquad (2\text{-}14)$$

where A" is a proportional constant. Since the first term in eq.(2-14) is determined solely by the photophysical properties of the solute, it is the second term that is affected by the thermal properties of the medium. If we can somehow modulate the properties (ρ, C_p, and dn/dT) and measure U_{slow}/U_{tot}, U_{PL} can be estimated from the plot of U_{slow}/U_{tot} vs. 1/nS. Changing the thermodynamical properties of medium is accomplished by changing the temperature or solvent.

The solvent dependence method is demonstrated in the case of anthracene by changing the solvent in a series of alkanes.[25] Figure 13 shows the plot of the U_{slow}/U_{tot} of anthracene in benzene probed at 458 nm. After subtracting the TA contribution by the CCL method, U_{slow}/U_{tot} depends on the thermal character, C/nS. This solvent dependent time profile can be explained by the solvent dependent efficiency of the thermal effect. The U_{slow}/U_{tot} from the pure thermal effect which is related to the photophysical character of the solute cannot be obtained unless the solvent dependent value is extrapolated to C/nS =0 ($S > \infty$). After this extrapolation, the true quantum yield of the triplet formation can be determined.[25]

A similar method is possible by changing the temperature in an aqueous solution case, because the thermal expansion coefficient, and thus the signal intensity due to the thermal effect is a strong function of temperature [26] as will be discussed later (section 2-5). When this method is to be used, one needs to keep in mind that other relevant photophysical properties of the solute could be sensitive to the temperature or the solvent too. In many cases, we have to *assume* that they are not changed by the variation of these parameters. If we want to be sure that the obtained value is reliable, we should also measure the photophysical properties against the thermal properties of the media as long as we can.

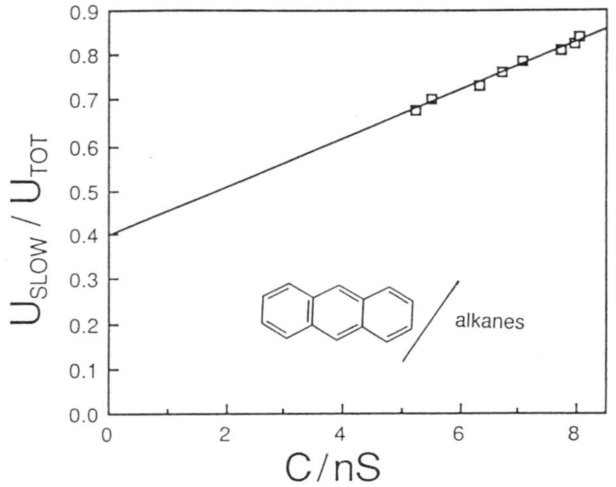

Fig.13 Plot of U_{slow}/U_{tot} of anthracene against C/nS in a series of alkanes

(c) Polarization dependence

If we monitor the transient absorption with linearly polarized light, we expect a linear dichroism in the signal within a time scale of the reorientational relaxation time in the solution. In a case that the direction of the transition dipole moment of the sample at the probe wavelength is parallel to that at the excitation wavelength, the transient absorption intensity for the parallel polarization ($E_{pump}//E_{probe}$) should be stronger than for the perpendicular polarization ($E_{pump}\perp E_{probe}$). In a case of the perpendicular transition dipole moment, the relative intensity should be opposite. A similar idea can be applied to the refractive index change. The signal intensity due to PL should depend on the relative orientation of the polarizations between the pump and probe light.

This is clearly observed in a case of Rhodamine 6G(R6G)/methanol sample. [27] After pumping R6G with a pulsed laser having 70 ps pulse width at 532 nm an expansion of the probe beam (640 nm with 200 fs pulse width) is observed which indicates that a concave transient lens is created by the photoexcitation.

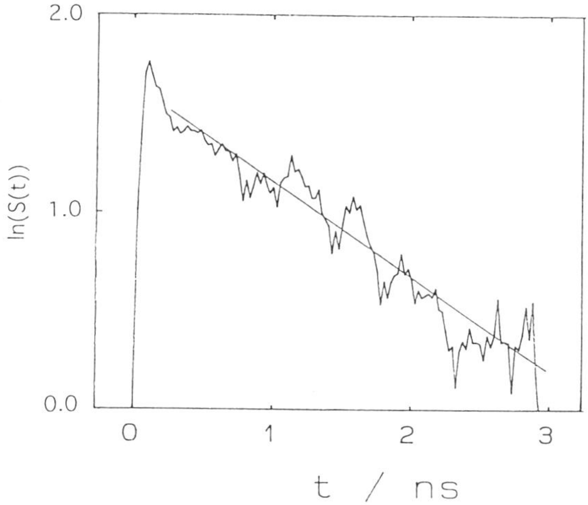

Fig. 14 Temporal profile of TrL (population lens) after the photoexcitation of R6G

The temporal profile of the TrL signal monitored by a photodetector through a pinhole is shown in Fig.14. The main part of the signal (after 200 ps from the pumping) decays with a lifetime of 2 ns. This lifetime is in good agreement with the lifetime of the excited singlet (S_1) state of R6G in methanol.[28] The signal can neither be due to the thermal origin (section 2-5) nor the Kerr response (section 2-6), because, for both cases, the decays should be the thermal diffusion time (ms order) and the reorientational time (120 ps), respectively, and both of which are very different from the observed 2 ns decay. On the basis of these considerations, the signal is attributed to PL.[27]

At the initial part of the profile, a sharp peak is observed when $E_{pump}//E_{probe}$, whereas a fast rise followed by a slower rise component observed under $E_{pump} \perp E_{probe}$. The characteristic time constant of the fast decay for $E_{pump}//E_{probe}$ and the slower rise for $E_{pump} \perp E_{probe}$ is around 120 ps, which is close to the reported orientational relaxation time of R6G in methanol.[29] This faster dynamics in the early time range can be explained by the reorientational relaxation. Since the pump beam preferentially excites R6G whose transition dipole moment is oriented parallel to the direction of the polarization, and the refractive index of PL is considered to be mainly determined by the absorption of the $S_1 \leftarrow S_0$ transition, the refractive index change along the pump polarization is expected to be larger than that of the perpendicular direction.

If the refractive index change comes only from several absorption bands which are accessible experimentally, the intensity of PL can be estimated from the dichroic ratio of the absorption bands. The limitation of this method is that the dichroic ratios of all the absorption bands which could contribute to PL are generally very difficult to be measured. Also this method can be applied only within the time range of the reorientational relaxation.

Recently the contribution of PL in the observed photothermal beam deflection after the photoexcitation of bacteriorhodopsin was estimated based on this idea in tens to hundreds microsecond time scale.[30] In this case, the sample is a protein in purple membrane, which makes the reorientational relaxation very slow even in an aqueous solution at room temperature. However we are not sure if the PL comes only from the absorption band they have measured.

(d) Time resolved method

If the metastable state or the intermediate species have a relatively long lifetime (longer than microseconds), the contribution of PL (as well as the volume lens signal discussed later) can be separated from the thermal effect by the time resolved manner.

The separation technique is based on the different time scale for the pure energy transport and mass transport processes. Since the thermal diffusion is generally much faster than the mass diffusion, the lens signal due to the chemical species (PL and/or volume lens) should last longer even after the decay of the thermal effect as long as the species lives longer than the thermal diffusion time. This method is easily applicable in the TG case, because the thermal diffusion time of the TG method is much shorter than those of the TrL case. The shorter thermal diffusion time in the TG case is due to the short length of the fringe distance, Λ, (section 3-2) than the relatively large beam radius, w, in the TrL case. In the TG signal, the thermal diffusion time is expressed by

$$I_{TG} \propto \exp(-2D_{th}q^2t)$$

where D_{th} is the thermal diffusion constant that is given by $\kappa/\rho C_p$ (κ;thermal conductivity, ρ density and C_p heat capacity at constant pressure) and q is the grating wave vector ($2\pi/\Lambda$), while the effect of the thermal diffusion is written as

$$I_{TrL} \propto \exp\{-2(r^2/w^2)/(1+2tk_{th})\}/(1+2k_{th}t)$$

for the TrL method ($k_{th}=4D_{th}/w^2$). They are microsecond orders in the TG case and tens of millisecond orders in the TrL case. On the other hand, the mass diffusion times for the both cases are given by replacing the D_{th} to the diffusion constant of the molecule and they are in orders of hundreds of microsecond (TG) and second (TrL). Recently this method is indeed applied to evaluating the volume and enthalpy changes in a photochemical reaction as described in section 2-7.[31]

(e) Acoustic measurement

Whenever heating occurs and/or molecular volume changes, the medium expands (or shrink) and acoustic signal appears. Even though this phenomena is well known in the TG case as the acoustic oscillation, a similar acoustic is observed in the TrL case, too (section 2-5-1). This acoustic signal can be easily separated from the other contribution by the characteristic time profile and this component is useful for the separation of the thermal effect or volume effect from PL. The origin of the acoustic signal is due to the expansion (or shrink) of the medium, it is expressed by [31]

$$I_{ac} = \{A(Q+\Delta H)/C_p\rho + B\Delta V\}^2 \tag{2-15}$$

where A is a constant, Q thermal energy released by the nonradiative transitions, ΔH is the enthalpy change which usually appear together with the volume change, and $B\Delta V$ is the contribution from the volume change. (Note that the above equation holds for if the heat releasing processes and the chemical reaction (ΔH and ΔV) occur promptly after the photoexcitation. If one of them is not fast enough compared with the acoustic transit time, the temporal evolution should be convoluted with the acoustic profile. Then the acoustic intensity is not simply related to these quantities as (2-15).) The remaining signal except the acoustic signal in this time region is due to PL (or the population grating in the TG case) (of course, if other signals without accompanying the medium expansion (such as the temperature lens (grating)) can be neglected).

2-5. Thermal effect

After a photon is absorbed by a molecule to create an excited state, the optical transition to the Franck-Condon state leads the molecule to a non-equilibrium configuration. For instance, if the excitation wavelength does not match the 0-0 transition, an excess energy in the vibrational manifold in the upper state is dispersed into all vibrational modes (intramolecular vibrational relaxation (IVR)) and then the energy flows into the bath modes (vibrational cooling) as phonons. In broad sense, the vibrational cooling could occur through the multiphonon process and the ladder process.[32] In the multiphonon process, the vibrations lose energy via higher order interactions with a bath of multiphonon states. The vibration would relax by emitting a burst of phonons. In the ladder process, vibrations lose energy via lower order interactions with neighboring vibrational state and phonons in energy. The excess vibrational energy descends the rungs of the vibrational ladder step by step with losing the energy to phonons. A distinction which mechanism is dominant in the energy relaxation process of a molecular system is a still very delicate and difficult problem. For that purpose, the energy relaxation processes have to be studied in detail, in particular in a ultrafast time scale. So far, this type of researches have been performed by monitoring the population of the vibrational states [32], but not from the translational manifold. All or a part of the energy is released to the medium ultimately as the translational energy, unless all of the input photon energy is released as luminescence or stored as chemical energy, both of which are very rare. Studying the dynamics of the translational energy; how does the translational mode of the bath gain the energy, will give an important clue to the mechanism of the vibrational relaxation mechanism.

The conventional TL experiment introduced in section 2-2 uses this heat energy as the origin of the signal in rather long time scale. Leaving the applications of the traditional TL effect to other reviews and original papers of this field [4,9-18], in this section I concentrate on a newly observed thermal effect, which will open a new field in a ultrafast detection of translational energy.

The refractive index change after the heat releasing can be expressed in terms of arbitrary chosen thermodynamical variables. One of the choice is the pressure, P and the entropy, S. In this case,

$$\delta n=\left(\frac{\partial n}{\partial p}\right)_S dP+\left(\frac{\partial n}{\partial S}\right)_p dS \tag{2-16}$$

This equation gives a direct decomposition of the scattered or diffracted light due to the refractive index fluctuation into a part connected with isobaric entropy fluctuation and with adiabatic pressure fluctuation.[33] Linearized hydrodynamics shows that the former one corresponds to a diffusive process while the latter one is related with acoustic oscillation. Although the choice of thermodynamical variable P and S allows a direct interpretation of these two contributions and frequently used in literatures, a different choice, density ρ and temperature T can be more useful for intuitive understanding. In this case, the refractive index change due to the temperature change is written as

$$\delta n=\{\left(\frac{\partial n}{\partial \rho}\right)_T \frac{\partial \rho}{\partial T}+\left(\frac{\partial n}{\partial T}\right)_\rho\}\delta T \qquad (2\text{-}17)$$

The first term represents the refractive index change through the change of the density and the second term is that due to the pure temperature rise without accompanying the density change. Since for normal organic solvents, the contribution of the first term is about ten times larger than that of the second term [34], the TL signal so far has been interpreted only in terms of the density change term.

The second term represents the pure thermal effect. Theoretically the second term in eq.(2-17) should always exist definitely, but there have been rather controversial debates for the non-negligible contribution in the refractive index change. For example, Eisenberg [35-a] expressed the temperature dependence of the refractive index of water as

$$n^2-1/n^2+2=A\rho^B\exp(-CT) \qquad (2\text{-}18)$$

where ρ is a density, and A, B, and C are constants. The constant C is related to the second term of eq.(2-17), and he attributed the C term as the temperature dependent structural change of water. However he could not observe clear evidence of the second term in other organic solvents such as ethanol, benzene, and carbon tetrachloride.[35-b] Although in 1967, Zaitsev et al. [36] attributed observed light scattering signal to the pure thermal scattering, Mack [37] in 1969 raised a question on their observation and reported a stimulated thermal light scattering signal which he assigned to the scattering due to the entropy fluctuation using a picosecond laser. However, Pohl [38] re-assigned the light scattering signal to the density fluctuation from theoretical and experimental studies. Scarlet [39] also studied the stimulated light scattering using a picosecond laser system and concluded that only the density fluctuation was observed. In 1986 Maison et al. [40] reported clear evidence for the existence of the second term in water using a precise determination of the Landau-Placzek ratio in light scattering experiments. However, Miller et al. [41] did not observed the existence of this component in the TG signals even at 4 °C or 0 °C of aqueous solutions. Generally the temperature dependence of the refractive index of the solutions can well expressed by the Lorentz-Lorenz relation, which is based on an assumption that the molecular polarizability is temperature independent; that is, the second term can be neglected.

Recently we have analyzed the temporal profile of the rise part of the TrL signal and found that the TrL signal from aqueous solution rises faster than that expected only from the density lens signal (Fig.15) (*vide infra*).[42] We have

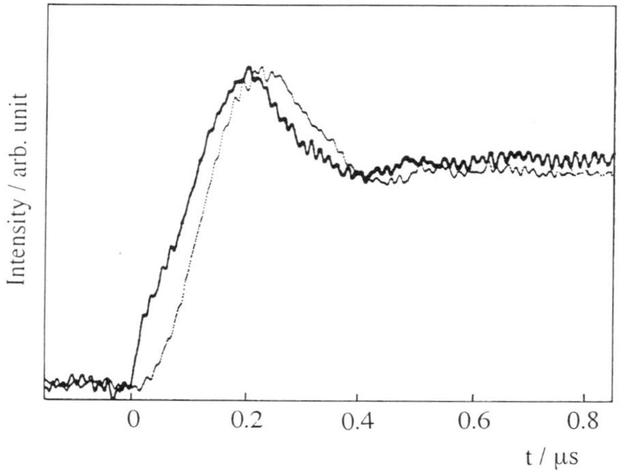

Fig.15 Temporal profiles of TrL signals of nitrobenzene in water (solid line) and in benzene (dotted line)

attributed the faster rise component to the contribution of the first term (we call it the temperature lens (Temp.L) signal). After this study, I have obtained clear evidence of the Temp.L component by changing the temperature of aqueous solutions.[43]

2-5-1 Density lens (Dens.L)

This is the lens signal that I have described in section 2-2. The signal builds up with a time constant of the acoustic transit time. More accurately the rise profile can be calculated from the continuum and Navirer Stokes equation and the energy transport equation. When we assume a Gaussian type spatial profile for the excitation beam, the density change, ρ_1, which is equivalent to the density lens signal can be calculated by

$$\rho_1 = -(A+B)\exp(-r^2/D)/D + (E-F)\times$$

$$\int_r^{\infty} \{\exp(-2(\frac{s-vt}{w})^2) - \exp(-2(\frac{s+vt}{w})^2)\}/(s^2-r^2)^{-1/2}ds \qquad (2\text{-}19)$$

where

$$A = \varepsilon\alpha\beta/\pi C_p$$

$$B = [\varepsilon(\gamma-1)\gamma^e/2\pi ncv^2]\ \partial/\partial t$$

$$D = w^2/2 + 4\lambda t/\rho_0 C_p$$

$$E = \varepsilon\alpha\beta/(2\pi^3)^{1/2} wvC_p \partial/\partial t$$

$$F = \varepsilon\gamma^e/2(2\pi^3)^{1/2} ncv^2\ \partial^2/\partial t^2$$

and α optical absorption coefficient, λ the thermal conductivity, $\gamma = C_p/C_v$, γ^e the electrostriction coupling coefficient and β the coefficient of thermal expansion. The first term of (2-19) represents the diffusive mode of the density variation ρ_1^{diff} and the second one is the acoustic mode (ρ_1^{ac}). The Dens.L signal is given by the spatial curvature of ρ_1, $\partial^2\rho_1/\partial r^2$, which is proportional to ρ_1 itself. If the temperature variation contributes to the TrL signal through the second term of eq.(2-17), the contribution can be calculated by $\partial^2 T_1/\partial r^2$, where T_1 is the temperature variation, and it is given by

$$T_1 = [-\rho_1^{diff} + (\gamma - 1)\rho_1^{ac}]/\beta\rho_0$$

Figure 16 shows the temporal profile of the TrL signal calculated with the thermodynamical parameters of water. The acoustic effect is clearly observed as the peak of the TrL signal just after the excitation (t=0) in the density variation. The calculated temporal profile of the density effect reproduces the TrL signal observed after the photoexcitation of nitrobenzene (which is an ideal molecular heater (instantaneous heat release) in a time scale longer than the nanosecond order as described later) in benzene almost perfectly, which means that the main part of the signal consists of Dens.L.[42]

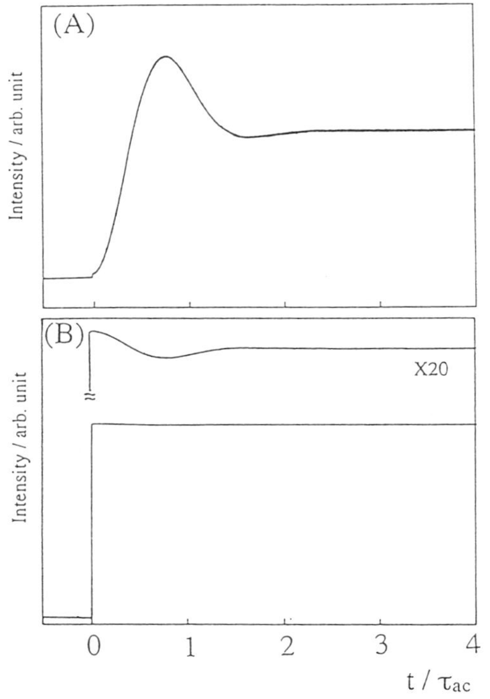

Fig.16 Calculated temporal profiles of Dens.L (A) and Temp.L (B) signals in water.

2-5-2 Temperature lens (Temp.L)

As shown in Fig.16, the acoustic effect in the calculated temporal profile of the Temp.L signal is negligible. The Temp.L signal rises immediately after the heat release and the signal is a near step function type. The small contribution of the acoustic effect is seen on the top of the signal as a negative peak. The weakness of the effect is due to a fact that γ of water is nearly unity.

Comparing the temporal profiles from the water and benzene solutions (Fig.15), we note that the signal from the aqueous solution rises faster. This faster rise component becomes more apparent as temperature decreases (Fig.17-A).[43] Around 4 °C, the signal consists of only a step function type in which the rise is determined only by the instrumental response (pulse width of the excitation beam and response of the photodetector). Below 4 °C, the signal shows a dip in the flat region and ultimately it becomes negative (negative indicates the increase of the probe light density (converging lens)).

Before going further, it is important to carefully consider the origin of the step function type signal; that is, this signal could be due to the population lens or other origins of lens signals. This possibility of PL can be denied based on the following reasons.

(a) From a picosecond TG experiment, we found that the lifetime of the excited singlet and triplet states of nitrobenzene is less than 10 ps and 1 ns, respectively. Therefore there should be no metastable excited states in the nanosecond order.

(b) Nitrobenzene is photochemically stable and no new chemical species should be created after the photoexcitation.

(c) A similar signal is observed after photoexcitations of $FeCl_3$ as well as 9,10-diazaphenanthrene in water. Both of these chemical species have short lifetimes of the excited states. It is very unlikely that these three chemicals with very different chemical nature provide similar PL signals.

(d) Most importantly, the decay of the step function type signal is the same as the decay of the Dens.L signal; that is, thermal diffusion time. If the signal is due to PL, the decay should be determined by the mass diffusion time, which is much slower than the thermal diffusion time.

The possibility of the other type signals (such as the volume lens and Kerr lens) can be also excluded from the above facts.

These temperature dependent signals can be explained in terms of the superposition of the Temp.L and Dens.L signals as seen in the theoretically simulated curves in Fig.17-B. According to the analysis, the signal intensity of Temp.L is almost constant within the temperature range we examine, while the

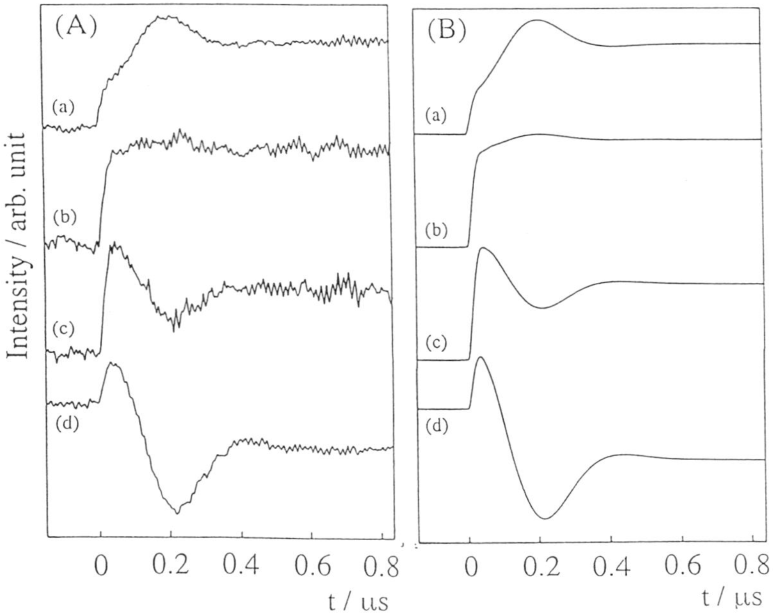

Fig. 17 (A) Observed temporal profile of the TrL signal after photoexcitation of nitrobenzene in water at various temperatures (a) 7.9 °C, (b) 4.0 °C, (c) 1.7 °C and (d) -3.5 °C. (B) Best fitted calculated signals with a superposition of the Temp.L and Dens.L components.

Dens.L term is sensitive to the temperature.[43] We found that the temperature dependence of the Dens.L signal intensity comes from the temperature dependent $d\rho/dT$.

The most important result of these studies is that it opens a new field in studying the fast dynamics of energy by the photothermal method. Until this finding, the fastest time resolution of the photothermal technique is achieved only by the TG method with a large crossing angle between two excitation beams (*vide infra*). However, even by the TG method, the acoustic oscillation makes the fast dynamics obscure and a few tens of picoseconds have been the fastest time resolution. Even by this time resolution, it is not fast enough to directly detect the build up of the translational temperature after the vibrational relaxation. Using the TrL (temperature lens) method, we can detect the heat releasing process faster than a few picoseconds.

For example, Fig.18 shows the time resolved TrL signal observed after the photoexcitation of Ni^{2+} in aqueous solution with a picosecond laser system (pulse width 30 ps).[44] The peak observed at the initial part of the signal has about the same time profile as the excitation pulse and it is attributed to the Kerr lens signal discussed in the next section. Subtracting the Kerr lens signal by the method explained in the next section, we find that the TrL signal due to the thermal effect rises faster than 5 ps, which means that we now have a technique to detect the temperature rise with a time constant faster than 5 ps. It is quite reasonable to think that the time resolution can be further improved by using a shorter laser pulse for the excitation.

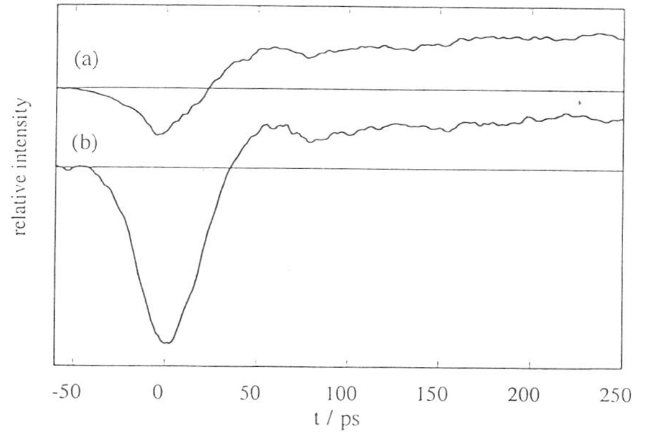

Fig.18 The TrL (Temp.L) signal observed in picosecond time scale (a) $E_{pump} \perp E_{probe}$ and (b) $E_{pump} // E_{probe}$. The Dens.L signal which will rise up with a time constant of 100 ns is not observed in this time scale.

2-6. Kerr lens

When strong electric field of light interacts with matter, it induces a distortion of the electronic charge distribution of the molecule and aligns the most-polarizable axis of the molecule along the direction of the field. These processes change the refractive index of the system anisotropically. This effect has been detected as the optically induced birefringence (Optical Kerr effect) by monitoring a rotation of the polarization plane of the light.[45] At the same time, it could be a part of the TrL signals (the Kerr lens).

A probe beam (640 nm, 200 fs pulse width) focusing is observed for CS_2 after an intense pump pulse (same character as the probe pulse but more intense) under the parallel polarization ($E_{pump}//E_{probe}$) ($I_{//}$ signal) (Fig.19(a)).[46] The created lens is a convex type in the entire time region. We could eliminate the contribution of transient absorption (, which could be due to multiphoton absorption) completely by the CCL method. Under the $E_{pump}\perp E_{probe}$ condition ($I_{\perp}$ signal) (Fig.19(b)), the lens signal is a convex type within the pump pulse, but it changes to a concave type after the pump pulse leaves from the sample. At an angle of 54.7° of the polarization planes between the pump and probe pulses, the signal consists of only the fast response convex lens (Fig.19(c)).

These observations can be explained in terms of the creation of the Kerr lens by the pump beam. Indeed, we have confirmed that the Kerr signal detected with a usual setup; the polarization of the pump pulse is rotated 45° from that of the probe pulse and the change in the polarization state of the probe beam is monitored through a crossed polarizer, shows in good agreement with the observed signal. A difference is the quadratic nature of the Kerr signal in the homodyne detection system. Under a weak-signal limit, the Kerr signal is given by $\{(n_2)_{//}-(n_2)_{\perp}\}^2$, whereas that of the TrL signal is just by $(n_2)_{//}$ and $(n_2)_{\perp}$.

According to a molecular orientational model, the polarization dependence of n_2 of the nuclear (n_2^{nuc}) and electronic (n_2^{ele}) contributions are predicted to be

$$(n_2^{nuc})_{xx}=-2(n_2^{nuc})_{xy} \tag{2-20a}$$

and

$$(n_2^{ele})_{xx}=3(n_2^{ele})_{xy}. \tag{2-20b}$$

Physically this relationship can be understood by the anisotropic polarizability of molecule. When anisotropic molecules with cylindrical symmetry (polarizabilities parallel to and perpendicular to the molecular axis are, respectively, $\alpha_{//}$ and $\alpha_{\perp}$ ($\alpha_{//}>\alpha_{\perp}$)) are aligned by the electric field of light, which is polarized along the x

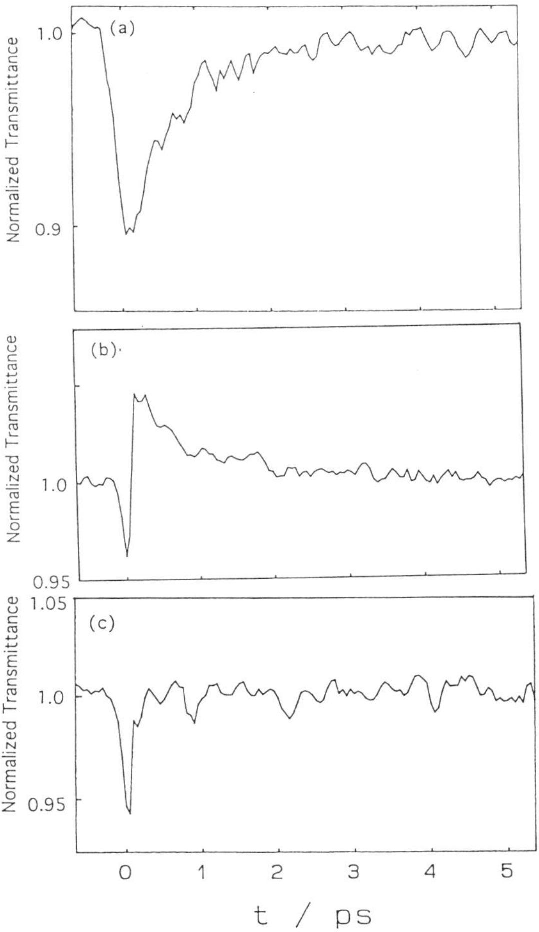

Fig.19 Temporal profiles of the TrL (Kerr lens) signals after pumped CS_2 for (a) $E_{pump}//E_{probe}$, (b) $E_{pump} \perp E_{probe}$, and (c) at an angle 54° .

axis, the optical susceptibility along the x axis becomes larger and is given by

$$\chi_{xx}=\chi_{av}+2N\delta\alpha g/3 \qquad (2\text{-}21)$$

where

$$\chi_{av}=N\alpha_{av},$$

$$\alpha_{av}=(1/3)(\alpha_{/\!/}+2\alpha_{\perp}), \quad \text{and} \quad \delta\alpha=\alpha_{/\!/}-\alpha_{\perp}$$

N is the number density, g represents the fraction of the aligned molecules. To the perpendicular direction, it is given by

$$\chi_{yy}=\chi_{zz}=\chi_{av}-N\delta\alpha g/3 \qquad (2\text{-}22)$$

We can verify eq.(2-20) as follows. Figure 20 (a) shows the time profile of $I_{/\!/}+2I_{\perp}$. In this superposition the slow (nuclear) response disappears and the

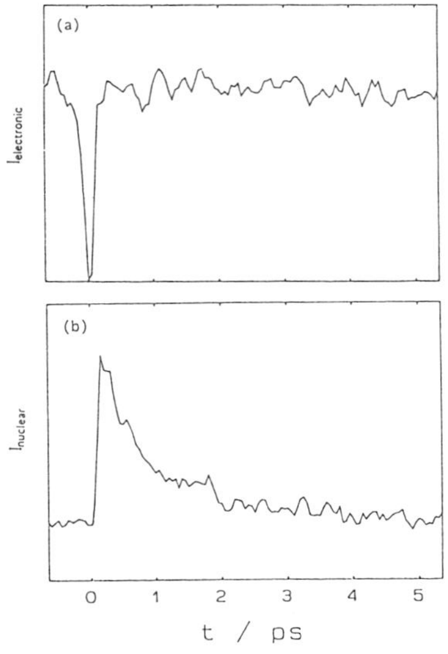

Fig.20 Temporal profiles of (a) $I_{/\!/}+2I_{\perp}$ ($=I_{electronic}$), and (b) $I_{/\!/}-3I_{\perp}$ ($=I_{nuclear}$)

signal consists of only the fast (electronic) response (the instrumental response). Figure 20 (b) shows the $I_{//}-3I_{\perp}$ signal, which gives rise to only the slow (pure nuclear) response. The disappearance of the slow component at the 54.7° angle is the consequence of $n_2^{nuc}=0$ at this special angle.

The temporal dependence of the nuclear response reflects the molecular reorientational relaxation in the solution and it is important to monitor the motion in a fast time scale. Utilizing this separation method, we now can trace the orientational motion even within the period of the pumping pulse.

The Kerr lens signal due to the OKE has been detected by the z scan method.[24] As mentioned before, however, our TrL method is more flexible technique than the z scan to study the molecular dynamics associated with the OKE through the temporal polarizations, and wavelengths dependencies.

2-7. Volume lens

When a molecular volume changes by photoexcitation, eventually the volume in the light irradiated region will change, which contributes to $\chi^{(3)}$. This phenomenon is more general than we might think because the molecular volume will change even just by creating the photoexcited state through the solvent-solute interaction. Even the solvent rearrangement around the excited molecule will induce the nonlinearity. The effect is pronounced when a photochemical reaction takes place. This signal component is very interesting and important in chemistry and molecular biology because sometimes the molecular volume changes in chemical reactions play a crucial role in understanding the reactions.

Experimentally pressure effects on the equilibrium or rate constants of chemical reactions have been used for obtaining the information on the volume changes (ΔV) or activation volume changes ($\Delta V^{\ddagger}$). However, experimentally, these methods are difficult to be applied to fast reactions or irreversible reactions. Theoretically, the relation between the pressure dependence and ΔV or $\Delta V^{\ddagger}$ is based on the thermodynamical relations. If the relation is not appropriate to describe the reaction system, the measured ΔV or $\Delta V^{\ddagger}$ cannot be directly related to the actual volume change and even the meaning of ΔV or $\Delta V^{\ddagger}$ becomes obscure. For instance, the lifetime of the precursor excited state by the nonradiative transition beside by the reaction can be pressure dependent through the variation of the phonon density in the system. In that case, the pressure dependence of the reaction rate cannot represent the volume change ($\Delta V^{\ddagger}$) of the reaction.

The photoacoustic method which detects the pressure wave induced by the volume change can be another technique and indeed frequently used for estimating ΔV. Still it has a serious demerit such that the temporal evolution of the reaction is hard to be traced from the signal.

Quantitative measurements of $\chi^{(3)}$ due to this volume change by the TrL and TG methods are direct methods for ΔV, and have less limitation in the time resolution.[30,31,41] A difficulty to be overcome for the measurement, however, is the separation of the volume component from the thermal effect, which generally contributes to the density change after photoexcitation. (One of the exceptional case is the density change that arises from a large amplitude protein motion. [41] In this case the density change due to this effect is dominant and indeed is so large that it is not necessary to be separated.) Sometimes, two strategies have been used; changing the temperature or solvent. Both of them are based on a modulation of the signal efficiency of the thermal effect as described in 2-4-b.

If the reaction takes place in aqueous solution, the temperature dependence method can be used. As stated in section 2-5-2, the thermal expansion coefficient of water is a strong function of temperature, and in particular, it vanishes at 4 °C. Therefore the signal observed at this temperature should come from the volume lens (of course, and from PL or other effects except the thermal effect). When organic solvents are the media, a series of solvents or binary mixtures of solvents which do not affect the other reaction parameters (such as the quantum yield, the volume change, the enthalpy change and so on) except the thermal expansion coefficient is used for modulating the efficiency. Then, by extrapolating to the zero thermal expansion limit, the magnitude of the volume change can be estimated.

An advantage of these methods is that ΔV induced by the photoexcitation can be quantitatively measured sensitively including the temporal dependence. If the reaction takes place stepwise, ΔV of each step can be separately monitored with the information of the reaction rate as well. There are two disadvantages in these methods. First, as stated above, we have to extrapolate the signal intensity to the zero thermal expansion limit. Sometimes it will induce a large experimental error. Second, we have to assume always that the modulated parameters, temperature or solvent, do not affect the photophysical properties or ΔV. It is of course desirable to verify that the assumption is correct, but it is usually very difficult to be performed.

Recently we proposed another way to overcome these limitations, time resolved method, that is related to the separation method of PL as described in the section 2-4-d. We applied this method to a photodissociation reaction of diphenyl-cyclopropenone (DPCP) (Scheme 2) using the TG method. (In principle, the same experiment should be possible by using the TrL method. The advantage of the TG method is the faster decay of the thermal contribution (cf.2-4-d)) After the photoexcitation of DPCP, the TG signal of the thermal origin appears. After that signal decays, there appears another weaker slowly decaying component. Apparently the signal is due to the presence of the chemical species created by the reaction. By comparing that signal intensity with the thermal signal, the volume change of the reaction can be determined.

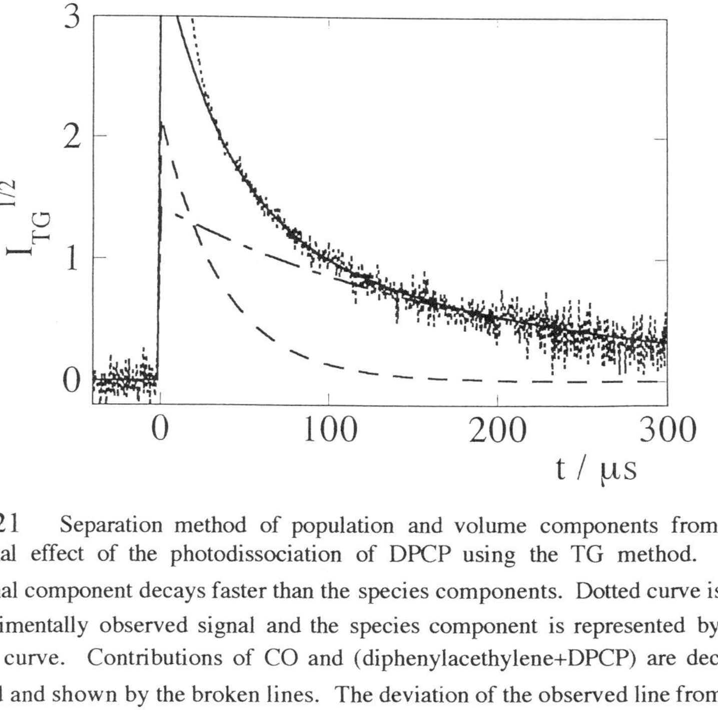

Fig.21 Separation method of population and volume components from the thermal effect of the photodissociation of DPCP using the TG method. The thermal component decays faster than the species components. Dotted curve is the experimentally observed signal and the species component is represented by the solid curve. Contributions of CO and (diphenylacethylene+DPCP) are decomposed and shown by the broken lines. The deviation of the observed line from the calculated on in t<20 μs is due to the thermal contribution.

The species grating or species lens component decays by the mass diffusion process in the solution. Interestingly, if several chemicals are involved in the reaction and the diffusion constant of them are different, not only the species component from the thermal effect but also the signal from each species can be separated by the time resolved manner. Then, not only the total volume change in the reaction but also the molecular volume (or partial molar volume) of each species can be measured. For example, in Fig.21, since the diffusion constant of CO is so different from those of DPCP and diphenylacethylene, the contribution of CO can be separated easily and the partial molar volume was determined. By using this time-resolved method, we do not need to vary the thermal properties of the matrix and naturally we are free from any extrapolation or the assumption mentioned above.

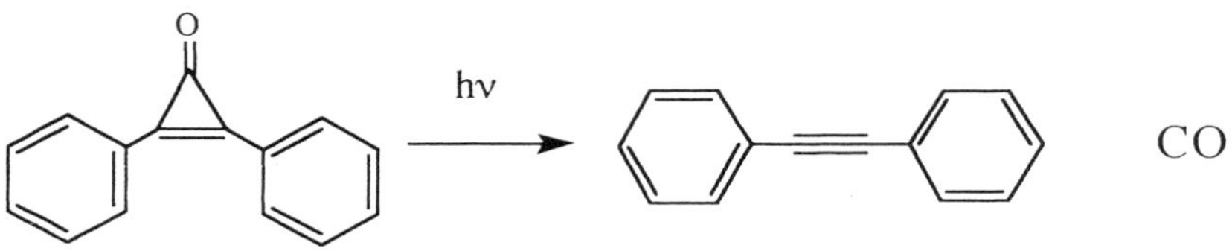

Scheme 2

3. Transient grating

3-1. Basic theory

Transient grating (TG) is a very general name for describing some four wave mixing techniques. Sometime it includes the degenerative four wave mixing, stimulated (Rayleigh or Brillouin) light scattering. In the TG experiment, two coherent excitation pulses are time coincident at the sample at $t'=t_1=t_2$ in eq.(1-7) and the created polarization emits radiation to the phase matching direction.

$$P^{(3)}_i(t)=\int dt'\chi^{(3)}_{ijkl}(t-t')E_j(t)\ E_k(t')E_l(t') \qquad (3\text{-}1)$$

To calculate the TG signal intensity measured by the probe light entered to the sample at t=t' with respect to the excitation pulses, it is given by [47]

$$I_{TG}(t)=\int_{-\infty}^{\infty} dt''\ I_j(t''-t)[\int_{-\infty}^{t''} dt'\ \chi^{(3)}_{ijkl}(t''-t')I_{kl}(t')]^2. \qquad (3\text{-}2)$$

By changing the delay between the probe and pump pulses or using a cw probe beam, the time dependence of the TG signal and as a consequence, kinetics of the molecular dynamics which contributes to $\chi^{(3)}$ can be monitored.

Intuitively, the TG signal can be understood as the diffracted light by the spatial sinusoidal modulation of the refractive index (phase) or absorption (amplitude) gratings, which is induced by the interference between the two coherent laser light (Fig.22). [48] Under assumptions of a plane wave for the excitation light and an equal light intensity for the two beams, the sinusoidal

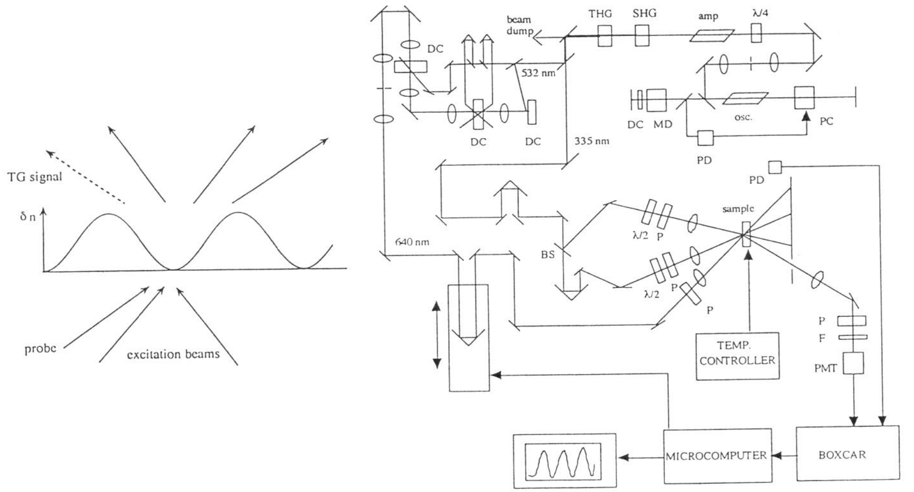

Fig.22 Transient grating experimental setup. (DC; dye cell, PD; photodiode, PC; Pockels cell, P; polarizer, λ/2; half wave plate, F; filter, BS; beam splitter) The setup is for a psec UV excitation (355 nm; THG of a Nd:YAG laser) and probed at various wavelengths by the dye laser. The polarizations of the three beams and the signal can be controlled independently.

optical interference pattern are described as

$$I(x,t)=2I_0f(t)[1+\cos(2\pi x/\Lambda)] \quad (3\text{-}3)$$

where I_0 is the intensity in each beam, and x is the coordinate which is transverse to the average direction of the excitation beams. The fringe spacing, Λ, is given by

$$\Lambda=\lambda_{ex}/2\sin(\theta/2) \quad (3\text{-}4)$$

using the wavelength of the excitation, λ_{ex} and crossing angle, θ, between the two beams. The light-matter interaction creates the sinusoidal modulation in n and k caused by various factors discussed in section 1-2. The grating is monitored by the diffraction efficiency of a cw probe beam or another time delayed laser pulse (Fig.22). For the thick grating, the probe is brought in at an appropriate angle to satisfy the phase matching condition (Bragg condition for the thick grating case). Under a weak diffraction condition, the diffraction efficiency, η, due to the grating is proportional to the square of the variations in the refractive index (Δn) and absorption (Δk),[48]

$$\eta=\exp(-2Kd/\cos\theta)[\sin^2(\pi\Delta nd/\lambda\cos\theta)+\sinh^2(\Delta kd/2\cos\theta)]$$
$$\cong a(\Delta n)^2+b(\Delta k)^2 \quad (3\text{-}5)$$

where K is an absorption coefficient, d is a thickness of the sample and λ is the wavelength of the probe. The first term describes the effect of the refractive index change (phase grating) and the second term is resulting the absorption change (amplitude grating).

The TG signal can originate from various $\chi^{(3)}$ sources as the TrL signal. The systems utilized in the TG experiment are generally carefully chosen to avoid the interference between these components. Otherwise the interference obscures the phenomena interest. Some of the identification or separation methods are the same or similar to those described in the previous sections. For example, by selecting the wavelengths for the pump and probe light, the phase and amplitude grating can be separated (section 2-3).[49] The separation of the volume effect from the thermal effect was demonstrated by Zimmt and co-workers.[50] They varied the thermodynamical properties of the matrix using a series of alkanes and determined the volume change associated with the photoisomerization of tetraphenylethene. (cf. section 2-7).

One of unique techniques in the TG method is the polarization selection. By the works originating from Hellwarth [51] it was shown that the different elements of the nonlinear susceptibility tensor possesses different polarization characters and different phenomena in the TG signal can be individually probed. Although polarization selection is possible even in the TrL case as shown in section 2-6, only the polarizations of the pump and probe beams can be selected. In the TG case, the polarizations of the two pump beams can be controlled independently beside that of the probe beam. Etchepare et al. first demonstrated how to fully exploit the polarization selective nature of the TG signal by separating the nuclear and electronic Kerr effects.[52] Deeg and Fayer [53] utilized this technique to separate the electronic OKE, nuclear OKE, excited state grating and acoustic wave (thermal contribution). It provides the determination of the anisotropy of the excited state grating and the geometrical relation between the excitation transition dipole and the probe transition dipole. The polarization selective TG can be understood by the grating decomposition method. [54] The ability to separate absorptive gratings with different anisotropy should enable successful application of this technique to separate out overlapping spectral components.[55]

Another unique TG method which could be used for the separation of the real and imaginary parts of $\chi^{(3)}$ is the optical heterodyne detection.[56,57] The optical heterodyne is achieved by providing a local oscillator filed in addition to the signal field at the detector, resulting in much higher sensitivity and at the same time in separation of the real imaginary parts. If the optical phase of the local oscillator is well defined with respect to that of the probe field, the material response can be in a linearized form thereby avoiding the cross term inherent to a mudulus-square detection. If the local field is in-quadrature with respect to the probe field, the signal becomes proportional to the real part, $\mathrm{Re}(\chi^{(3)})$ representing the transient birefringence. By adjusting the local oscillator to be in-phase with respect to the probe field, the imaginary part of $\chi^{(3)}$ (transient dichroism) is measured.

There are extensive studies on various dynamics using the TG method [56]. For examples, chemical reactions in those we are interested in this review were studied by the TG method by many groups. [57] Here I review some examples of the studies on the thermal effect and on the molecular dynamics.

3-2. Thermal effect

The TG method is increasingly being used for the studies on the excited state dynamics ever since Salcedo et al. demonstrated the usefulness in their paper on energy transfer in molecular crystal.[58] Further, the energy transfer process of anthracene crystal [59] as well as in solution phase [60] have been studied. However these studies utilized the population grating component. Compared to the studies using the population grating, studies on excited state dynamics from the heating effect have been rather limited. Probably the reason is that the time resolution of the signal is determined not only by the pulse width but also by the acoustic transit time, τ_{ac}, as described in 2-5.

Nevertheless, detection of the heat energy from the radiationless transition provides unique information which cannot be detected by monitoring other properties, (such as the population of the excited states). As stated in section 2-5, studying the energy dynamics from the translational temperature is very interesting and important not only because the nonradiative transition is the most commonly observed phenomenon after the photoexcitation of any molecules or materials (and these energies ultimately flow into the translational freedom) but also because to elucidate the very initial and fundamental step of the nonradiative transition and vibrational relaxation. Energy relaxation from the electronic to vibrational or from vibrational to (rotational-) translational freedoms have been studied extensively in gas phase. In condensed phase, a lot of questions still remain unresolved due to the short time scales involved. These ultrafast detection of the translational energy in condensed phase will shed a light on the elementary step of the energy transfer processes, which will eventually flow into the energy of the translational freedom.

When the matrix is heated by the nonradiative transition and the temperature of the matrix is increased by δT, the variation of the temperature is described by the heat diffusion equation. If the thermal diffusion coefficient (D_{th}) is temperature independent (, which is usually correct under a very small temperature variation), it is given by

$$d\delta T(x,t)/dt = D_{th}\, \partial^2 \delta T(x,t)/\partial t^2 + F(x,t)/\rho C_p \qquad (3\text{-}6)$$

where F(x,t) is the rate of the heat generation through nonradiative transition and

$$F(x,t)=\varepsilon\, \phi I(x,t) \otimes f(t) \qquad (3\text{-}7)$$

where $\otimes$ denotes the convolution, ε is the absorption coefficient at the excitation wavelength, I(x,t) is the excitation laser intensity given by eq.(3 3), ϕ the quantum yield of the heat releasing step, and f(t) is the rate of the heat-releasing process. (The previous heat energy Q in (2-13) and (2-15) is given by the integration of F(t) in time.) This spatially modulated (time dependent) temperature change induces the modulation of the refractive index of the medium by the relation of eqs.(2-16) or (2-17). This modulation can be a cause of the TG signal (eq.(3-5)).

Decomposition of the refractive index change by eq.(2-16) gives an isobaric entropy fluctuation (Rayleigh wave or diffusive wave) and an adiabatic pressure fluctuation (Brillouin wave or acoustic oscillation). On the other hand, if we choose the second decomposition way based on eq.(2-18), there are two components in the TG signal due to the thermal effect; temperature grating (Temp.G) and density grating (Dens.G) which are similar to the thermal effect in the TrL case.

3-2-1 Density grating (Dens.G)

Dens. G is the one usually observed in most of the TG studies. The thermal effect after the photoexcitation was detected by Mack,[37] and Key et al. [61] as the periodic variation of the refractive index. Soon after that, it has been applied to the measurements of thermal diffusivities in solutions [62], liquid crystals [63], solids [64], gases [65,a-d] and supercritical fluids [65,e]. Also, since the TG method is the background free detection, it can detect a very small temperature rise. Utilizing this excellent sensitivity, there has been a number of applications to detect weak (one or two photon) absorptions.[66,67] In particular, when nonlinear processes such as multiphoton absorption, optical saturation, self-quenching and so on are involved in the photoexcitation or deactivation processes, the grating pattern is distorted from the sinusoidal curve and second or higher harmonic spatial component is created to produce the anharmonic (higher order) signals. The detection of such a higher order signal is a prominent way to study the nonlinear processes selectively.[68] The thermal effect can be detected in a gas phase at rather high pressure (the TG signal is dominated by the population grating signal at low pressure.).[65,69]

Concerning the thermal detection, the TG method can be categorized in the photothermal techniques such as the photoacoustic and thermal lens method. The most notable merit, if we compare the TG method with the other ones, may be the fast time resolution. In fact, until we have developed the TrL method (Temp.L), TG is the fastest response technique for detecting the temperature rise. This fast response comes from the fact that the acoustic transit time of the grating case is given by $\tau_{ac}=\Lambda/v$, which is much shorter than that of the TrL case. By taking this advantage, fast excited state dynamics have been detected.[70] As clearly seen from eq.(3-7), the signal intensity depends on the temporal profile of the deactivation processes and the quantum yield of the nonradiative transition. The energy deposition quantum yield was measured for iron(III)tetra-phenylporp hyrin chloride as well as the thermal diffusion.[71]

We have applied this technique to study the photophysical processes involved in the excited triplet states [72] as well as the fluorescence states.[73] After the photoexcitation of photochemically stable organic molecules in solution under air saturated condition, generally strong thermal grating signals, which decay to the baseline with the thermal diffusion time were observed. At the initial part of the signal, we found two different rises which can be analyzed by

$$I_{TG}(t)^{1/2}=C\{A+B-A\exp(-k_f t)-B\exp(-k_s t)\}\otimes\exp(-D_{th}q^2t)\otimes I(t) \quad (3\text{-}8)$$

where A, B, and C are constants, $\otimes$ represents convolution, and I(t) is the temporal profile of the pump beam. It should be noted that the Brillouin wave (acoustic oscillation) generally observed in a fast time scale of the TG signal is smeared out by the broader laser pulse width than the acoustic period and only the Rayleigh component of Dens.G is observed in this study. The faster rise component comes from the heating effect due to the nonradiative transitions from the S_1 state and the instrumental response. The slower rise comes from the heating effect from the deactivation of the T_1 state. From the time constant of the rise curve, the lifetimes of the triplet states can be determined.

Under the air saturated condition, the T_1 lifetimes are generally shortened by the quenching due to oxygen dissolved in solution. Therefore the quenching rate of the T_1 state by oxygen can be sensitively and conveniently studied from the rise part of the TG signal. An alternative way for the quenching study is to use the transient absorption technique, but the application is always limited by the probe wavelength which should be located within a T-T absorption band. Photophysically interesting observation was obtained in the photoexcited dynamics of pyridazine. A remarkably shortened triplet lifetime ca.100 ns was measured in solution at room temperature (at lower temperature the lifetime is ca. 3 μs from the TL studies [21]). The short lifetime is not due to the quenching by oxygen but is intrinsic. The quantum yield of the triplet formation was determined from the relative intensity of the two rising component, A/B, and the energy balance relation similar to eq.(2-6) to be 0.1, which is much smaller than that at the low temperature. These temperature dependent lifetime and ϕ_{isc} support the model we proposed for the nonradiative transition of ortho-diazaarom atic molecules (Fig.7).

Harris and co-workers have reported applications of rise parts of the density grating signal to study a photoisomerization reaction, and photodissociation reactions of I_2, Cl_2 and Br_2.[74] The photodissociation and recombination of these halogens are directly monitored from tens of nanosecond time scale. The quantum yield of the photodissociation was estimated by using the excitation dependence of the diffraction efficiency from the anharmonic thermal grating created by saturated absorption. The rate constant of non-geminate recombination was extracted from the temporal profile of the TG signal. Other spectroscopic techniques, EPR or the transient absorption methods, could trace such reaction kinetics of the intermediate species. However, the sensitivity of the transient absorption method is rather low compared with the TG detection and it is difficult to monitor only a trace of intermediate species. A conventional time-resolved EPR (TREPR) method based on a cw-EPR spectrometer and a laser system cannot possess a tens nanosecond time resolution. Recently TREPR based on well developed FT-EPR can have such a fast time resolution. However considering the energetic information as well as the quantum yield determination of the reactions, we must admit that the Dens.G (thermal grating) method provides very unique data which are hardly accessed by other spectroscopies.

Not only the dynamics of the lowest excited states but also those of the higher excited states can be studied by the TG method with a slight modi fication.[75] A photoexcited metastable state was created by a pre-pulse before the grating pulses and the wavelength of the grating light was tuned to the transient absorption band. The temporal profile of the two step excited TG signal reflects the dynamics of the higher excited states. By this method, the intermolecular energy transfer process from the higher excited triplet state was studied.

In the picosecond time scale, not only the Rayleigh component but also Brillouin wave (acoustic oscillation) appears. Dynamics of the excited state should be extracted from the acoustic curve by a theoretical fitting procedure. For example, vibrational energy relaxation in heme proteins has been studied in relation to the protein conformational change [76] and the very short lifetime of vibrational modes and rapid vibrational energy distribution were found. Other applications to the studies of the energetics of photoisomerization have been reported.[77]

As long as the TG signal comes from only the density change contribution, the time resolution is limited by the acoustic signal. A faster time resolution can be achieved by making Λ small; that is, by a large crossing angle with a short excitation wavelength. For instance, Fig.23 shows the acoustic grating signal detected after the TG excitation with a crossing angle 30° at 355 nm. The acoustic period is ca. 500 ps and by a simulation, about 50 ps time resolution for the heat releasing process can be achieved.

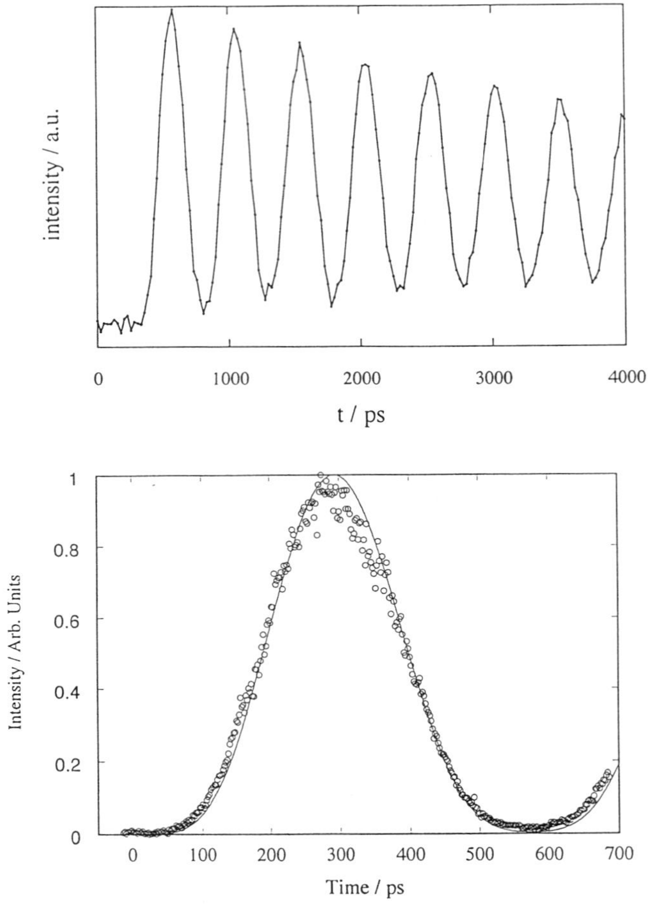

Fig.23 An example of an acoustic TG signal. We can obtain 50 ps time resolution by deconvolution by the acoustic oscillation (lower part).

Since the time resolution of this signal is determined by the acoustic transit time, and that time is shorter as the fringe spacing becomes shorter, the higher time resolution can be achieved by making Λ small. Theoretically, adopting counter-propagation beam geometry gives the shortest fringe spacing (eq.(3-4)) and, hence, the fastest time resolution. Indeed, Genberg et al. have reported that the rise time of the density grating signal under that geometry with 366 nm excitation provides less than 20 ps.[71,76] However for the counterpropagation geometry, the wavelength of the probe light should be the same as that of the pump light to satisfy the phase matching condition (degenerative four-wave mixing), and as a consequence, contribution of the population grating, which is overlapped to the acoustic signal and makes the thermal effect obscure is essentially unavoidable. Furthermore, even under this condition, the time resolution is not enough to trace the elementary step of the vibrational energy relaxation.[76] In order to investigate the energy transfer processes to the translational freedom from the vibrational freedom, we need further fast detection system such as the temperature grating as I will discuss next.

3-2-2 Temperature grating (Temp.G)

A theoretical consideration as described in 2-5, a similar type of TG signal as Temp.L is expected from the second term of eq.(2-7). This component can be called temperature grating (Temp.G). The density grating signal gives familiar acoustic signal (Brillouin wave) (Fig.24-A) and 'thermal grating' signal (Rayleigh wave) that is described in the previous section, whereas Temp.G is of a step function type with a weak acoustic oscillation on it (Fig.24-B)[43], which means that the Rayleigh wave is dominant over the Brillouin wave. The weak Brillouin wave in the Temp.G signal is a consequence of nearly equal values between C_p and C_v (heat capacities with a constant pressure and a constant volume).

These components are observed clearly by the same strategy used in section 2-5. Above 4 °C of aqueous solution, the dominant signal is due to the Dens.G. Around 4 °C, the pure Temp.G signal is observed dominantly (Fig.25). As discussed in section 2-5-1, the rise time of the Temp.G signal should be very fast and, in this case, it is determined by the pulse width of the pump laser. This Temp.G signal can be used for ultrafast (faster than the rise time of the acoustic signal) detection of the temperature rise (Fig.25).

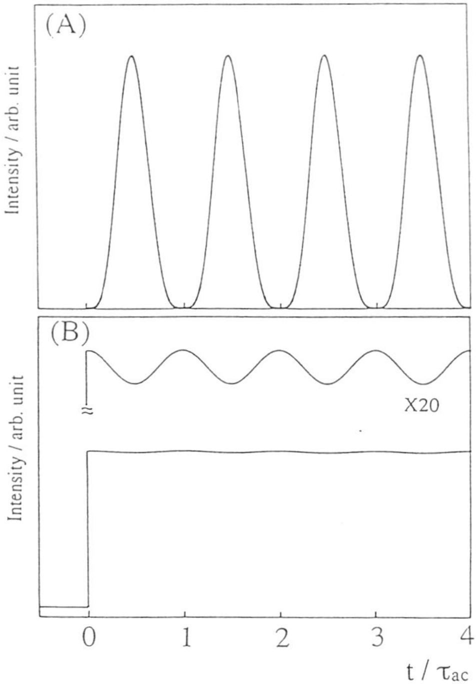

Fig.24 Calculated temporal dependence of Dens.G (A) and Temp.G (B) using the thermodynamical parameters of water.

This component can be seen not only from aqueous solution. Recently we observed the Temp.G component from many organic solvents, such as benzene, alkanes, alcohols etc.[44] Interestingly, their observed values of $(\partial n/\partial T)_\rho$ is larger than that of water. Therefore, we conclude that the manifested Temp.G signal from the aqueous solution is the result of the unique temperature dependent $\partial\rho/\partial T$ of water not from a large value of $(\partial n/\partial T)_\rho$. From these observations, the origin of the refractive index change by the temperature rise with a constant density is suggested to be the interaction induced polarizability of the solvent molecules. If this is actually the origin, the ultimate time resolution of this signal can be a time period of the molecular collision, which is less than 100 fs in a condensed phase. The nature of this component is now being investigated in detail using the TG method.

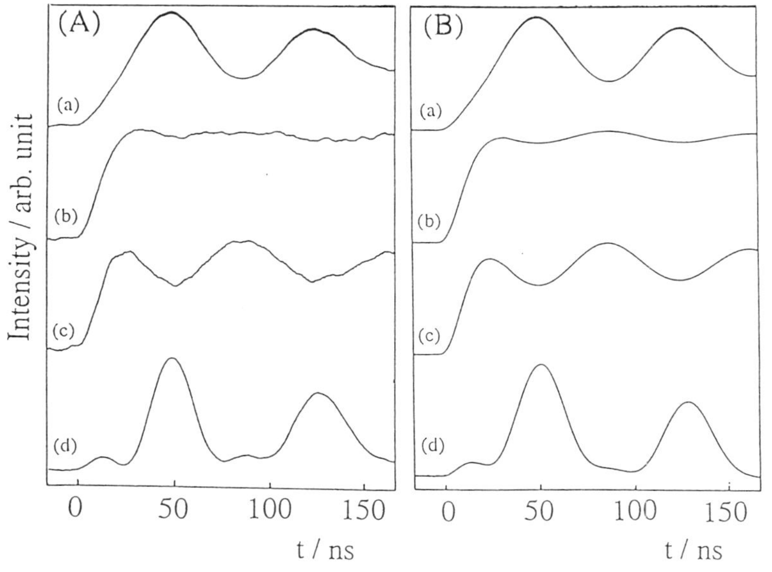

Fig.25 (A) Observed temporal profiles of the TG signal in water at various temperatures (a) 5.2 °C, (b) 4.0 °C, (c) 3.0 °C and (d) -2.1 °C. (B) Best fitted calculated signals with a superposition of the Temp.G and Dens.G components.

3-3. Reorientational relaxation in solution

Since chemical reactions usually take place in flexible system such as liquid phase, they are more or less affected by the dynamics of the surrounding matrix. Barrier heights along the reaction coordinate may be altered by the solvent dielectric constant or polarizability whereas the dynamics such as reorientational and translational motion essentially control the kinetics of the reaction. The rotational motions of molecules in solution have been studied extensively also in order to probe the nature and dynamics of solvent-solute interactions. While the orientational dynamics can be examined indirectly using frequency domain method such as Rayleigh-wing spectra, these dynamics give rise to a very low

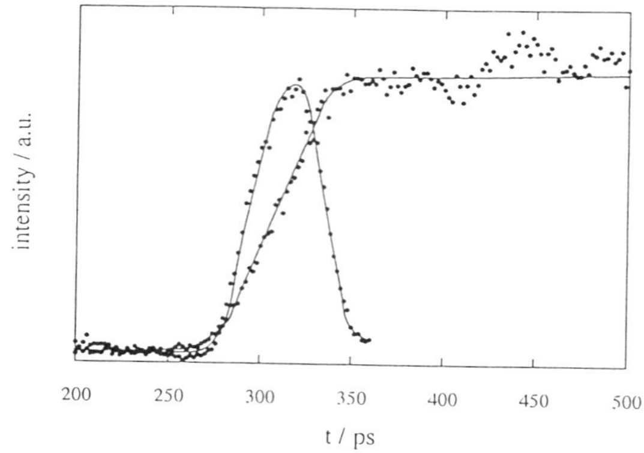

Fig.26 An example of the Temp.G signal in picosecond time scale observed at 4 °C in aqueous solution with the experimental setup shown in Fig.22. The pulse shown at 300 ps is the response function of the system (pulse widths of the excitation and probe lasers).

intensity pedestal resulting from collision and interaction induced local motions with a superimposed sharp Lorentzian peak of high intensity, time resolved techniques allow direct measurement. Fluorescence or transient absorption polarization has been widely used for exploring the molecular motion. In the fluorescence method, two strategies have been mainly used, the depolarization measurement of the emission in time domain or under steady state, and the time dependent (dynamic) Stokes shift measurement.[78-a] However the fluorescence method can probe only emissive excited states and the time resolution of such experiment may be limited by the detection system rather than the excitation pulse width. (If the up-pumping technique is used, the limitation of the low time resolution can be removed. However for that, fluorescence should be rather strong.) The transient absorption technique has an advantage in the time resolution but still it cannot be applied to any systems because the optical transition is essential in the detection and also suffered from the low sensitivity. By the TG method based on a similar idea (using the population grating), Phillion et al. have measured the orientational relaxation times of R6G in various alcohols in picosecond time scale by the TG method in 1975.[29] Owing to the excellent sensitivity and time resolution, this TG method have been applied to many systems so far.

Development of intense pulsed laser opened a new field in these researches; that is, ultrafast nonlinear spectroscopies enabled us to observe the ultrafast reorientational motion directly without real excitation (OKE). As shown in eq.(3-5), the TG signal is directly related to $\chi^{(3)}$, which is proportional to the peak-null difference of the refractive index. Since the molecular motion is related to the refractive index (dielectric constant), the autocorrelation function of the dielectric constant ($C^{\varepsilon\varepsilon}(t)$) is related to the dielectric response function $G^{\varepsilon\varepsilon}(t)$ as [78-80]

$$I(q,\omega) \propto (kT/\omega)\mathrm{Im}[G^{\varepsilon\varepsilon}(q,\omega)] \qquad (3\text{-}9)$$

and

$$C^{\varepsilon\varepsilon}{}_{ijkl}(t)=<\varepsilon_{ij}(0)\varepsilon^*{}_{kl}(t)>/<\varepsilon_{ij}\varepsilon^*{}_{kl}>$$
$$=-kT\int G^{\varepsilon\varepsilon}(t')dt'$$

where $G^{\varepsilon\varepsilon}(q,\omega)$ is related to $G^{\varepsilon\varepsilon}(t)$ by the Fourier transfer. $C^{\varepsilon\varepsilon}(t)$ can be projected out in terms of the reorientational contributions as

$$C^{\varepsilon\varepsilon}(t) = C^{OR}(t)+C^{II}(t)+C^{CT}(t)$$

The first term is the orientational contribution, the second one interaction induced term, and the last one the cross term. Here the correlation between the orientation and translation motions is neglected. When the pair correlation is neglected, C^{OR} is now simply the single-molecule polarizability anisotropy

$$C^{OR}(t) \approx C_2(t)=<P_2[u_i(0)u_j(t)]> = (3/2)<\cos^2\theta(t)-1/3> \qquad (3\text{-}10)$$

P_2 is the second order Legendre polynominal and $u_i(t)$ is a unit vector along the axis of the i-th molecule. As a result, the orientational dynamics can be extracted from the dynamics of the TG signals.

Dynamics of liquid crystals [79] and many simple solutions [80] have been studied using this techniques. Deeg and Fayer have beautifully demonstrated the polarization selection method in studying the molecular reorientation without disturbed by the thermal effect which could be involved in the signal by multiphoton absorptions (Fig.27). A different version of the grating technique in a frequency domain can be also used for the studies of ultrafast motions.[81]

Sengupta et al. have investigated local dynamics of a polymer (poly(2-vinylnaphthalene)) using the femtosecond TG method with the polarization selection (Fig.28).[82] In the solid phase, the nuclear response was fitted well by

$$G^{\varepsilon\varepsilon}(t)=C_1\{\exp(-k_b t)-\exp(-k_c t)\}$$

where k_b and k_c are the rise and decay rate constants of the nuclear response and the k_b^{-1}=140 fs decay is observed. The decay is attributed to librational damping

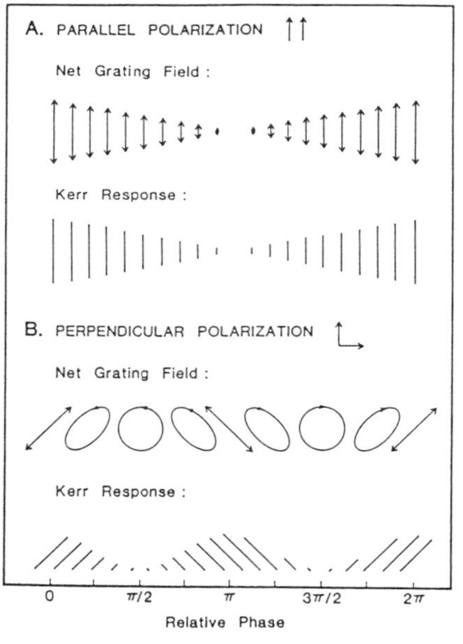

Fig.27 An example of polarization selection of the excitation beams.[79,a] (A) Polarization of both beams are parallel. In this case, the intensity grating could be induced if the materials absorb the light. Then the acoustic oscillation and thermal grating signal appear. (B) Perpendicular case. Although the direction of the polarization sinusoidally varies depending on the phase between two light waves, the overall light intensity remains constant. Therefore the TG signal under this condition should be free from the thermal effect, while the Kerr response can be probed as the TG signal.

of the naphthyl side groups, whose local structure is precisely determined by the packing constrains (no room for orientational freedom). In the CCl_4 solution, there are two other slower orientational relaxation components besides the librational decay.

$$G^{\varepsilon\varepsilon}(t)=C_l\{\exp(-k_b t)-\exp(-k_c t)\}+C_r\{1-\exp(-k_b t)\}\{\exp(-k_1 t)+\exp(-k_2 t)\}$$

Interestingly, these decays are temperature- independent in the range of -3 to 55 °C, although the bulk sheer viscosity of the solution and even the viscosity of pure CCl_4 change considerably. This observation indicates that the dynamics of the naphthyl groups are controlled by the local intramolecular constrains, namely, the orientational anisotropy does not decay by the diffusive motion but rather by

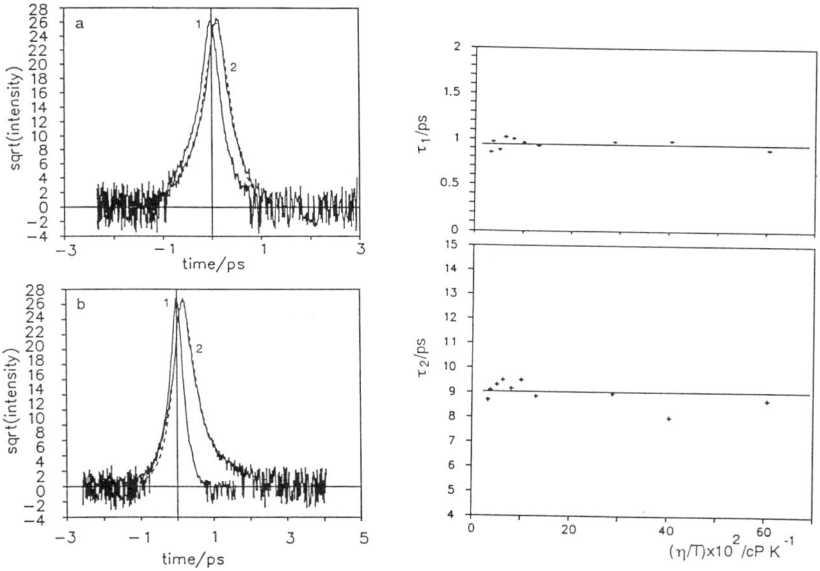

Fig.28 (left) The TG signal due to the Optical Kerr effect of poly(2-vinylnaphthal ene) (P2VN). The curves 1 and 2 are the electronic and nuclear responses, respectively, selectively probed by controlling the polarization of the pump and probe beams. The upper figure represents the response of the P2VN glass and the lower one P2VN/CCl_4 solution at 25 °C. (Right) Nonhydrodynamic behavior of the reorientational decay in P2VN/CCl_4 solution from -3 to 55 °C of the (a) fast component and (b) slow component.

relaxation back to the initial local geometry. A similar phenomenon was observed in the orientational dynamics of phenyl side groups in poly(methyl-phenylsiloxan e).[83] After the librational relaxation, phenyl side group orientational dynamics are temperature- and viscosity-independent. Again the lack of the η/T dependence (η; viscosity) is attributed to the well defined local structures that exist for a time much longer than the time scale of the observed relaxation.

3-4. Translational diffusion of chemically active species

One of the most notable achievements accomplished by the TG method is the detection of the translational movement of molecules. Traditionally, the diffusion constant (D) has been measured by various techniques, such as the tracer diffusion, NMR, a capillary method, and so on.[84] A difficulty in measurements by these techniques comes from a fact that generally translational diffusions in condensed phase are very slow processes. For example, a typical D of a medium size organic molecule in nonviscous liquid is ca. 10^{-9} m^2s^{-1}. If one tries to measure D by detecting the spatial movement with an accuracy 1 mm, the measurement will require ca. 16 min. It will be no problem waiting this period as long as the molecule is stable and it will not be so difficult to keep the experimental conditions (such as temperature) constant. These traditional methods, however, cannot be applied to short-lived species. Furthermore if a solution is very viscous, or a diffusing species is very big (like a macromolecule), the period for the measurement could be too long to keep the experimental conditions constant and to be free from any mechanical vibrations. Several methods have been developed for aiming to overcome these difficulties; such as those based on the analysis of the transient effect of fluorescence [85], based on the analysis of magnetic field or chemically induced electron (nuclear) spin polarization (CIDE (N)P) effect [86,87], so called photochemically space-intermittency method [88,89], and so on. In these methods, a response after photoexcitation (fluorescence, CIDE(N)P, product yield) is detected and the response is analyzed by appropriate theories using a diffusion constant as a fitting parameter. These methods will work fine for determination of D, when and only when the theories can exactly describe the response quantitatively and the other parameters except D are already known.

On the other hand, as shown below, the TG method for the measurement of D not only requires micro or milliseconds in time but also it can overcome many difficulties in the traditional methods. It enables us to study what was impossible to be performed so far.

The principle for the measurement can be described again by eq.(3-2). In this case, the response function, $\chi^{(3)}$, is governed by the spatial position of chemical species created by a photoexcitation (by photochromic processes or photochemical reactions) and also depleted by the reaction. The sinusoidal pattern of the excitation light induces the spatial modulation of the chemical species, which should have their own characteristic optical properties. Then the optical properties are the sinusoidal function of the position and $\chi^{(3)}$ is proportional to the peak-null difference of the refractive index, δn, and optical density, δk.

$$I_{TG}(t) \propto \alpha\{\delta n_{th}(t)+\delta n_i(t)-\delta n_j(t)\}^2+\beta\{\delta k_i(t)-\delta k_j(t)\}^2 \qquad (3\text{-}11)$$

where i and j represent the intermediate species (or product) and precursor , respectively, and δn_{th} term is due to the thermal effect which generally appears after photoexcitation and also contains the reaction enthalpy change. Since δn and δk are functions of the spatial distribution of the species and governed by the translational diffusion, the TG signal reflects the translational movement in the solution. The time profile of the grating can be monitored by the diffraction efficiency of the probe beam.

Generally the magnitudes of δn and δk terms due to the chemical species are not large (even compared with the δn_{th} term). In fact, they should not be large in order to prevent any secondary effect of the solute. Therefore, experimentally a weak TG signal should be detected with a nano to microsecond time resolution. For that purpose, we preferably use a pulsed laser system with a pulse width 10 ns for the excitation and a cw laser for the probe light. The excited wavelength should be tunable to select a desirable absorption band of the reactant. The TG signal is averaged by a digital oscilloscope for improving the S/N ratio.

This TG method for measuring D has the following merits over the conventional techniques.

(a) Since the TG method can monitor the spatial movement among the sinusoidal fringes whose distance is about a few micrometer, it takes only a short time to complete one experiment. As a result, a wide range of diffusion constant, D (10^{-7} to 10^{-15} m^2s^{-1}) can be measured without difficulty.

(b) The TG signal is detected without background light and, hence, it possesses high sensitivity, which enables us to study the solution dynamics with a very dilute concentration of the tracer. The dilute concentration of solute is favorable to ensure that the observed dynamics is truly controlled by the solvents.

(c) Since the grating can be created in a very small space, only a small amount of sample is required. It is important for a study of such as biological molecules.

(d) By choosing the direction of the fringe against a specific axis of the material, anisotropy of the diffusion can be investigated.

(e) If we vary the fringe spacing, Λ, the species should move by various distances before smearing out the grating. This procedure can be used for discriminating the diffusion process from a natural decaying process of unstable species. This Λ dependent TG measurement will give an opportunity to find an anomalous diffusion process, which shows a distance dependent diffusion coefficient.

(f) For extracting D from the data, it does not require any assumptions on the reaction, such as the quenching rate or mechanism, creating mechanism of spin polarization etc.

At first, this notable merit of the TG method is applied to the measurement of diffusion constants of tracers (small molecules), which should be photochromic, in very viscous solutions such as a liquid crystal [90] or polymer solutions [91,92]. Successively the application has been extended to many liquid crystals[93], gels [94], a biomolecule [95], glasses [96] and other simple solutions[97]. These results could be interpreted within the framework of the traditional theories of diffusions and the results have provided sound grounds for elucidation of the mass transport processes in these systems. The photochromisms in the above experiment occur by the (reversible) cis-trans photoisomerization or reversible bond breaking. For instance, Methyl Red, a derivative of azobenzene, is frequently used as the photochromic tracer molecule (Scheme 3). Their main interest was the dynamics of the matrix but not the dynamics of the diffusing species.

Recently, however, totally unexpected results, at least on the basis of the conventional theories of diffusion have been obtaining for unstable (chemically active) molecules by applying the TG method as shown below.

Scheme 3 Photoisomerization of Methyl Red

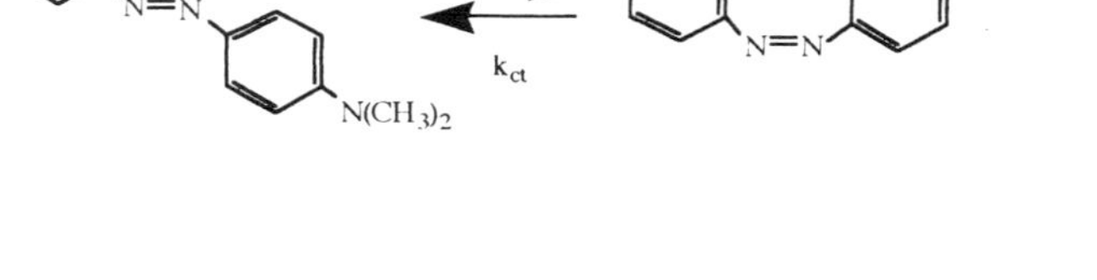

3-4-1. Measurement of D of transient species

In chemistry, the translational diffusion of chemically reactive species (such as radicals) plays an essential role in the whole reaction systems. Indeed, most of theories treating chemical reactions such as the reaction kinetics [98], electron transfer, energy transfer, magnetic field effect, CIDE(N)P and so on[86,87] contain D of intermediate species of the reaction. Nevertheless, due to the short-lifetime and the long period requirement for the measurement, it has been impossible to even estimate D of such species (except only a few examples [88,89]), and there was almost no data on that. Then how one has been able to use the theories which contain D of the intermediate species? Implicitly or explicitly, most of them have used an assumption that D of an intermediate species is the same as (or close to) that of a stable molecule with a similar size and shape, and used the data on such a stable one. In that case, another question arises; is that assumption correct?

It is true that all of the mass transport theories that have been developed till now do not concern if the species is radical or stable species.[84] For example, the Stokes-Einstein relation, one of the most frequently used and also the basis for many modified theories, is written as

$$D_{SE}=kT/f \qquad (3\text{-}12)$$

$$f= 6\pi\eta r\{(1+2\eta/\beta r)/(1+3\eta/\beta r)\}$$

where f is the friction coefficient, η; viscosity, r; radius of the solute, and β; coefficient for boundary condition (, which is 0 for slip boundary and ∞ for stick boundary condition). The shape of the diffusing species can be taken into a consideration by Perrin.[99] Many modifications based on this relation have been proposed.[100] After extensive efforts, the theoretically calculated D seems to reproduce the experimentally observed D for stable molecules in many cases. However, still we cannot answer to the question raised at the end of the previous paragraph; how one can know if these theories can be applied to unstable radicals without experimentally measuring D? We used the TG method by taking the advantage of the remarkable reduction of the required time to access D of unstable species for the first time.[101]

The time evolution of these species grating depends on the diffusion process of the species and concentration change due to chemical reactions. If the concentrations of the chemical species are constant during the measurement, the time dependencies of them are expressed by the diffusion equation.

$$dC(x,t)/dt = D\, \partial^2 C(x,t)/\partial x^2 \qquad (3\text{-}13)$$

where C(x,t) is the concentration of the species. The resultant TG signal is expressed by

$$I_{TG}(t) = \alpha\{\delta n_{th}\exp(-D_{th}q^2t) + \delta n_i \exp(-D_i q^2 t) - \delta n_j \exp(-D_j q^2 t)\}^2 + \beta\{\delta k_i \exp(-D_i q^2 t) - \delta k_j \exp(-D_j q^2 t)\}^2 \qquad (3\text{-}14)$$

where subscripts i and j, respectively represent the intermediate species (or product) created by the photoexcitation and the reactant, $D_{i(j)}$ is the diffusion constant of the i(j)-species in the solution.

The assumption of the constant concentration of species i (reactant j should be stable without the excitation light) can be experimentally checked by the measurement of the decay rate constant of the TG signal (k) as a function of q^2 or by directly monitoring the concentration from, for example, the transient absorption or EPR. If a chemical reaction from the intermediate species proceeds non-negligibly during the period of the TG measurement, the temporal profile of the TG signal deviates from eq.(3-14) and the plot of k vs. q^2 should not be linear or should possess non-zero intercept with the ordinate. It is important to stress that D can be determined from the slope of the decay rate constants against q^2 plot even if the intercept has a finite value as long as the plot shows linear dependence. If the reaction proceeds by the second order kinetics with a rate comparable to the diffusion, the temporal profile can no longer be expressed by eq.(3-14) and we cannot determine the decay rate constant in a usual sense.

The relative intensities of these terms in eq.(3-14) are determined by the magnitudes of $\delta n_{i(j)}$ and $\delta k_{i(j)}$. Since both of them are related to each other through the Kramers-Kronig relation, the species that possess strong absorption bands near the probe wavelength give stronger TG signals. If the absorption bands of the species are located in the UV region with small extinction coefficients, the contribution of the TG signal would not be significant or probably not be observed. This is an important criterion for the assignment of the chemical species in the TG signal.

3-4-2. Anomalous diffusion of radicals

A typical example of the TG signal in the chemical reaction system is shown in Fig.29 for pyrazine/2-propanol.[101] First, a strong TG signal which is assigned to the thermal grating is observed and it decays to the baseline with a time constant of the thermal diffusion among the fringes. After reaching to the baseline it rises and then decays again. Apparently the slow time development is due to the mass diffusion. The dip between the thermal grating and the population grating is a consequence of the interference effect between these components. Since the thermal grating is the pure phase grating with negative sign, the species grating should consist of the phase grating with positive sign for the stronger component. After the thermal grating, the square root of the TG signal is well expressed by two exponentials with different signs of the pre-exponential factors.

$$I_{TG}(t)^{1/2} = a_s \exp(-k_s t) + a_f \exp(-k_f t) \qquad (3\text{-}15)$$

where subscripts s and f stand for the slow and fast components, respectively and $a_f < 0 < a_s$, $|a_s| > |a_f|$. Now the assignment of the chemical species appeared in the TG signal is very important for further discussions on the diffusing process in solution. We have shown that the chemical species which gives rise to the rise and decay components are assigned to pyrazine and pyrazinyl radical created by the photoinduced hydrogen abstraction reaction based on the following grounds.[101]

(a) Pyrazine is already known to create pyrazinyl radical in 2-propanol solution efficiently by light irradiation from the excited triplet state (Scheme 4).[102]

Scheme 4 The hydrogen abstraction reaction of pyrazine in 2-propanol. The $^3n\pi^*$ state of pyrazine is the precursor state.

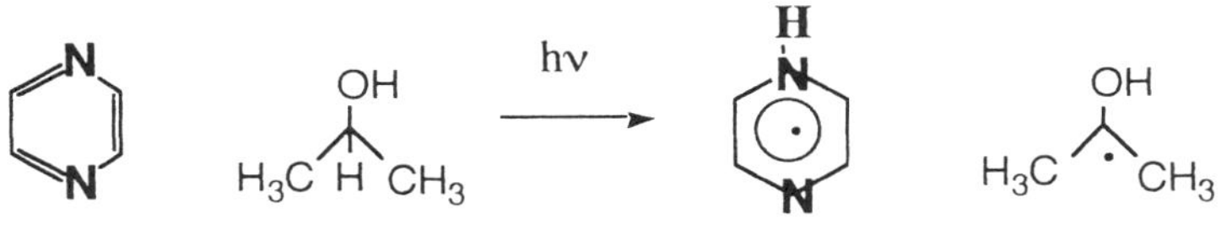

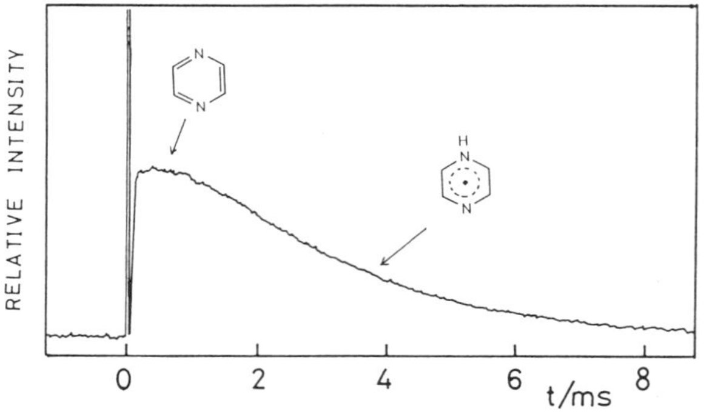

Fig.29 Typical temporal dependence of the TG signal which represents the mass diffusion of pyrazine/2-propanol at room temperature. The initial spike-like signal is the thermal grating signal.

(b) The slow decay component becomes weaker and decays faster under the air saturated condition. Since it is unlikely that dissolved nitrogen or oxygen changes the diffusion process in solution, the slower decay must be due to species which is quenched by oxygen and created from the T_1 state of the precursor molecule.

(c) The created radicals as well as the parent molecules (pyrazine and 2-propanol) possess their absorption bands at shorter wavelength than that of the He-Ne laser. Therefore all of δn_i and δn_j should be positive.

(d) Pyrazine has much stronger absorption bands in the near UV region compared with that of 2-propanol. Therefore the depletion of pyrazine should form a larger phase grating than 2-propanol.

(e) The transient absorption measurement reveals a strong absorption band near 350 nm which lasts in tens of ms order. The spectrum is similar to that of the pyrazinyl radical and there is no trace of absorption band which resembles that of 2-hydroxypropyl radical.

(f) The TG experiment on acetone/2-propanol, which produces the 2-hydroxypro pyl radical exclusively shows that the TG signal is relatively weak and the diffusional motion of this radical is much faster than the species which shows the slow decay.

(g) More directly, we could confirm that D of pyrazine determined by an independent method, NMR, agrees well with D of the rise component from the q^2 plot described below.

We believe that 2-propanol and 2-hydroxypropyl radical do not appear in the TG signal because of the smaller extinction coefficients in shorter wavelength and also possible reverse hydrogen abstraction reaction,[103] The 2-hydroxyprop yl radical could be extinguished by this mechanism efficiently before the TG signal appears.

Scheme 5 Reverse hydrogen abstraction reaction.

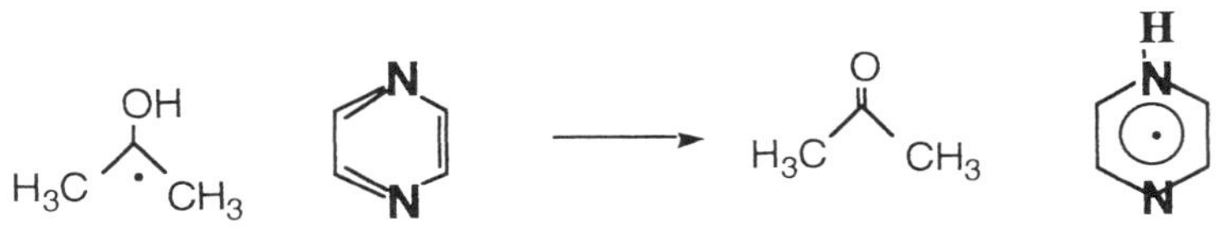

Figure 30 shows the q^2 dependence of k_s and k_f. The slopes for k_f and k_s represent D of pyrazine and pyrazinyl radical, respectively.

A striking feature of the results is the very small D of the transient radical compared with the parent molecule (pyrazine) even though the molecular volume and shape are not different each other so much (the difference in volume is the only one hydrogen atom in pyrazine).

In other hydrogen abstraction reaction systems as well as other photochemical reaction systems shown in Scheme 6,(a)-(d) as examples, similar TG signals are observed.[104] Based on the same analysis as that in the pyrazine case, we find that *the radical diffusions are generally slower than the parent molecules with similar volumes* at least in the hydrogen abstraction reactions. Studies on various solutes reveal that the ratios of D's of the parent molecules (D_{par}) to those of the radicals (D_{rad}) depend on the molecular size; for example, D_{par}/D_{rad} is about 4 for pyrazine (volume 0.13 nm^3) and about 1.3 for biquinoline (volume 0.61 nm^3) (Fig.31).[104]

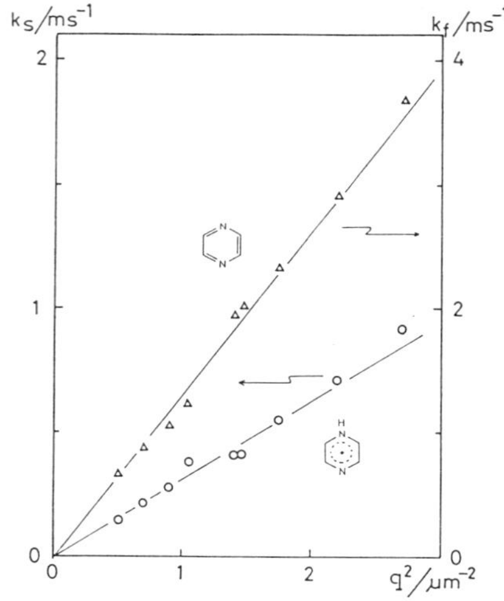

Fig.30 Plot of the decay rate constants of k_s (circles) and k_f (triangles) vs. q^2

Scheme 6 Several examples of photochemical reactions to create intermediate species for the measurement of D by the TG method. In (d), the lifetime of the first reactive species (diphenylcarbene) is too short to be detected and D of the secondary radical, diphenylmethyl radical, was measured in cyclohexane.

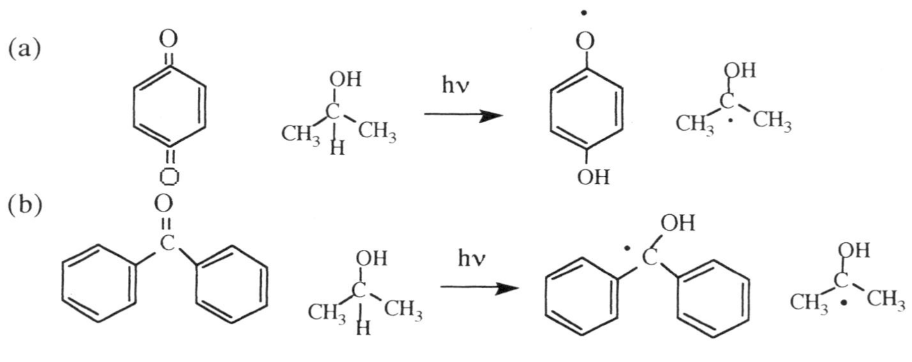

(c)

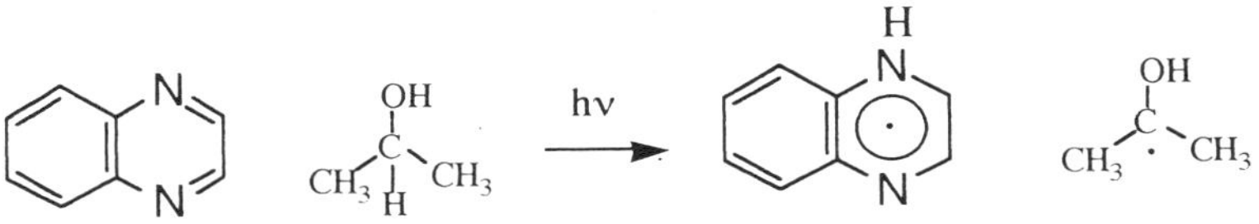

(d)

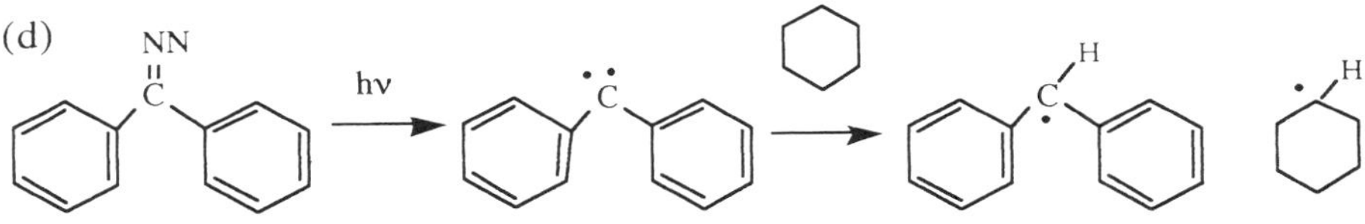

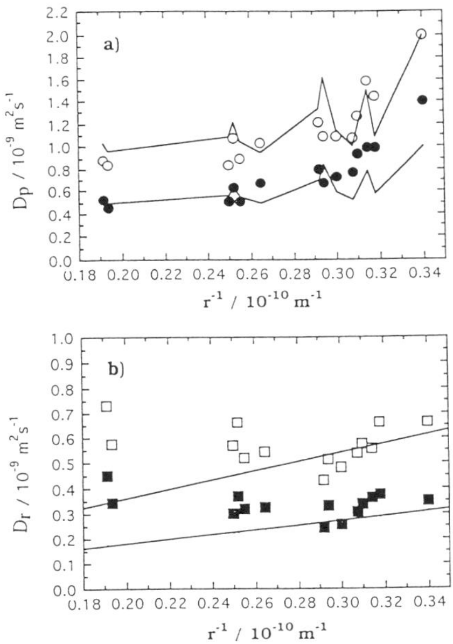

Fig.31 Molecular size dependence of D of the parent molecules (a) and those of the radicals (b) in 2-propanol (closed circles) and in ethanol (open circles). The solid lines represent calculated D's against r^{-1} from a modified SE equation in (a) (ref.104 in detail) and D_{SE} in (b).

This dependence results in an anomalous molecular size dependence of D_{rad}; D_{rad} increases (moves faster) with increasing the molecular size in a certain molecular size range. Neither this size dependence nor the smaller D's of the radicals compared with those of the non- radical molecules can be explained from any conventional theories of the diffusion even qualitatively.

There are several possibilities for explaining the anomalously slow diffusion of radicals. First, the slow diffusion could be due to an aggregate of radicals. However considering the dilute concentration of the radicals (10^{-4} to 10^{-6} M), we can exclude the aggregation process of the radicals during the observation time period. Second, the slow diffusion can be interpreted in terms of a specific radical-solvent (or -other solute) interaction. In the presence of such an interaction, the radicals may be accompanied by several solvent molecules during the movement. The increase of the apparent size of the radical should slow down the diffusion. One possible candidate for the origin of the attractive interaction is the hydrogen bonding interaction created by the formation of -OH or -NH groups in the radicals. However we have found that the radicals diffuse slower than the parent molecules even in nonpolar solvents, which hardly make the hydrogen bonding with the radicals. Furthermore the radicals which do not possess any -OH group such as shown in Scheme 4-d, show slower diffusion. Results of similar measurements at various temperatures indicate that the activation energies of D_{rad} as well as D_{par} are close to the activation energies of the viscosities.[105] This fact means that the diffusions of the parent molecules as well as the radicals are mainly governed by the hydrodynamic force of the solvent.

The above anomalous molecular size dependence can be explained in terms of the friction of the radicals and parent molecules. The friction coefficient of a species i, f_i, is defined by

$$f_i = kT/D_i \tag{3-16}$$

For the case of the SE equation, the Stokes friction coefficient is given by $f_{SE}=6\pi\eta r$ with the stick boundary condition. An excess friction, Δf, of the radical gained by the conversion from the parent molecule to the corresponding radical is given by

$$\Delta f = f_{rad} - f_{par} . \tag{3-17}$$

Here, we consider that the radical is diffusing accompanied by several molecules and that the effective molecular volume is defined by including the attached molecules around the radical. We further assume that one π electron has a capability of increasing the effective molecular volume only by a certain amount, V_0, because the unpaired π electron-solvent interaction may be constant even with increasing the molecular volume. Then the additional increase of the radius from the initial value, r, is given by

$$\Delta r \approx V_0/4\pi r^2 \tag{3-18}$$

If a relation $\Delta f \propto \Delta r$ is used, we obtain a simple relation between Δf and r as

$$\Delta f \propto 1/r^2$$

This relation is actually observed experimentally (Fig.32).

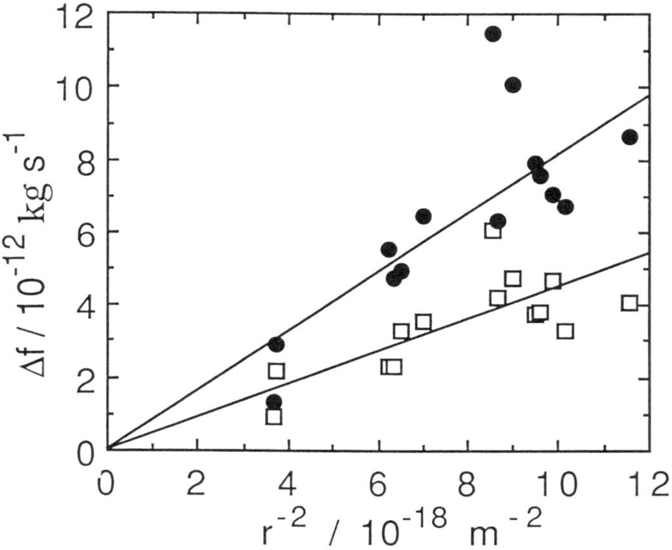

Fig.32 Plots of the excess friction Δf vs. r^{-2}.

So far, anomalously small D's of the radicals, the above mentioned molecular size dependence, and the temperature dependence of D's have been consistently explained in terms of an apparent radical volume expansion which is caused by the attractive intermolecular interaction between the radical and the surrounding molecules.

Slow diffusion was also observed for ionic radicals [106]. For instance, N,N,N',N'-tetramethyl-p-phenylenediamine (TMPD) is ionized by one photon process to produce a geminate pair of the TMPD cation radical and an electron (Scheme 7). After a geminate recombination process in the nanosecond range, the solvated electron is scavenged by a solvent and/or an impurity in the solution. The TMPD free cation survives for a rather long time in alcoholic solvents. The TG signal after photo-excitation of TMPD in ethanol is shown in Fig.33.

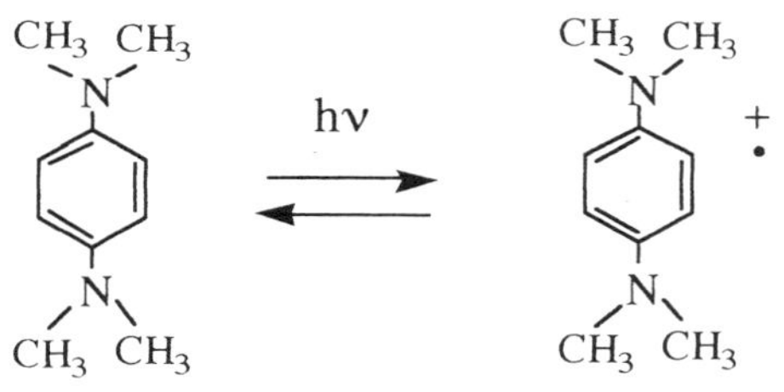

Scheme 7 Photoionization of TMPD

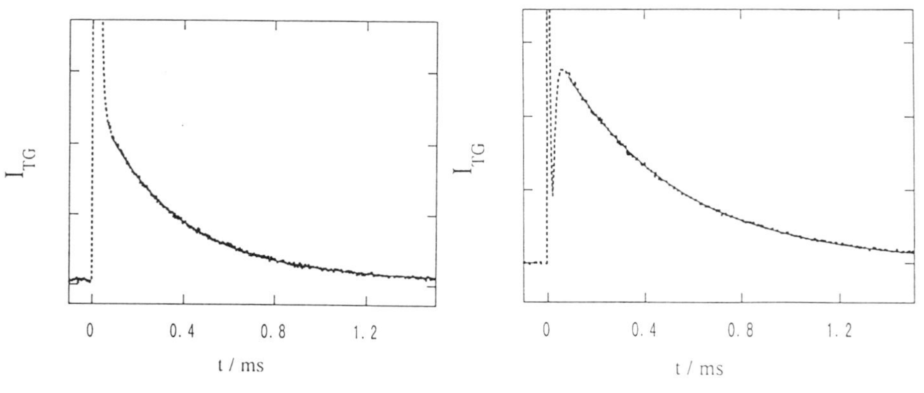

Fig.33 The TG signals (broken lines) after photoexcitation of TMPD in ethanol probed at (left) 633 nm and (right) 515 nm and the calculated lines (solid lines) with a single exponential function (left) and bi-exponential function (right).

The TG signal probed at 633 nm shows the thermal grating signal and species grating without an interference dip between them. This is explained by the dominant contribution of the amplitude grating (δk term in eq.(3-14)), because the TMPD cation possesses a strong absorption band around 600 nm.[107] The temporal profile of the species grating component can be analyzed by a single exponential function. Considering the strong absorption at the probe wavelength, we assigned the signal to the species grating of the TMPD cation. Indeed, when the wavelength of the probe light is shifted to shorter, the interference dip between the thermal and species gratings appears and the decay constant of the species grating is changed. These observations can be explained by the increase of the phase grating and, at the same time, by the increase of the TMPD contribution in the signal. From the slopes of the k vs. q^2 plots, diffusion constants of TMPD and TMPD cation radical are determined to be 1.8 m^2s^{-1} and 0.91 m^2s^{-1}, respectively. Again, the diffusion of the cation radical is slower than that of the parent molecule. Comparing these D's with those of the neutral radicals we presented above, we concluded that the slower diffusion of the TMPD cation is due to the unpaired electron.

Recently the stronger intermolecular interaction between the radicals and the surrounding molecules were suggested from an independent spectroscopic method; Raman scattering spectra.[108]

This anomalously slow radical diffusion must cast a serious problem to many relevant fields on chemical reactions because there is no reason to believe that interpretations on the kinetics of chemical reactions which have ever made without knowing the diffusion constant of intermediate species are correct.

4. Summary

In this review, I present chemical applications of the transient lens and transient grating spectroscopies, which belong to the four wave mixing techniques, for aiming to understand the intramolecular energy transfer processes and molecular dynamics in solution. In section 2, following the general description of the transient lens signal, I categorize the origins of the lens signal as population lens, density lens, temperature lens, Kerr lens, and volume lens based on our studies. For each component, methods for discriminating and for separation from other sources are presented. Furthermore, each lens signal is used for elucidating various photophysical properties after the light-matter interaction. This is a rather newly developed and very convenient method with a few simple optics and it still provides valuable information of $\chi^{(3)}$ as much as those from other techniques. I hope that this review will provide a trigger and a basis for further studies using this method in different fields of nonlinear optics.

The TG method is one of the frequently used four wave mixing techniques. It has been applied to investigations of various dynamics and characters of materials. In this review, I present only on topics of ultrafast temperature rise detection and investigations of molecular dynamics in solution. The ultrafast temperature rise detection by this method or by the TrL method will be used for the study of the intramolecular energy dynamics in fast time scale as well as the intermolecular energy transfer processes including the electronic-vibrational-(rotational-)translational transfer. In the molecular dynamics topic, two recently observed unusual phenomena were discussed; molecular motions which are independent of the bulk viscosity in ultrafast time scale and translational diffusion of the chemical active species. In particular, although the exact mechanism is not established yet, the result of the anomalous slow radical diffusion must be relevant in many related fields. A theory which incorporates the molecular character (such as radical) should be constructed. Apparently such a theory should be based on the experimental results on D's of the radicals, excited state, ions and so on. Since the diffusion processes of the chemically active species are involved in many theoretical studies on chemical reaction, the fact of the slow diffusion of the radicals, which is revealed for the first time, may change the traditional interpretations.

Finally a temporal behavior of the TG signal dealt in this review is schematically illustrated in Fig.34. Just after the photoexcitation, the temperature is raised by the vibrational relaxation in the upper state and the raise is appeared as the Temp.G signal. Since the acoustic (Brillouin) wave is relatively weak than the Rayleigh wave in Temp.G, the signal rises with the time constant of the deactivation (heat releasing) processes in the upper state. Then the deactivations of the lowest excited singlet state followed by the triplet state appear as the Temp.G and Dens.G signals. First the acoustic oscillation is dominant for the Dens.G signal. From the period and the damping rate, the acoustic velocity and the viscoelastic parameters can be measured. After the complete damping of the acoustic wave, the Rayleigh wave (diffusive mode) component remains and it decays with a time constant of the thermal diffusion time. If a photochemical reaction takes place, the enthalpy change can be another source of the thermal signal.

Besides the thermal effect, various molecular dynamics contribute to the signal (Fig.34(b)). In the fastest time region, the electronic response of OKE appears. It is followed by the nuclear response OKE, which reflects the reorientational motions. The population grating signal due to the excited singlet and triplet state also contribute to the signal if the pumping is under the resonant condition. In a later time region, the translational diffusion of the metastable excited states or new species created by a photochemical reaction is observed.

A similar feature is expected from the TrL signal. In a wide time range, a variety of characteristic dynamics appear in these TrL and TG signals and from the dynamics we can study many fundamental rules which are important in chemistry (and of course in many scientific fields).

Acknowledgement

Author is indebted to Prof.N.Hirota for helpful discussions on the contents in this review.

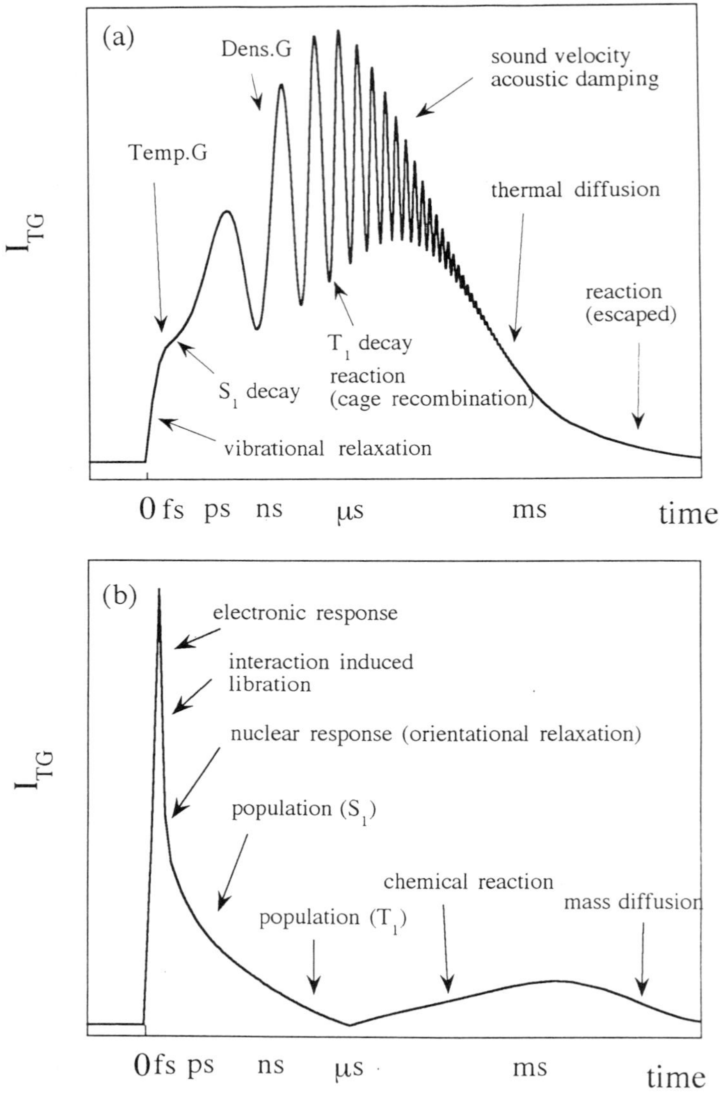

Fig.34 Temporal behavior of the TG signal in a wide time scale due to (a) the thermal effect and (b) non-thermal effect. The time axis is arbitrary scaled so that the temporal profile can be clearly seen. Just an approximate order of time is shown.

References

1. R.W.Boyd,"Nonlinear Optics",Academic Press, London,(1992).
   Y.R.Shen,"The principle of nonlinear optics", John-Wiley&Sons, NewYork(1984).
2. P.N.Prasad,D.J.Williams,'Introduction to nonlinear effects in molecules and polmers' ,Wiley, New York, 1991.
3. "Phenomena induced by intermolecular interactions",ed.by G.Birnbaum,Plenum Press,New York,(1985).
   L.C.Geiger,B.M.Ladanyi,J.Chem.Phys.,**87**,191(1987).
4. J.R.Whinnery,Acc.Chem.Res.,**7**,225(1974).
   H.L.Fang,R.L.Swofford, in "Ultrasensitive Laser Spectroscopy", ed. by D.S. Kliger,Academic Press, New York (1983)
   D.S.Kliger, Acc.Chem.Res.,**13**,129(1980).
   S.E.Braslavsky,G.H.Heibel, Chem.Rev.,**92**,1381(1992).
5. M.Terazima, T.Hara, N.Hirota, J.Phys.Chem.,**97**,10554(1993).
6. J.F.Power,Appl.Opt.,**29**,52, and 841(1990).
   S.E.Bialkowski, Appl.Opt.,**24**,2792(1985).
7. A.Yariv, "Quantum Electronics",John-Wiley, NewYork,(1989).
8. J.P.Gordon, R.C.C.Leite, R.S.Moore, S.P.S.Porto, J.R.Whinnery, Bull.Am.Phys.Soc.,**9**,501(1964).
9. J.P.Gordon, R.C.C.Leite, R.S.Moore, S.P.S.Porto, J.R.Whinnery, J.Appl.Phys.,**36**,3(1965).
10. P.R.Longaker, M.M.Litvak, J.Appl.Phys.,**40**,4033(1969).
    C.Hu, J.R.Whinnery, Appl.Opt.,**12**,72(1973).
    R.T.Bailey,F.R.Cruickshank,D.Pugh,W.Johnstone, J.Chem.Soc.,Farad, II,**76**,633(1980).
    S.J.Sheldon, L.V.Knight, J.M.Thorne,Appl.Opt.,**21**,1663(1982).
    R.T.Bailey,F.R.Cruickshank,R.Guthrie,D.Pugh,I.J.M.Weir, Mol.Phys., **48**,81(1983).
    S.J.Jacobs, Chem.Phys.,**132**,71(1989).
11. R.T.Bailey,F.R.Cruickshank,D.Pugh,W.Johnstone,Chem.Phys.Lett., **59**,324(1978).
    A.Salazar,A.Sanchez-Lavega,J.Fernandez,J.Appl.Phys., **65**,4105(1989).
12. (a) K.Daree, Opt.Comm.,**4**,238(1971).
    (b) K.Fuke,A.Hasegawa,M.Ueda,M.Itoh,Chem.Phys.Lett.,**84**,176(1981).
    (c) S.F.Lebedkin,A.D.Klimov,Chem.Phys.Lett.,**190**,313(1992).
    (d) R.T.Cambron,J.M.Harris, J.Phys.Chem.,**97**,13598(1993).
13. D.R.Siebert,F.R.Grabiner,G.W.Flynn,J.Chem.Phys.,**60**,1564(1974).

J.R.Barker.T.Rothem,Chem.Phys.,**68**,331(1982).
P.L.Trevor.T.Rothem,J.R.Barker,Chem.Phys.,**68**,341(1982).
W.Braun,T.J.Wallington, Chem.Phys.Lett.,**140**,441(1987).
M.Xing-Xiao,X.Zhu-De, Chem.Phys.Lett.,**98**,563(1983).
B.M.Toselli,T.L.Walunas,J.R.Backer,J.Chem.Phys.,**92**,4793(1990).

14. (a) M.Scotoni,M.Zen,D.Bassi,A.Boschetti,M.Ebben,Chem.Phys.Lett., **155**, 233(1989).
(b) R.W.Redmond,K.Heihoff,S.E.Braslavsky,T.George,Photochem.Photobio, **45**,209(1987).
(c) F.R.Grabiner,D.R.Siebert,G.W.Flynn,Chem.Phys.Lett.,**17**,189(1972).
(d) K.Fuke, M.Ueda, M.Itoh, J.Am.Chem.Soc.,**105**,1091(1983).
(e) M.Terazima, T.Azumi, Chem.Phys.Lett.,**153**,27(1988).
(f) M.Terazima, T.Azumi, Appl.Phys.Lett.,**54**,2398(1989).
(g) M.Terazima, T.Azumi, J.Am.Chem.Soc.,**111**,3824(1988).
(h) M.Terazima, T.Azumi, J.Phys.Chem.,**94**,4775(1990).
(i) M.Terazima, T.Azumi, Appl.Phys.Lett.,**57**,1485(1990).
(j) M.Terazima, H.Kanno, T.Azumi, Chem.Phys.Lett.,**173**,327(1990).
(k) M.Terazima, N.Hirota, H.Shinohara, Y.Saito, J.Phys.Chem.,**95**,9080(1991).
(l) M.Terazima, T.Azumi, N.Hirota, J.Photochem.Photobio.,**65**,21(1992).
(m) X.Song, J.F.Endicott, Chem.Phys.Lett.,**204**,400(1993).

15. J.H.Brannor,D.Magde, J.Phys.Chem.,**82**,705(1978).
A.J.Ranponi,J.A.Caird,J.Appl.Phys.,**63**,5476(1988).
J.Shen, R.D.Snook, Chem.Phys.Lett.,**155**,583(1989).
J.Degan,K.Reinecke,H.-H.Schmidtke, Chem.Phys.,**162**,419(1992).
M.L.Lesiecki,J.M.Drake, Appl.Opt.,**21**,557(1992).

16. J.Stone, J.Opt.Soc.Am.,**62**,327(1972).
M.E.Long,R.L.Swofford,A.C.Albrecht,Science, **191**,183(1976).
R.L.Swofford,M.E.Long,M.S.Burberry,A.C.Albrecht, J.Chem.Phys., **66**,664(1977).
R.L.Swofford,M.E.Long,A.C.Albrecht, J.Chem.Phys.,**65**,179(1976).
Y.Mizugai,F.Takimoto,M.Katayama,Chem.Phys.Lett.,**76**,615(1980).
W.B.Jackson,N.M.Amer,A.C.Boccara,D.Fournier,Appl.Opt.,**20**, 1333(1981).

17. R.A.Leach,J.M.Harris, Anal.Chem.,**56**,2801(1984).
K.Miyaishi,T.Imasaka,N.Ishibashi, Anal.Chem.,**54**,2039(1982).
K.L.Jansen, J.M.Harris, Anal.Chem.,**57**,1698(1985).
C.D.Tran,Appl.Opt.,**40**,1108(1986).
S.E.Bialkowski, Z.F.He, Anal.Chem.,**60**,2674(1988).
A.I.Larkin,V.M.Petropavlovskii,M.Yu.Cherkasov,Opt.Spectrosc. **68**,630(1990).

18. A.J.Twarowski,D.S.Kliger, Chem.Phys.,**20**,253 and 259(1977).
L.D.Ziegler,B.S.Hudson, Chem.Phys.Lett.,**71**,113(1980).
K.Fuke,S.Sakamoto,M.Ueda,M.Itoh,Chem.Phys.Lett.,**74**,546(1980).
H.L.Fang,T.L.Gustafson,R.L.Swofford, J.Chem.Phys., **78**,1663(1983).
J.K.Rice,R.Wanderson, J.Phys.Chem.,**90**,6793(1986).
P.R.Salvi,P.Foggi,R.Bini,E.Castellucci,Chem.Phys.Lett.,**141**, 417(1987).
J.R.Guckert,R.W.Carr, J.Phys.Chem.,**90**,4286(1986).
19. (a) M.Terazima, T.Azumi, Chem.Phys.Lett.,**141**,237(1987).
(b) M.Terazima,S.Yamauchi,N.Hirota,J.Chem.Phys.,**83**,3234(1985).
20. R.M.Hochstrasser,D.A.Wiersman,J.Chem.Phys.,**56**,528(1972).
E.C.Lim,J.Stanislaus,J.Chem.Phys.,**53**,2096(1970).
T.Takemura,K.Yamamoto,I.Yamazaki,H.Baba,Bull.Chem.Soc.Jpn., **45**,1839(1972).
S.Yamauchi,T.Ueno,N.Hirota,Mol.Phys.,**47**,1333(1982).
21. M.Terazima, S.Yamauchi, N.Hirota, J.Chem.Phys.,**84**,3679(1986).
M.Terazima, S.Yamauchi, N.Hirota, O.Kitao, H.Nakatsuji, Chem.Phys., **107**,81(1986).
M.Terazima, T.Azumi, Chem.Phys.Lett.,**145**,286(1988).
22. M.Terazima, N.Hirota, J.Phys.Chem.,**96**,7147(1992).
23. C.A.Carter,J.M.Harris,Appl.Opt.,**23**,476(1984).
S.Wu, N.J.Dovichi, J.Appl.Phys.,**67**,1170(1990).
24. M.Sheik-Bahae,A.A.Said,E.W.VanStryland, Opt.Lett.,**14**,955(1989).
M.Sheik-Bahae,A.A.Said,T.-H.Wei,D.J.Hagan, E.W.VanStryland IEEE.J.Quantum Electron.,**26**,760(1990).
H.Ma,A.S.L.Gomes,C.B.deAraujo,Appl.Phys.Lett.,**59**,2666(1991).
A.Agnesi,G.C.Reali,A.Tomaselli,Opt.Lett.,**17**,1764(1992).
25. M.Terazima, T.Hara, N.Hirota, J.Phys.Chem.,**97**,13668(1993).
26. M.Franko, C.D.Tran, Anal.Chem.,**61**,1660(1989).;Chem.Phys.Lett., **158**,31(1989).
27. M.Terazima, Chem.Phys.Lett.,**230**,87(1994).
28. K.A.Selanger, J.Falnes, T.Sikkeland, J.Phys.Chem., **81**,1960(1977).
D.W.Phillion, D.J.Kuizenga, A.E.Siegman, Appl.Phys.Lett., **27**,85(1975).
29. D.W.Phillion, D.J.Kuizenga, A.E.Siegman, Appl.Phys.Lett., **27**,85(1975).
30. P.J.Schulenberg,W.Gartner,S.E.Braslavsky,J.Phys.Chem., **99**,9617(1995).
31. M.Terazima, T.Hara, N.Hirota, to be published.
32. S.Chen, W.A.Tolbert, D.D.Dlott, J.Phys.Chem.,98,7759(1994).
J.R.Hill, D.D.Dlott, J.Chem.Phys.,89,830,842(1988).
H.Kim, D.D.Dlott, J.Chem.Phys.,93,1695(1990).
33. B.J.Berne, R.Pecora, "Dynamics light scattering", John-Wiley&Sons, New York (1976).

34. M.Terazima, to be published.

35. (a) H.Eisenberg, J.Chem.Phys.,**43**,3887(1965).
 (b) E.Reisler,H.Eisenberg, J.Chem.Phys.,**43**,3875(1965).

36. G.I.Zaistev, Yu.I.Kyzylasov, V.S.Starunov, I.L.Fabelinskii, JETP Lett., **6**,255(1967).

37. M.E.Mack, Phys.Rev.Lett.,**22**,13(1969),

38. D.Pohl, Phys.Rev.Lett.,**23**,711(1969).

39. R.I.Scarlet,Phys.Rev.Lett.,**26**,364(1971)., Phys.Rev., **6**,2281(1972).

40. G.Maisano,D.Majolino, F.Mallamace, P.Migliardo, F.Aliotta,C.Vasi, F.Wanderlingh, Mol.Phys.,**57**,1083(1986).

41. R.J.D.Miller, Ann.Rev.Phys.Chem.,**42**,581(1991).
 L.Genberg,L.Richard, G.McLendon, R.J.D.Miller,Science,, **251**,1051(1991).

42. M.Terazima, N.Hirota, J.Chem.Phys.,**100**,2481(1994).

43. M.Terazima, Chem.Phys.,**189**,793(1994).

44. M.Terazima, to be published.

45. J.Prost, J.R.Lalanne, Phys.Rev.,**8A**,2090(1973).
 C.Flytzanis, Y.R.Shen., Phys.Rev.Lett.,**33**,14(1974).
 G.K.L.Wong, Y.R.Shen, Phys.Rev., **10A**,1277(1974).
 K.Sala, M.C.Richardson, Phys.Rev.,**12A**,1036(1975).
 J.Etchepare, G.Grillon, I.Thomazeau, A.Migus, A.Antonetti, J.Opt.Soc.Am., **2B**,649(1985).
 P.P.Ho, R.R.Alfano, Phys.Rev.,**20A**,2170(1979).

46. M.Terazima, Opt.Lett.,**20**,25(1995).

47. J.G.Fujimoto,T.K.Yee, IEEE,J.Quant.Electro.,**QE-22**,1215(1986).

48. H.Kogelnik, Bell.System Tech.J.,**48**,2909(1969).
 A.E.Siegman, J.Opt.Soc.Am.,**67**,545(1977).
 W.J.Tomlinson,G.D.Aumiller, Appl.Opt.,**14**,1100(1975).

49. K.A.Nelson, R.Casalegno, R.J.D.Miller, M.D.Fayer,J.Chem.Phys.,**77**,1144(1982).
 M.D.Fayer, Annu.Rev.Phys.Chem.,**33**,63(1982).
 K.A.Nelson, D.R.Lutz, M.D.Fayer, Phys.Rev.,**24B**,3261(1981).

50. J.Morais, J.Ma, M.B.Zimmt, J.Phys.Chem.,**95**,3885(1991).

51. R.W.Hellwarth, Prog.Quantum.Electron.,**5**,1,(1977).

52. J.Etchepare, G.Grillon, J.P.Chambaret, G.Hamoniaux, A.Orszag, Opt.Commun.,**63**,329(1987).

53. F.W.Deeg, M.D.Fayer, J.Chem.Phys.,**91**,2269(1989).

54. J.T.Fourkas, R.Trebino, M.D.Fayer, J.Chem.Phys.,**97**,69(1992).

55. G.P.Wiederrecht, W.A.Svec,M.P.Niemczyk, M.R.Wasielewski, J.Phys.Chem., **99**,8918(1995).,

56. W.Kohler, P.Rossmanith, J.Phys.Chem.,**99**,5838(1995).
P.Vohringer, N.F.Scherer, J.Phys.Chem.,**99**,2684(1995).
57. (a)H.J.Eichler, P.Gunter, D.W.Pohl, "Laser dynamic grating", Springer-Verlag, Berlin(1986).
(b) F.W.Deeg,J.Pinsl,C.Brauchle, J.Phys.Chem.,**90**,5719(1986).
(c) R.K.Grygier,P.A.Brugger,D.M.Burland, J.Phys.Chem.,**89**,112(1985).
(d) F.W.Deeg,J.Pinsl,Chr.Brauchle,J.Phys.Chem.,**90**,5715(1986).
(e) U.Schmitt,D.M.Burland, J.Phys.Chem.,**87**,729(1983).
(f) F.W.Deeg,J.Pinsl,C.Brauchle, J.Chem.Phys.,**76**,4502(1982).
(h) D.M.Burland,C.Brauchle, J.Chem.Phys.,**76**,4502(1982).
(i) G.C.Bjorklund, D.M.Burland,D.C.Alvarez,J.Chem.Phys., **73**,4321(1980).
58. J.R.Salcedo, A.E.Siegman, D.D.Dlott, M.D.Fayer, Phys.Rev.Lett., **41**,131(1978).
59. T.S.Rose, R.Righini, M.D.Fayer, Chem.Phys.Lett.,**106**,13(1984).
60. L.Gomez-Jahn, J.Kasinski, R.J.D.Miller,Chem.Phys.Lett., **125**,500(1986).
S.Scheider, R.Bierl, M.Seischab, Chem.Phys.Lett.,**230**,343(1994).
61. P.Y.Key, R.G.Harrison, V.I.Little, J.Katzenstein, IEEE,J.Quant.Electron., **QE-6**,641(1970).
62. H.Eichler, G.Salje, H.Stahl, J.Appl.Phys.,**44**,5383(1973).
K.Thyagarajan, P.Lallemand, Opt.Comm.,**26**,54(1978).
R.Yano,Y.Fukuda, T.Hashi,Chem.Phys.,**124**,315(1988).
63. W.Urbach,H.Hervet,F.Rondelez,Mol.Cryst.Liq.Cryst., **46**,209(1978).; and J.Chem.Phys.,**78**,5113(1983).
64. H.J.Eichler, G.Salje, H.Stahl, J.Appl.Phys.,**44**,5383(1973).
D.W.Phol, S.E.Schwarz, V.Irniger, Phys.Rev.Lett.,**31**,32(1973).
O.W.Kading,E.Matthias, R.Zachai, H.J.Fusser, P.Munzinger, Diamond and relat.Mater.,**2**,1185(1993).
A.Harata, H.Nishimura,T.Sawada,Appl.Phys.Lett.,**61**,132(1990).
C.D.Marshall,I.M.Fishman,R.C.Dorfman,C.B.Eom,M.D.Fayer, Phys.Rev.,**45B**,10009(1992).
I.M.Fishman,C.D.Marshall,J.S.Meth,M.D.Fayer,J.Opt.Soc.Am.,**8B**, 1880(1991).
65. (a) E.B.Cummings, Opt.Lett.,**19**,1361(1994).
(b) P.H.Paul, R.L.Farrow, P.M.Danehy, J.Opt.Soc.Am.,**12B**,384(1995).
(c) D.E.Govoni, J.A.Booze, A.Sinha, F.F.Crim, Chem.Phys.Lett.,**216**,525(1991).
(d) E.B.Cummings, I.A.Leyva, H.G.Hornung, Appl.Opt.,**34**,3290(1995).
(e) Y.Kimura,Kanda, M.Terazima, N.Hirota,Ber.Bunsenges.Phys.Chem., **99**,196(1995).
66. M.A.Buntine, D.Wchandler, C.C.Hayden,J.Chem.Phys., **102**,2718(1995).
67. R.L.Sutherland,E.Rea,L.V.Natarajan, T.Pottenger, P.A.Fleitz,

J.Chem.Phys.,**98**,2593(1993).

M.Zhao, Y.Cui,M.Samoc,P.N.Prasad, M.Runroe, B.A.Reinhard, J.Chem.Phys., **95**,3991(1991).

E.B.Cummings, Opt.Lett.,**19**,1361(1994).

D.J.McGraw, J.M.Harris, J.Opt.Soc.Am.,**B2**,1471(1985).

68. D.J.McGraw, J.M.Harris, Phys.Rev.,**34A**,4829(1986).

69. A.Dreizler,T.Dreier,J.Wolfrum, Chem.Phys.Lett.,**233**,525(1995).

70. R.J.D.Miller, in "Time resolved spectroscopy", ed. By R.J.H.Clark, R.E.Hester, John-Wiley&Sons, (1989).

L.Genberg, F.Heisel, G.McLendon, R.J.D.Miller, J.Phys.Chem., **91**,5521(1987).

L.Genberg, Q.Bao, S.Gracewski, R.J.D.Miller, Chem.Phys., **131**,81(1989).

71. J.R.Andrews, R.M.Hochstrasser. Chem.Phys.Lett.,**76**,207(1980).

72. M.Terazima, N.Hirota, J.Chem.Phys.,**95**,6490(1991).

M.Terazima, N.Hirota, Chem.Phys.Lett, **189**,560(1992).

73. M.Terazima, N.Hirota, J.Appl.Phys.,**73**,7672(1993).

74. X.R.Zhu, J.H.Harris,Chem.Phys.,**142**,301(1990).; ibid.,**124**,321(1988).

X.R.Zhu, J.H.Harris,Chem.Phys.Lett.,**186**,183(1991).;J.Phys.Chem., **97**,6650(1993). ;Chem.Phys.,**157**,409(1991).

R.T.Cambron, J.M.harris, J.Phys.Chem.,**99**,695(1995).; Anal.Chem., **67**,365(1995).

75. M.Terazima, N.Hirota, Chem.Phys.Lett.,**214**,541(1993).

76. R.J.D.Miller, Ann.Rev.Phys.Chem.,**42**,581(1991).

77. M.D.Zimmt, Chem.Phys.Lett.,**160**,564(1989).

J.Ma,G.B.Dutt,D.H.Waldeck,M.B.Zimmt, J.Am.Chem.Soc., **116**,10619(1994).

78. (a) M.Cho, S.J.Rosenthal, N.F.Scherer, L.D.Ziegler, G.R.Fleming, J.Chem.Phys., **96**,5033(1992).

(b) A.Waldman, U.Banin, E.Rabani, S.Ruhman, J.Phys.Chem., **96**,10842(1992).

(c) S.Ruhman, K.A.Nelson, J.Chem.Phys.,**94**,859(1991).

79. (a) G.Eyring, M.D.Fayer, J.Chem.Phys.,**81**,4314(1984).

(b) F.W.Deeg,S.R.Greenfield, J.J.Stankus, V.J.Newell, M.D.Fayer,J.Chem.Phys., **93**,3503(1990).

80. F.W.Deeg,J.J.Stankus,S.R.Greenfield,V.J.Newell, M.D.Fayer, J.Chem.Phys., **90**,6893(1989).

S.R.Greenfield, A.Sengupta, J.J.Stankus, M.D.Fayer, Chem.Phys.Lett., **193**,49(1992).

A.Henseler, E.Vauthey, Chem.Phys.Lett.,**228**,66(1994).

S.Ruhman,K.A.Nelson, J.Chem.Phys.,**94**,859(1991).

A.Waldman, U.Banin, E.Rabani, S.Ruhman, J.Phys.Chem., **96**,10842(1992).

Y-X.Yan, L-T.Cheng, K.A.Nelson,in "Advanced Nonlinear Spectroscopies" ed. by R.J.H.Clark and R.E.Hester, John-Wiley&Sons, (1988).

C.Kalpouzos, W.T.Lotshaw, D.McMorrow, G.A.Kenny-Wallace, J.Phys.Chem., **91**,2082(1987).

Y.J.Chang, E.W.Castner,Jr., J.Chem.Phys.,**99**,113(1993).

W.T.Lotshaw, D.McMorrow, C.Kalpouzos, G.A.Kenny-Wallace, Chem.Phys.Lett., **136**,323(1987).

81. R.Trebino, C.E.Barker, A.E.Siegman, IEEE,J.Quant.Electron., **QE-22**,1413(1986).
C.E.Barker, R.Trebino, A.G.Kostenbauder, A.E.Siegman, J.Chem.Phys., **92**,4740(1990).

82. A.Sengupta, M.Terazima, M.D.Fayer, J.Phys.Chem.,**96**,8619(1992).

83. A.Sengupta, M.D.Fayer,J.Chem.Phys.,**100**,1673(1994).

84. E.L.Cussler, "Diffusion", Cambridge University,Cambridge(1984).
H.J.V.Tyrrell, K.R.Harris, "Diffusion in liquids", Butterworths,London,(1984).
A.Abragam, 'The principle of nuclear magnetism', Oxford,(1961).

85. T.L.Nemzek, W.R.Ware, J.Chem.Phys.,**62**,477(1975).
R.Das, N.Periasamy, Chem.Phys.,**136**,361(1989).
A.D.Scully, S.Hirayama, D.hachisu, T.Tominaga, J.Phys.Chem.,**96**,7333(1992).

86. "Polarization and magnetic effects in radical reactions", ed.by Yu.N.Molin, Elsevier, Amsterdam (1984).

87. F.J.Adrian,J.Chem.Phys.,**94**,6251(1972).
J.B.Pedersen, J.H.Freed, J.Chem.Phys.,**58**,2746(1973).
H.J.Werner, Z.Schulten, K.Schulten, J.Chem.Phys.,**67**,646(1977).

88. R.M.Noyes,J.Am.Chem.Soc.,**81**,566(1959).
R.D.Burkhart, R.J.Wong,J.Am.Chem.Soc.,**95**,7203(1973).

89. S.A.Levison, R.M.Noyes,J.Am.Chem.Soc.,**86**,4525(1964).
R.D.Burkhart, R.F.Boynton, J.C.Merrill, J.Am.Chem.Soc.,**93**,5013(1971).

90. F.Rondelez, H.Hervet, W.Urbach, Chem.Phys.Lett.,**53**,138(1978).
H.Hervet, W.Urbach, F.Rondelez,J.Chem.Phys.,**68**,2725(1978).
W.Urbach, H.Hervet,F.Rondelez, J.Chem.Phys.,**83**,1877(1985).

91. H.Hervet,L.Leger,F.Rondelez,Phys.Rev.Lett.,**42**,1681(1979).
J.Zhang,B.K.Yu,C.H.Wang,J.Phys.Chem.,**90**,1299(1986).
A.V.Veniaminov,G.I.Lashkov,O.B.Ratner,N.S.Shelekhov,O.V.Bandyuk, Opt.Spectrosc.,**60**,87(1986).
J.A.Lee, T.P.Lodge, J.Phys.Chem.,**91**,5546(1987).
S.S.Gong,D.Christensen,J.Zhang,C.H.Wang, J.Phys.Chem.,**91**,4504(1987).

92. J.L.Xia, C.H.Wang, J.Chem.Phys.,**88**,5211(1988).
J.M.Widmaier,T.E.Ouriaghli,L.Leger,M.F.Marmonier, Polymer,**30**, 549(1989).

T.P.Lodge, J.A.Lee, T.S.Frick, J.Polym.Science,**B28**,2607(1990).
C.H.Wang,J.L.Xia, J.Chem.Phys.,**92**,2603(1990).
T.S.Frick,W.J.Huang,M.Tirrell,T.P.Lodge, J.Polym.Science, **B28**,2629(1990).
C.H.Wang,J.L.Xia, J.Phys.Chem.,**96**,190(1992).
93. H.Takezoe,S.Ichikawa,A.Fukuda,E.Kuze, Jpn.J.Appl.Phys.,**23**, L78(1984).
M.Hara,S.Ichikawa,H.Takezoe,A.Fukuda,Jpn.J.Appl.Phys.,**23**, 1420(1984).
W.Urbach,H.Hervet,F.Rondelez, J.Chem.Phys.,**83**,1877(1985).
M.Hara, H.Takezoe,A.Fukuda, Jpn.J.Appl.Phys.,**25**,1756(1986).
J.A.Wesson, H.Takezoe, H.Yu, S.P.Chen, J.Appl.Phys.,**53**,6513(1082).
F.Rondelez, Sol.State Commun.,**14**,815(1974).
K.Ohta, M.Terazima, N.Hirota, Bull.Chem.Soc.Jpn., in press.
94. J.A.Wesson, H.Takezoe, H.Yu,S.P.Chen, J.Appl.Phys., **53**,6513(1982).
95. K.W.Rhee,D.A.Gabriel,C.S.Johnson,Jr,J.Phys.Chem., **88**,4010(1984).
96. W.D.Dozier,J.M.Drake,J.Klafter, Phys.Rev.Lett.,**56**,197(1986).
97. T.Chang,H.Kim,H.Yu,Chem.Phys.Lett.,**111**,64(1984).
D.G.Miles,Jr.,P.D.Lamb,K.W.Rhee,C.S.Johnson,Jr, J.Phys.Chem.,**87**,4815(1983).
M.Terazima, K.Okamoto, N.Hirota, J.Phys.Chem.,**97**,5188(1993).
98. H.Schuh, E.J.Hamilton,jr.,H.Paul, H.Fisher, Helv.Chim.Acta, **57**,2011(1974).
H.Schuh, H.Fisher, Int.J.Chem.Kinet.,**8**,341(1976).
J.J.Tria, R.H.Johnsen, J.Phys.Chem.,**81**,1274(1977).
M.Sitarski, Int.J.Chem.Kinet., **8**,125(1981).
M.Lehni, H.Schuh, H.Fisher, Int.J.Chem.Kinet.,**11**,705(1979).
H.Paul, C.Segaud, Int.J.Chem.Kinet.,**7**,637(1980).
M.Lehni, H.Fisher, Int.J.Chem.Kinet.,**15**,733(1983).
S.Sridharan, J.A.McCammon, J.B.Hubbard, J.Chem.Phys.,**90**,237(1989).
99. F.Perrin, J.Phys.Radium.,**7**,1(1936).
100. A.Spernol, K.Wirtz, Z.Naturforsch.Teil.**A89**,522(1953).
H.J.Sandhu, J.Mag.Reson.,**17**,34(1975).
101. M.Terazima, N.Hirota, J.Chem.Phys.,**98**,6257(1993).
102. P.N.Moorthy, E.Hayon, J.Phys.Chem.,**78**,2615(1974).
D.V.Bent, E.Hayon, P.N.Moorthy,J.Am.Chem.Soc.,**97**,5065(1975).
S.Basu, K.A.McLauchlan, G.R.Sealy, Chem.Phys.Lett.,**88**,4631(1984).
103. T.Ohara, N.Hirota, to be published.
104. M.Terazima, K.Okamoto, N.Hirota, J.Phys.Chem.,**97**,13387(1993).
M.Terazima, K.Okamoto, N.Hirota, J.Chem.Phys.,**102**,2506(1995).
105. K.Okamoto, M.Terazima, N.Hirota, to be submitted.
106. M.Terazima, T.Okazaki, N.Hirota, Photochem.Photobio., in press.
107. N.Yamamoto, Y.Nakato, H.Tsubomura, Bull.Chem.Soc.Jpn.,**39**,2603(1966).

A.C.Albrecht, Acc.Chem.Res.,**3**,238(1970).
Y.Hirata, N.Mataga, J.Phys.Chem.,**88**,3091(1984).; *ibid.* ,**87**,3190(1083).; *ibid.* ,**87**,1680(1983).;*ibid.*, **89**,159(1992).
108. M.Terazima, H.Hamaguchi, J.Phys.Chem.,**99**,7891(1995).

PART TWO

Electronic Interactions & Interchromophore Energy Transfer

Electronic Interactions & Interchromophore Energy Transfer[†]

Gregory D. Scholes

and

Kenneth P. Ghiggino

School of Chemistry
The University of Melbourne, Parkville 3052 Australia

ABSTRACT: Electronic energy transfer (EET) from an initially excited donor chromophore to a suitable acceptor chromophore is a fundamental process in photochemistry, and plays a key role in a diversity of natural and synthetic light harvesting systems. To unravel the often complex entanglement of processes occurring in such systems, the mechanisms promoting EET must be well understood. In this work a detailed quantum mechanical model for the rate of EET is developed and analyzed, especially for the phenomena of superexchange-mediated EET and energy migration. The contributions by various EET mechanisms are investigated experimentally for a series of rigidly-linked bichromophores.

† A Ph.D. thesis written by G. D. Scholes describing work carried out under the supervision of K. P. Ghiggino at the University of Melbourne.

Contents

"But when it was seen in experiments, that's when it really became exciting. The miracle is that, in systems that are interesting, you can still understand behaviour in detail by a model with a small number of degrees of freedom."

It was Hohenberg, in the end, who brought the theorist and the experimenter together. ... Feigenbaum paid a call on Libchaber in Paris. They stood amid the scattered parts and instruments of Libchaber's laboratory. Libchaber proudly displayed his tiny cell and let Feigenbaum explain his latest theory. Then they walked through the Paris streets looking for the best possible cup of coffee.

Chaos, James Gleick.

Acknowledgments

GDS would especially like to thank Dr Ken Ghiggino for his generous support and assistance throughout my Ph.D. candidature. I also wish to thank Prof. Mike Paddon-Row for a very edifying introduction to "through-bond interactions" and for the "Paddon-Row molecules"; studied in Chapter 4. I thank Prof. Wilse Robinson for many enlightening discussions re radiationless transitions and the like (hence the evolution of Chapter 2), and Dr Dick Harcourt for his input into the development of the theory presented in Chapter 3, and for for teaching me, in a refreshingly enthusiastic spirit, some very useful quantum chemistry. Thanks also to Mr Andrew Clayton for many, many enjoyable and fruitful hours of discussions. Figures 3.1, 3.8 and 3.9 are reprinted with permission from Ref. 17 (Chapter 3); copyright (1994) American Chemical Society.

CHAPTER ONE

Introduction and Perspective

I. Proem

Interactions between chromophores (i.e., the light-absorbing group in a molecular system), when one is electronically excited, have import in a variety of instances. An excited state complex formed between two identical chromophores (which would in the ground state, and in the absence of external restraints, dissociate) - an excimer - is an example of a strongly-interacting system. Singlet state excimers play an important role in many areas of photochemistry and have been the subject of several recent investigations.[1,2] Related are exciplexes[3,4] (an excited hetero-complex), triplet state excimers[5] and so-called transannular interactions.[6-8] Transannular interactions are typified by molecular exciton interactions; the perturbation of the energy levels of a molecule in an array as compared to the lone moiety.[9-11] These are manifested as strong spectral shifts or splittings.

A further example of the impact of such interactions upon molecular spectroscopy and photochemistry is the phenomenon of interchromophore electronic energy (excitation) transfer, EET. Here, the electronic interactions introduced above act to promote the transfer of electronic energy from an initially-excited chromophore (the donor) to a proximate chromophore (the acceptor). Both electronic and dynamic aspects of this ubiquitous process, important in natural as well as synthetic systems, are the focus of this thesis.

Non-Radiative Processes. EET is a special instance of a non-radiative deactivation process, although, interestingly, a simple but predictive theory for EET was developed over 15 years before *intramolecular* non-radiative processes were understood. Much of the work presented in this thesis is based on a knowledge of non-radiative processes and related fundamental physical theories. It will therefore be relevant to commence by pedagogically reviewing some of the vast breadth of work in this area.

The mixing of an electronic configuration belonging to a discrete state of an atom or molecule with continuous spectrum configurations gives rise to autoionization, with this exact coincidence of energies rendering a treatment by ordinary perturbation theory inadequate. The basic approach to this problem was developed by Rice in 1933.[12,13] Such fundamental radiationless processes were examined further in the 1960s; paralleling and then merging with the development of molecular radiationless processes.[14,15] There were many other examples of such work appearing in the physics literature at this time which would later help form the mathematical basis for radiationless transition theory work, a noteworthy example being the paper by Kubo and Toyozawa investigating radiative and non-radiative transitions of a trapped electron in a crystal.[16,17]

To elucidate the dynamics of a quantum mechanical process (relevant to an observation that reflects behaviour on a microscopic scale, such as spectroscopic measurements) one usually tries to describe the time-evolution of an, initially, "stationary" system after application of a perturbation. One may proceed to calculate the probability that the system undergoes a transition from this initial state to any other state of the system by the action of the perturbation.*

A transition probability per unit time may then be determined for a perturbation which is independent of time, the case of physical interest being that where there is a (quasi-) continuous range of final state energy levels passing through the energy of

---

* See Heitler[17] p. 136 or Dirac[18] p.172.

the initial state.† The resultant fundamental rate expression, made famous by Enrico Fermi's lecture series[19] is the Fermi Golden Rule.

Once again, fundamental advances provide the foundation for description of complex molecular systems, but only after a physical interpretation has been elucidated. Much of the work presented in this thesis is based upon some basic equations of scattering theory,# together with an interpretation of molecular radiationless processes first presented by Robinson and Frosch.[22-27]

The basic model used in radiationless transition theory stems from the intuitive insight and reasoning of Robinson and Frosch.[22-24] It generally involves discrete states (primary levels), which carry all the oscillator strength, interacting with a quasi-continuum of states (the secondary levels) which may be, for example, high vibrational levels of a lower excited state. The model has since been the foundation for work explaining the lifetimes of molecules in rare gas matrices,[22] describing processes such as quantum beats,[28] electron transfer,[29-31] and now *supramolecular* energy transfer.

The key to the model is that the discrete states are *non-stationary*. As stated by Robinson and Frosch: the matrix elements for the electronic relaxation problem in the limit of weak coupling are determined by the presence of perturbations between electronic states which are almost, but not quite, stationary. This is because a stationary state may undergo transitions to other stationary states only by coupling with the radiation field, so that all time-dependent transitions between stationary states are radiative in nature. However, if the system is prepared in a non-stationary state of the *total* Hamiltonian, non-radiative "transitions" will occur. An important

---

† See Dirac[18] p. 178.

e.g. the work of Lippmann and Schwinger,[20] and of Gell-Mann and Goldberger[21]

feature of the model[%] is that Fermi's Golden Rule may be employed to find a rate for the non-radiative transition.

Several detailed physical elaborations of this basic model ensued, starting with the important work of Lin,[32,33] who further formalized the problem and attributed the perturbation (which acts to induce the non-radiative transition) to arise from deviations from the Born-Oppenheimer approximation, bringing about a coupling between vibrational and electronic motion. A series of papers followed (the key ones being by Bixon and Jortner,[32-36] and Chock, Jortner and Rice[37]) which, through elegant physics, provided insight into the phenomenon and provided new avenues for quantifying the theory. An interesting example of application of the theory is in the untangling of the higher vibrational levels of the naphthalene first singlet transition from the origin of the second singlet transition.[26,38-40]

EET Theory. A quantitative theory for the EET process was first elucidated by Förster in 1948.[41] Interestingly, although EET is a special case of a non-radiative process, this development came 15 years before the work of Robinson and Frosch. The reason for this is that EET is an intermolecular process, so it makes sense to think of preparing a *non-stationary* initial state of the system, which would be the stationary states of an isolated molecule (i.e., the perturbation is "turned on"), which, when perturbed by the presence of a second molecule; become non-stationary. The correct stationary states of the dimer then refer to a delocalization over both components of the system. However, whether such thinking is applicable to a *single molecule* was hotly debated. In the end it was the Robinson and Frosch approach that won general acceptance.

% Since it describes discrete states interacting with,or via, a (quasi-) continuum.

One of the key themes of Försters thinking was that of a *non-additivity of excited state properties*, meaning that the photophysical properties of a system may be quite different from those of its components.[42] It will be seen that such a theme also underlies the present work. Thus Förster distinguished, from a phenomenological viewpoint, three instances of this concept: (A) the strong coupling case, where major alterations in the absorption spectrum are observed; (B) the weak coupling case, where individual vibronic lines are perturbed; and (C) the very weak coupling case, where there are little or no alterations in the absorption spectrum.

In this latter case of very weak coupling it is unambivalent to consider the excitation as being temporarily localized then subsequently transferred to another component of the system. These ideas will also be examined and further explicated in Chapter 2.

In a system of N *noninteracting* (identical) chromophores, the energy of the system would be unchanged by transferring excitation from one chromophore to any other in the system:

i.e., $$A_1A_2^* \ldots A_N = A_1A_2 \ldots A_i^* \ldots A_N$$

However, in a real system, small interchromophore interactions remove this N-fold degeneracy. The correct stationary states for the excited system correspond to a delocalization of the energy over *all* chromophores,[9,11]

i.e., $$(A_1A_2 \ldots A_N)^*$$

Such a concept was first introduced in 1931 by Frenkel[43] in connection with the theory of transformation of electromagnetic radiation into heat in crystalline argon. The so-called oriented gas model was also discussed by Peierls.[44] In 1948 Davydov[45] developed the Frenkel-Peierls method for the treatment of the molecular exciton, in particular the problem of the electronic excited states of crystalline naphthalene.

It is also such interactions which promote the transfer of excitation between non-identical chromophores. From this starting point Förster showed that, in the very weak coupling case, the rate of EET could be given by a Fermi Golden Rule rate expression involving this interaction matrix element. The final expression was cast in terms of absorption and emission spectra of the acceptor and donor chromophores respectively.

The Förster expression for EET, then, is ideally suited for quantitative interpretation and prediction of experimental observations. It is appropriate for situations where the donor and acceptor chromophores have allowed singlet-singlet transitions, and are at reasonably large separations (ca. greater than 8 to 12 Å).

As such, the Förster theory is employed in almost every experimental study of EET; in systems ranging from model bichromophores to aromatic polymers to solutions of free dye molecules to monolayers and Langmuir-Blodgett films.

What happens, however, when the donor and acceptor chromophores do not obey these criteria demanded by the Förster theory? Dexter was concerned with the problem of sensitized luminescence in the solid state,[46] in particular, the situation where an impurity atom in a crystalline host may be sensitized by energy transfer (from another impurity atom or from the host) even though its electronic transition is electric dipole-forbidden. The "many-body" nature of solid state problems was emphasized, but Dexter's interest was in the mechanism of the energy transfer process in such situations.

Dexter's approach was to write out properly antisymmetrized initial and final state wavefunctions of the donor-acceptor pair. He then considered only two electron integrals, and showed that the Coulomb interactions (incorporated in the dipole approximation in Förster's theory) could be obtained without antisymmetry requirements. However, with antisymmetry considerations another interaction, the exchange integral, was recovered. The magnitude of this interaction is critically dependent upon the degree of overlap between donor and acceptor wavefunctions (i.e., they must be closely separated) and it was stated that exchange effects become important for magnetic dipole-allowed (and higher order) transitions.

The most important result of Dexter's work is that the exchange interaction (or more generally, as will be seen from this work, interactions depending upon orbital overlap between donor and acceptor chromophores) are attenuated approximately exponentially with increasing separation. Experimentally, the Dexter exchange interaction is, however, difficult to quantify.

What is seldom recognized is that when Dexter's work is used to interpret interactions between chromophores with *allowed* electronic transitions (remember, his paper discussed the sensitization of *forbidden* transitions) is that the exchange and Coulombic interactions act in conjuction, *not* exclusively. In some cases (due to the opposite signs of the respective interactions) this may appear to lead to a paradox - where the interaction vanishes at a certain close separation between donor and acceptor. However, as will be demonstrated by the new theory given in Chapter 3, such a situation would be exceptional.

Lin and coworkers have established a firm connection between the Förster theory for EET and non-raidative rate theory.[47-49] This is discussed in Chapter 2.

Craig and Thirunamchandran[50-54] as well as Andrews and co-workers[55-58] have employed a quantum electrodynamical (QED) model to describe Coulombic (specifically dipole-dipole) EET. Molecular QED deals with the dynamics of molecules interacting with an electromagnetic field, the key aspect being that the electromagnetic field is quantized and forms part of the entire dynamical system. In the case of EET, the molecules (or chromophores) are the sources of the field, which acts to mediate the electronic coupling through creation of a virtual photon. The mechanism, then, by which excitation energy is transferred from one molecule to another becomes (in a sense) more lucid.

The contribution of this work certainly does not just lie in esoterics, or physical appeal. In fact, some important advances have been made. For example, Andrews has employed a relatively simple approach to generalize the mechanisms of intermediate- and long-range EET as limits of a unified theory involving virtual photon coupling.[56] The intermediate-range mechanism (the near-zone limit of the theory) is equivalent to the theory of Forster, whilst the long-range (wave-zone limit) explicates the modification of the distance-dependence of the EET rate from R^{-6} to R^{-2} as relativistic retardation effects invoke a two-step trivial mechanism (uncorrelated emission and absorption events). Continuing from this work, Andrews and Juzeliūnas[57] examined the residual fluorescence anisotropy following EET over arbitrary interchromophore separations; establishing a connection with the depolarization associated with reabsorption.

Superexchange (Relayed) EET. Electronic coupling need not be realized soley by a direct interaction between donor and acceptor chromophores. If there are additional components present in the system (i.e. we are considering a *supramolecular* system) which couple electronically to the donor and acceptor chromophores, then EET may be mediated indirectly - by a multi-step virtual process, being relayed by these intervening moieties. Such a mechanism is known to promote long-range electron transfer, and recent evidence suggests that it may act to considerably enhance rates of EET, where typical relay units may include σ bond "spacer" groups, intervening chromophoric substituents, or a lipid bilayer.

The first evidence that a relay-type mechanism could significantly enhance rates of energy migration over large donor-acceptor separations was reported by El-Sayed, et al. for dilute mixed aromatic crystals;[59] their observations being later explained by Sternlicht, Nieman and Robinson.[60] These authors considered a mixed crystal consisting of host molecules doped with a small concentration of isotopic guest molecules. They postulated that the most important mechanism for promoting long-range triplet-triplet energy migration amongst the traps (isotopic guests) involved indirect trap-to-trap interactions through the virtual states of the host. This work was later extended by Colson and Robinson[61] who employed two different isotopic guests, producing traps of different depths (a trap and a supertrap).

Closs et al. first demonstrated quantitatively the utility of a σ-bond "spacer" in promoting efficient long-distance intramolecular EET.[62,63] They compared the rates of both triplet-triplet energy transfer and electron transfer in a series of compounds consisting of a 4-biphenylyl donor, a 2-naphthyl acceptor, and cyclohexane or decalin spacers. It was shown that the rate of triplet-triplet energy transfer followed the same trends as electron transfer for different stereoisomers, with the rate decreasing by approximately one order of magnitude in going from equatorial-equatorial substitution to equatorial-axial. This marked rate dependence on the conformation of connecting bonds is known to be indicative of through-bond-mediated electronic coupling.

Recently, Kroon et al.,[64] building upon an earlier communication,[65] examined intramolecular singlet-singlet energy transfer in a series of bichromophoric molecules containing rigid polynorbornyl spacers. It was similarly concluded that the EET was primarily mediated by through-bond electronic coupling, with an order of magnitude decrease in the rate being observed by adding a kink to the "all-trans" six-bond bridge.

The possible importance of superexchange interactions in many facets of supramolecular photochemistry is recognized, however their mechanism of action and significance are not fully appreciated. These aspects of superexchange mediated EET in supramolecular systems are investigated theoretically and experimentally in this thesis.

A superexchange mechanism was first suggested by Kramers to account for magnetic interactions between transition metal ions in oxide crystals.[66] Later, Halpern and Orgel examined its utility in mediating electron transfer between metal ions via π-bridges.[67] Soon after this, in an important paper, McConnell reported a theoretical analysis of the rate of intramolecular electron transfer in the mononegative ions of α,ω-diphenylalkanes.[68] He showed that the rate of electron transfer mediated through these σ-bridges should decrease approximately exponentially with its length (assuming a tight-binding approximation).

Orbital interactions "through bonds" were first examined in detail from a chemical perspective by Hoffmann et al.[69,70] Hoffmann's empirical observations of through-space and through-bond interference effects and "parity rule" were later discussed by Paddon-Row,[71] and are still the focus of much attention (both theoretical and experimental) by Paddon-Row and co-workers.[72-75] Photoelectron spectrotroscopic studies of through-space and through-bond interactions following a similar tack were reported by Martin and Mayer,[76] and by Heilbronner and Yang[77] at this time also.

Succeeding this work a plethora of studies of long-range electron transfer have ensued. Kuznetsov and Ulstrup[31] formulated a dynamical model for long-range intramolecular electron transfer which incorporated nuclear motions and compared direct, superexchange via high energy intermediate states, and superexchange via low energy intermediates. Larsson,[78] on the other hand, looked at electronic aspects of the process and used the Lowdin partitioning method[79] to reduce any problem to an effective two-state system. Significant and unambiguous experimental investigations in 1986 by Closs et al[80] and in 1987 by Oevering et al.[81] provided convincing evidence of the theories established over the previous years, including information on the the distance-dependence and the *cis-trans* (linking chain configuration) rule.

Since then, a vast number of papers have appeared in the literature which focus on electronic aspects of long-range electron transfer.[74,82-89] A key review of such work is that of Newton.[90]

Energy Migration. Energy migration is a process by which the excitation localized at one chromophore in a multichromophoric system is relocated to another chromophore of the same type, elsewhere in the system by a multistep process. For a polymer a different chromophore (with a lower excitation energy) may be tagged onto the chain, in which case the energy migration process will be terminated when excitation is transferred to this "trap". Observation of the time-dependence of fluorescence from the trap chromophore, subsequent to initial excitation of the polymer, can provide information on the energy migration process [91].

II. Relevance and Applications

The EET process is of fundamental interest in photochemistry. It plays a key role in the photosynthetic unit where special antennae molecules promote a vectorial transport of energy to the reaction centre, and its relevance extends to a myriad of synthetic systems including polymer photophysics, surface photochemistry and molecular crystals.

EET is also commonly utilized in, for example, photosensitization, enabling indirect excitation of specific chromophores, or in the operation of synthetic light antennae systems,[92] which are sometimes designed to mimic the natural photosynthetic unit, or may be vinyl aromatic polymer systems.[93,94] In any case, these synthetic "light harvesting" systems are formulated to promote the efficient vectorial transport of light energy from the "antennae" to an energy trap, for utilization or storage. They may provide valuable insight into the natural photosynthetic process, or may be studied with the goal of realizing efficient conversion of solar energy into useable potential energy. As such, the EET process has inspired a variety of theoretical and experimental studies.

Photosynthesis. The natural photosynthetic unit in higher plants and algae is thought to consist mainly of about 300 chlorophyll molecules plus other pigments. This complex serves as an energy-gathering "antenna" which promotes the transport of light energy to the reaction centre, where the energy is trapped and used to drive chemical reactions.[95] One such example is the main light-harvesting antenna complexes of green photosynthetic bacterium (which consist of supramolecular aggregate structures formed by self-organization of bacteriochlorophylls).[96,97] If such a system is to operate efficiently then excitation energy must be transferred from one pigment molecule to another with high efficiency. Furthermore, the overall mechanism of energy transport from the initially-excited pigment site to the reaction centre trap must be highly efficacious.

Such efficacy may be achieved firstly by an optimal arrangement of the different pigments radially outward from the reaction centre, which can be understood from classical considerations of the process. That is, if the energy migration is assumed to occur as a random walk, migrating amongst the pigments until it encounters a trap site, or is lost via pigment fluorescence, then such an optimal pigment configuration would "funnel" the energy toward the reaction centre (by spectral overlap considerations). Secondly, and complementary to the first point, it is proposed here (for the first time) that the rate of such a highly efficient process may be understood in terms of a quantum mechanical model describing a coherent vectorial transport of energy from the initially-excited "donor" pigment to the trap.

Synthetic Light Harvesting Systems. An escalating number and diversity of novel synthetic systems which utilize EET are arising. They range from model bichromophoric systems[98-102] to aromatic polymers[91,103-106] to multichromophoric β-cyclodextrins[107] to Langmuir-Blodgett films[108] to photomolecular devices[92,109,110] (that is, supramolecular structures capable of performing valuable light-induced functions, such as vectorial energy transport or remote photosensitization).

Some of the novel light-harvesting systems investigated by Balzani et al.[100,109-111] provide the opportunity to direct EET vectorially inwards or outwards between transition metal chelated ligands within an ordered supramolecular framework. However, whilst such systems are highly interesting from empirical and asthaetic viewpoints, the complexity of such an energy transport prohibits any detailed elucidation of the mechanism at this stage. Correlations and patterns observed along a series of related molecules have, however, allowed these workers to obtain considerable insight into the functionality of these basic photochemical molecular devices. This is achieved, for example, by varying the type of metal ion, the type of bridging ligand, the type of terminal ligand, and the position of various "components" in the supramolecular structure. The aims of such studies are to design artificial photochemical molecular devices capable of performing light-induced energy migration and charge separation processes.

It is known that energy migration amongst aromatic pendant chromophores on a polymer backbone is an important feature of polymer photophysics. About 20 years ago it was recognized that a lower-energy "trap" chromophore could be covalently attached to such a polymer, thereby allowing one to study the efficiency of trap sensitization after excitation of the main polymer component.[106,112-115] These were the first examples of "photon-harvesting polymers",[94] or the "antenna effect".[105]

Quite a number and diversity of investigations into this topic have since ensued, particularly with a view to understanding the dynamics of the energy migration and trapping process. Typically, either "hopping" or "diffusion" models form the basis of such analyses.[116-118]

The present work also addresses the energy migration to a trap process, but from a quantum mechanical viewpoint. It is seen that trap sensitization can be achieved via a coherent process related to superexchange. The high efficiency of trap sensitization which is often observed experimentally can thus be rationalized.

III. Current Status and Understanding of the Topic

Electronic Coupling - Förster versus Dexter. At interchromophore separations of approximately 10 to 150 Å, Förster's theory [41,42,119] provides a satisfactory description of the rate of excitation transfer. At larger separations the distance-dependence of the electronic coupling is modified by effects arising as a consequence of the finite speed of light [56,58]. As described here, and shown quantitatively for the first time in Refs. [120-122], at separations of less than ca. 10 Å the form of the distance dependence of electronic coupling becomes more complex. As a result, Förster's Equation becomes unreliable. This is important, since such close interchromophore separations are characteristic of those found in a plethora of topical systems, including aromatic polymers and some biological systems.

When molecular orbitals localized on donor and acceptor chromophores overlap, additional EET mechanisms are possible. It is shown that interchromophore interactions resulting from direct overlap of electron density consist of more than just an exchange interaction. Thus, it is shown that at short separations Dexter's treatment of excitation transfer [46] (via an exchange mechanism) is incomplete and close interchromophore separations are characteristic of those found in a plethora of topical systems, including aromatic polymers and some biological systems.

When molecular orbitals localized on donor and acceptor chromophores overlap, additional EET mechanisms are possible. It is shown that interchromophore interactions resulting from direct overlap of electron density consist of more than just an exchange interaction. Thus, it is shown that at short separations Dexter's treatment of excitation transfer [46] (via an exchange mechanism) is incomplete and not appropriate for describing allowed singlet-singlet EET in molecular systems. For this reason the author prefers to describe interchromophore interactions depending upon orbital overlap as "short-range" interactions. Evaluation of interchromophore electronic coupling, valid over all separations is presented in Chapter 3.

In 1980 Zimmermann et al.[123] examined the EET process between donor and acceptor chromophores attached to either end of rigid "organic rods" consisting of bicyclo[2.2.2]octane moieties linked bridgehead to bridgehead. They aimed to systematically study such "molecular light pipes". They did not observe a simple dipole-dipole coupling between donor and acceptor chromophores, but considered the possibility of a modification of the distance dependence of EET by an exchange interaction (interfering with the Coulombic interaction). They also conjecture that such a "short-range" interaction may be mediated through-bond, which in the light of recent work (cf. Chapter 4) appears to be the most likely interpretation.

An extensive and systematic study of through-space exchange-mediated EET (based on the Dexter model[46]) has been undertaken by Speiser et al.[98,124,125] By investigating EET in a series of macrocyclic α-diketones incorporating an aryl group (specifically *p*- and *m*-cyclophanes and the ortho analogues, benzocycloalkenediones) they obtained experimentally, for the first time, the parameters incorporated in Dexter's theory.[98] This was made possible by firstly ensuring poor spectral overlap between donor and acceptor chromophores, and that the electronic transition of the donor was forbidden, (so that the dipole-dipole interaction is negligible) and secondly the semi-rigid geometries of the bichromophores allowed distance dependencies to be elucidated. A parallel study on triplet-triplet EET in these molecules was unable to provide quantitative correlations between EET efficiency and interchromophore separation, although the efficiency was shown to be sensitive to molecular geometry.

In a more recent study Levy et al.[124] developed a model to account for the observed structural and temperature dependencies on the rate of Dexter-type intramolecular EET. The temperature effect was attributed to conformational changes induced by temperature variations. This study, together with a cognate, primarily computational, investigation[125] demonstrated that it is important to account for orientation effects on the magnitude of the short-range interactions, although the simple Dexter distance-dependence relation holds for symmetric systems. The possibility that the short-range interactions may be also propagated via the connecting linkages (a through-bond interaction) was not considered.

A further example of EET in a series of bichromophores is the work of Valeur et al. They use EET as a probe to gain experimental proof for calculated statistical distributions of conformations of the flexible polymethylene chains linking donor and acceptor.[99] Another recent study of interest is that of Osuka et al.[126] who have reported observations of EET from a zinc porphyrin to a free-base porphyrin in a series of conformationally-restricted hybrid diporphyrins. Gust et al. have investigated the role of linkage bonds in triplet and singlet EET in carotene-porphyrin dyads.[101] They conclude that a through-bond superexchange mechanism (involving the π-electrons of the linkage bonds) is significant. Further examples of interest are given in Chapter 6 of the book by Balzani and Scandola.[92]

In passing, it may be worthwhile to note one recent article which is in disagreement with the conclusions of Chapter 3. EET in lanthanide dinuclear complexes with *p-tert*-butylcalix[8]arene is found to occur between two types of interacting donor-acceptor pairs; one closely interacting (3.7 Å) and the other in the intermediate separation regime (11 Å). The authors conclude that both EET processes are mediated by a dipole-dipole mechanism.[127] It appears that this conclusion is based on an erroneous conception of the role of short range interactions in mediating allowed singlet-singlet EET. As mentioned above, the Coulombic and exchange (and superexchange-mediated interactions) operate simultaneously, *not exclusively to one another*.

Superexchange. Energy transfer between two chromophores may occur over large separations via a Coulombic interaction between transition dipoles of the excited donor and ground state acceptor.[42,54-57] It has been suggested that this interaction may also be propagated indirectly, via intervening molecules (e.g. solvent) between the donor and acceptor chromophores.[53] In situations where there is some degree of interchromophore orbital overlap, EET can also be mediated by short-range, overlap-dependent electronic coupling, one of which is the quantum mechanical exchange interaction.[46,128] Indeed, there is substantial experimental evidence to suggest that such intramolecular EET can even occur over distances greatly exceeding the sum of the van der Waals radii of the donor and acceptor groups, in rigidly linked donor–bridge–acceptor systems. In such an instance the connecting bridge is no longer simply a molecular spacing unit, but can effect orbital overlap between distant donor and acceptor groups. In such cases the "short-range" interaction is promoted by through–bond coupling between the chromophore orbitals with those of the bridge.[62-64,90,101,102,129]

Energy Migration. Most work to date on energy migration in polymers has concentrated on dipole-dipole energy transfer hops between nearest neighbour chromophores.[91] Interchromophore separations in aromatic polymers are relatively small (ca. 4 to 8 Å), so the dipole approximation may not provide a good approximation to the actual electronic coupling between chromophores.[120] The nature of this electronic coupling at such separations is pursued in detail in this work, including the role of "superexchange" interactions via intervening σ bonds or chromophores. Superexchange via chromophores identical with the donor (on which the initial excitation is localized) is, of course, the energy migration process.

In general, at short to intermediate interchromophore separations, measured rates of EET in multichromophoric systems (such as bichromophores, aromatic polymers, or biological systems) can sometimes be substantially larger than predicted by simple dipole-dipole theory[120, 128]. The aims of the present work are to: establish the distance and orientation dependencies of the electronic coupling at short and intermediate separations; extend this to the description of relayed excitation transfer; and to examine the consequences of these developments on the rate of EET - particularly when applied to aromatic polymer systems.

Unresolved Questions. As femtosecond spectroscopy becomes a reality further questions are arising. For example, in conclusion to a study in which ultrafast EET in the photosystem I core antenna was directly observed using the femtosecond fluorescence upconversion technique,[130] the authors noted that one may ask: "Does EET involve strong, intermediate or weak coupling (or the combination of all three)? What are the roles of inhomogenous broadening and/or electronic and vibrational dephasing in the EET process?" These quite fundamental questions will be addressed by the theory presented in Chapter 2.

Most of the past studies concerning the nature of the interactions responsible for mediating EET at short-range (i.e., where there is some degree of overlap between donor and acceptor orbitals) have been based on the work of Dexter.[46] Hence the process is often said to be mediated by the "Dexter exchange mechanism". It is shown here that such a treatment is specious. In addition, a further complication at close separations, the possible importance of contributions from higher multipole terms in the expansion of the Coulombic interaction, is addressed here, particularly for dipole-allowed transitions.

The significance of a superexchange mechanism in promoting EET is just being realized. Some previous studies have elucidated its import, but not its detailed mechanism. The present work addresses both issues and uses the results as a stepping stone to a quantum theory for energy migration.

The reason that energy migration to a trap can be so efficient remains an important unanswered question. The present work takes a quantum mechanical viewpoint so as to more rigorously analyse the phenomenon. As in electron transfer (in, for example, bacterial photosynthetic reaction centres[131-133]) there are two types of mechanism. One is a hopping mechanism, such that the energy is transferred from donor to acceptor, via the intermediate chromophores in an incoherent, stepwise manner, such that the intermediate chromophores may be regarded as genuine kinetic intermediates. A superexchange mechanism, on the other hand, effects EET from donor to acceptor via a coherent mixing of donor, acceptor *and* intermediate state wavefunctions. The role of random hops versus coherent this tunnelling contributions to energy migration are discussed in Chapter 2.

IV. Synopsis and Aims of the Thesis

The general goals of this thesis are (i) to gain a detailed insight into the microscopic mechanisms underlying EET phenomena in supramolecular systems; (ii) to develop a general understanding and theoretical formulation of the problem, which may be cast in a form related to experiment; and (iii) to undertake experimental studies of EET and interchromophore interactions in several rigid bichromophoric molecules.

Topics to Address. To unravel the often complex entanglement of processes occurring in supramolecular systems, the mechanism promoting a single EET step must be well understood so that rates may be estimated reliably and accurately. The present works aims to ascertain a detailed description of electronic coupling in bichromophoric systems by elucidating and comparing singlet-singlet and triplet-triplet EET mechanisms.

This work is extended to supramolecular systems via a dynamics-oriented approach. The significance and mechanisms of superexchange and energy migration are investigated and unravelled from the same perspective. An approach is developed and applied to the energy migration process, thus providing considerable insight into the phenomenon.

The Approach. As already discussed, the general energy transfer phenomenon is to be addressed quantum mechanically. This is presented in Chapter 2 from the viewpoint of radiationless transition theory. After discussion of the basic donor-acceptor interaction, theories for superexchange-mediated EET and energy migration are developed simultaneously. This highlights their semblances and shows clearly how they differ.

As is already well-known, a key parameter for determining the rate of such non-radiative processes is the electronic coupling. The nature and properties of the electronic coupling in relation to EET is not completely comprehended for many common situations. Chapter 3 aims to dissect the matrix element for donor-acceptor interactions in EET, thus detailing its characteristics in many practical instances.

In Chapter 4, experimental measurements on a novel series of rigidly-linked bichromophores are reported and related to the theory presented in Chapters 2 and 3. The major conclusions are reviewed in Chapter 5.

Thus, the general organization of the thesis is as follows:

- Chapter 1 - Introduction to the topic plus a summary and brief history of our current understanding.
- Chapter 2 - Detailed quantum mechanical development of the general rate equations suitable for the correct description of EET, together with a discussion of their interpretation and consequences.
- Chapter 3 - A theoretical study of the electronic interaction matrix element which promotes EET. Both general and computational investigations are reported.
- Chapter 4 - An experimental study of several novel rigidly-linked bichromophores illustrates the significance and applicability of the theoretical work presented in previous chapters.
- Chapter 5 - Précis of the major conclusions and ramifications of the work.

References

(1) H. Saigusa, S. Sun and E. C. Lim, J. Phys. Chem. **96**, 2083 (1992).

(2) H. Saigusa, S. Sun and E. C. Lim, J. Chem. Phys. **97**, 9072 (1992).

(3) J. Michl and V. Bonačić-Koutecký, *Electronic Aspects of Organic Photochemistry* (Wiley, 1990).

(4) J. Cai and E. C. Lim, J. Chem. Phys. **96**, 8265 (1992).

(5) J. Cai and E. C. Lim, J. Chem. Phys. **97**, 3892 (1992).

(6) D. S. McClure, Can. J. Chem. **36**, 59 (1958).

(7) R. Polák and J. Paldus, Theoret. Chim. Acta (Berl.) **4**, 37 (1966).

(8) J. Koutecký and J. Paldus, Coll. Czech. Chem. Commun. **27**, 599 (1962).

(9) D. S. McClure, *Electronic Spectra of Molecules and Ions in Crystals* (Academic Press, New York, 1959).

(10) M. Kasha, H. R. Rawls and M. A. El-Bayoumi, Pure Appl. Chem. **11**, 371 (1966).

(11) D. P. Craig and S. H. Walmsley, *Excitons in Molecular Crystals* (Benjamin, NY, 1968).

(12) O. K. Rice, Phys. Rev. **33**, 748 (1929).

(13) O. K. Rice, Phys. Rev. **34**, 1451 (1929).

(14) U. Fano, Phys. Rev. **124**, 1866 (1961).

(15) R. A. Harris, J. Chem. Phys. **39**, 978 (1963).

(16) (a) R. Kubo, Phys. Rev. **86**, 929 (1952).

(b) R. Kubo and Y. Toyozawa, Prog. Theor. Phys. **13**, 160 (1955).

(17) W. Heitler, *The Quantum Theory of Radiation* (Clarendon, Oxford, 1954).

(18) P. A. M. Dirac, *The Principles of Quantum Mechanics;* 4th ed. (Clarendon, Oxford, 1958).

(19) E. Fermi, *Notes on Quantum Mechanics* (University of Chicago Press, Illinois, 1961).

(20) B. A. Lippmann and J. Schwinger, Phys. Rev. **79**, 469 (1950).

(21) M. Gell-Mann and M. L. Goldberger, Phys. Rev. **91**, 398 (1953).

(22) G. W. Robinson, J. Molec. Spectrosc. **6**, 58, 509 (1961).

(23) G. W. Robinson and R. P. Frosch, J. Chem. Phys. **37**, 1962 (1962).

(24) G. W. Robinson and R. P. Frosch, J. Chem. Phys. **38**, 1187 (1963).

(25) G. W. Robinson, in *Excited States*, edited by E. C. Lim (Academic Press, N.Y., 1974) Vol. 1.

(26) G. W. Robinson and C. A. Langhoff, Chem. Phys. **5**, 1 (1974).

(27) J. Jortner, S. A. Rice and R. M. Hochstrasser, Adv. Photochem. **7**, 149 (1969).

(28) J. Jortner and R. S. Berry, J. Chem. Phys. **48**, 2757 (1968).

(29) J. Ulstrup and J. Jortner, J. Chem. Phys. **63**, 4358 (1975).

(30) M. Bixon and J. Jortner, Faraday Disc. Chem. Soc. **74**, 17 (1982).

(31) A. M. Kuznetsov and J. Ulsrtup, J. Chem. Phys. **75**, 2047 (1981).

(32) S. H. Lin and R. Bersohn, J. Chem. Phys. **44**, 3768 (1966).

(33) S. H. Lin, J. Chem. Phys. **44**, 3759 (1966).

(34) M. Bixon and J. Jortner, J. Chem. Phys. **48**, 715 (1968).

(35) M. Bixon and J. Jortner, J. Chem. Phys. **50**, 4061 (1969).

(36) M. Bixon and J. Jortner, J. Chem. Phys. **50**, 3284 (1969).

(37) D. P. Chock, J. Jortner and S. A. Rice, J. Chem. Phys. **49**, 610 (1968).

(38) C. A. Langhoff and G. W. Robinson, Chem. Phys. **6**, 34 (1974).

(39) G. W. Robinson and J. O. Berg, Can. J. Phys. **53**, 2068 (1975).

(40) J. O. Berg, Chem. Phys. Letts. **41**, 547 (1976).

(41) Th. Förster, Ann. Physik **2**, 55 (1948).

(42) Th. Förster, in *Modern Quantum Chemistry*, edited by O. Sinanoğlu (Academic Press, 1965) Vol. III.

(43) J. I. Frenkel, Phys. Rev. **37**, 17, 1276 (1931).

(44) R. E. Peierls, Ann. Physik **13**, 905 (1932).

(45) A. S. Davydov, J. Exptl. Theoret. Phys. (USSR) **18**, 210 (1948).

(46) D. L. Dexter, J. Chem. Phys. **21**, 836 (1953).

(47) S. H. Lin, Mol. Phys. **21**, 853 (1971).

(48) S. H. Lin, Proc. Royal Soc. London A **335**, 51 (1973).

(49) S. H. Lin, W. Z. Xiao and W. Dietz, Phys. Rev. E **47**, 3698 (1993).

(50) D. P. Craig and T. Thirunamachandran, *Molecular Quantum Electrodynamics* (Academic Press, New York, 1984).

(51) D. P. Craig and T. Thirunamachandran, Adv. Quantum Chem. **16**, 97 (1982).

(52) D. P. Craig and T. Thirunamachandran, Acc. Chem. Res. **19**, 10 (1986).

(53) D. P. Craig and T. Thirunamachandran, Chem. Phys. **135**, 37 (1989).

(54) D. P. Craig and T. Thirunamachandran, Chem. Phys. **167**, 229 (1992).

(55) D. L. Andrews and B. S. Sherborne, J. Chem. Phys. **86**, 4011 (1987).

(56) D. L. Andrews, Chem. Phys. **135**, 195 (1989).

(57) D. L. Andrews and G. Juzeliūnas, J. Chem. Phys. **95**, 5513 (1991).

(58) D. L. Andrews and G. Juzeliūnas, J. Chem. Phys. **96**, 6606 (1992).

(59) M. A. El-Sayed, M. T. Wauk and G. W. Robinson, Mol. Phys. **5**, 205 (1962).

(60) H. Sternlicht, G. C. Nieman and G. W. Robinson, J. Chem. Phys. **38**, 1326 (1963).

(61) S. D. Colson and G. W. Robinson, J. Chem. Phys. **48**, 2550 (1968).

(62) G. L. Closs, P. Piotrowiak, J. M. MacInnis and G. R. Fleming, J. Am. Chem. Soc. **110**, 2652 (1988).

(63) G. L. Closs, M. D. Johnson, J. R. Miller and P. Piotrowiak, J. Am. Chem. Soc. **111**, 3751 (1989).

(64) J. Kroon, A. M. Oliver, M. N. Paddon-Row and J. W. Verhoeven, J. Am. Chem. Soc. **112**, 4868 (1990).

(65) H. Oevering, J. W. Verhoeven, M. N. Paddon-Row and E. Cotsaris, Chem. Phys. Letts. **143**, 488 (1988).

(66) H. A. Kramers, Physica **1**, 182 (1934).

(67) J. Halpern and L. E. Orgel, 1960 **29**, 32 (1960).

(68) H. M. McConnell, J. Chem. Phys. **35**, 508 (1961).

(69) R. Hoffmann, A. Imamura and W. J. Hehre, J. Am. Chem. Soc. **90**, 1499 (1967).

(70) R. Hoffmann, Acc. Chem. Res. **4**, 1 (1971).

(71) M. N. Paddon-Row, Acc. Chem. Res. **15**, 245 (1982).

(72) M. N. Paddon-Row, S. S. Wong and K. D. Jordan, J. Am. Chem. Soc. **112**, 1710 (1990).

(73) M. N. Paddon-Row and K. D. Jordan, in *Modern Models of Bonding and Delocalization*, edited by J. F. Liebman and A. Greenberg (VCH, New York, 1988) p 115.

(74) K. D. Jordan and M. N. Paddon-Row, Chem. Rev. **92**, 395 (1992).

(75) M. J. Sheppard, M. N. Paddon-Row and K. D. Jordan, J. Am. Chem. Soc. **116**, 5328 (1994).

(76) H.-D. Martin and B. Mayer, Angew. Chem. Int. Ed. Engl. **22**, 283 (1983).

(77) E. Heilbronner and Z.-Z. Yang, Top. Curr. Chem. **115**, 1 (1983).

(78) S. Larsson, J. Am. Chem. Soc. **103**, 4034 (1981).

(79) P.-O. Löwdin, J. Mol. Spectrosc. **10**, 12 (1963).

(80) G. L. Closs, L. T. Calcaterra, N. J. Green, K. W. Penfield and J. R. Miller, J. Phys. Chem. **90**, 3673 (1986).

(81) H. Oevering, M. N. Paddon-Row, M. Heppener, A. M. Oliver, E. Cotsaris, J. W. Verhoeven and N. S. Hush, J. Am. Chem. Soc. **109**, 3258 (1987).

(82) P. Bertrand, Chem. Phys. Lett. **140**, 57 (1987).

(83) J. R. Reimers and N. S. Hush, Chem. Phys. **134**, 323 (1989).

(84) M. A. Ratner, J. Phys. Chem. **94**, 4877 (1990).

(85) A. Kuki, Structure and Bonding **75**, 49 (1991).

(86) C. A. Naleway, L. A. Curtiss and J. R. Miller, J. Phys. Chem. **95**, 8434 (1991).

(87) R. Gleiter and W. Schäfer, Acc. Chem. Res. **23**, 369 (1990).

(88) P. Siddarth and R. A. Marcus, J. Phys. Chem. **94**, 2985 (1990).

(89) C. Liang and M. D. Newton, J. Phys. Chem. **97**, 3199 (1993).

(90) M. D. Newton, Chem. Rev. **91**, 767 (1991).

(91) K. P. Ghiggino and T. A. Smith, Prog. Reaction Kinetics **18**, 375 (1993).

(92) V. Balzani and F. Scandola, *Supramolecular Photochemistry* (Ellis Horwood Limited, 1991).

(93) J. Guillet, *Polymer Photophysics and Photochemistry* (Cambridge University Press, 1985).

(94) S. E. Webber, Chem. Rev. **90**, 1469 (1990).

(95) D. Gust and T. A. Moore, Adv. Photochem. **16**, 1 (1991).

(96) A. R. Holzwarth, M. G. Müller and K. Griebenow, J. Photochem. Photobiol. **B 5**, 457 (1990).

(97) A. R. Holzwarth, K. Griebenow and K. Schaffner, Z. Naturforsch **45 C**, 203 (1990).

(98) S. Hassoon, H. Lustig, M. B. Rubin and S. Speiser, J. Phys. Chem. **88**, 6367 (1984).

(99) M. Kaschke, N. P. Ernsting, B. Valeur and J. Bourson, J. Phys. Chem. **94**, 5757 (1990).

(100) J.-P. Collin, S. Guillerez, J.-P. Sauvage, F. Barigelletti, L. DeCola, L. Flamigni and V. Balzani, Inorg. Chem. **31**, 4112 (1992).

(101) D. Gust, T. A. Moore, A. L. Moore, C. Devadoss, P. A. Liddell, R. Hermant, R. A. Nieman, L. J. Demanche, J. M. DeGraziano and I. Gouni, J. Am. Chem. Soc. **114**, 3590 (1992).

(102) G. D. Scholes, K. P. Ghiggino, A. M. Oliver and M. N. Paddon-Row, J. Phys. Chem. **97**, 11871 (1993).

(103) K. P. Ghiggino, A. J. Roberts and D. Phillips, Adv. Polym. Sci. **40**, 69 (1981).

(104) D. Phillips and G. Rumbles, Polymer Photochemistry **5**, 153 (1984).

(105) J. E. Guillet, *Polymer Photophysics and Photochemistry* (Cambridge University Press, Cambridge, 1985).

(106) J. E. Guillet and W. A. Rendall, Macromolecules **19**, 224 (1985).

(107) M. N. Berberan-Santos, J. Canceill, J.-C. Brochon, L. Jullien, J.-M. Lehn, J. Pouget, P. Tauc and B. Valeur, J. Am. Chem. Soc. **114**, 6427 (1992).

(108) M. Yamamoto, S. Ito and S. Ohmori, in *Photochemical Processes in Organized Molecular Systems*, edited by K. Honda (Elsevier, 1991).

(109) P. Belser, A. von Zelewsky, M. Frank, C. Seel, F. Vögtle, L. DeCola, F. Barigelletti and V. Balzani, J. Am. Chem. Soc. **115**, 4076 (1993).

(110) G. Denti, S. Campagna, S. Serroni, M. Ciano and V. Balzani, J. Am. Chem. Soc. **114**, 2944 (1992).

(111) G. Denti, S. Serroni, S. Campagna, V. Ricevuto, A. Juris, M. Ciano and V. Balzani, Inorg. Chim. Acta **198-200**, 507 (1992).

(112) R. F. Cozzens and R. B. Fox, J. Chem. Phys. **50**, 1532 (1969).

(113) R. B. Fox, T. R. Price, R. F. Cozzens and W. H. Echols, Macromol. **7**, 937 (1974).

(114) J. S. Aspler, C. E. Hoyle and J. E. Guillet, Macromol. **11**, 925 (1978).

(115) X.-X. Ren and J. E. Guillet, Macromolecules **18**, 2012 (1985).

(116) A. I. Burshtein, Sov. JETP Phys. **35**, 882 (1972).

(117) S. G. Fedorenko and A. I. Burshtein, Chem. Phys. **98**, 341 (1985).

(118) (a) V. M. Kenkre and R. S. Knox, Phys. Rev. B **9**, 5279 (1974).

(b) H. Haken and P. Reineker, Z. Physik, **249**, 253 (1972).

(c) M. Grover and R. Silbey, J. Chem. Phys., **54**, 4843 (1971).

(119) Th. Förster, Z. Naturforsch **4a**, 321 (1949).

(120) G. D. Scholes and K. P. Ghiggino, J. Phys. Chem. **98**, 4580 (1994).

(121) G. D. Scholes and K. P. Ghiggino, J. Photochem. Photobiol. A: Chem., **80**, 355 (1994).

(122) (a) R. D. Harcourt, G. D. Scholes and K. P. Ghiggino, J. Chem. Phys. (in press, December 1, 1994).

(b) G. D. Scholes, R. D. Harcourt and K. P. Ghiggino, J. Chem. Phys. (submitted).

(123) H. E. Zimmerman, T. D. Goldman, T. K. Hirtzel and S. P. Schmidt, J. Org. Chem. **45**, 3933 (1980).

(124) S.-T. Levy, M. D. Rubin and S. Speiser, J. Am. Chem. Soc. **114**, 10747 (1992).

(125) S.-T. Levy and S. Speiser, J. Chem. Phys. **96**, 3585 (1992).

(126) A. Osuka, K. Maruyama, I. Yamazaki and N. Tamai, Chem. Phys. Lett. **165**, 392 (1990).

(127) P. Froidevaux and J.-C. G. Bünzli, J. Phys. Chem. **98**, 532 (1994).

(128) G. D. Scholes, A. H. A. Clayton and K. P. Ghiggino, J. Chem. Phys. **97**, 7405 (1992).

(129) G. D. Scholes, K. P. Ghiggino, A. M. Oliver and M. N. Paddon-Row, J. Am. Chem. Soc. **115**, 4345 (1993).

(130) M. Du, X. Xie, Y. Jia, L. Mets and G. R. Fleming, Chem. Phys. Lett. **201**, 535 (1993).

(131) H. Guo, L. Liu and K.-Q. Lao, Chem. Phys. Lett. **218**, 212 (1994).

(132) R. A. Marcus, Chem. Phys. Lett. **133**, 471 (1987).

(133) R. A. Marcus, Chem. Phys. Lett. **146**, 13 (1988).

CHAPTER TWO

Dynamics of Electronic Energy Transfer

I. Preamble

This Chapter is concerned with the dynamics of interchromophore electronic excitation transfer (EET), including the possible influence of intermediate virtual relay configurations. The study was undertaken to provide a detailed understanding of the microscopic EET process; primarily to enable rates of superexchange-mediated EET and energy migration to be determined.

Firstly, the meaning of a rate for EET and the limits within which such a definition is valid are examined; particularly when the donor, A, and acceptor, B, chromophores are identical. The impetus for this study is that in many instances Förster's Equation[1-3] for the rate of EET is applied without first considering whether or not the approximations made during its development are reasonable for the situation under consideration.[4] For instance, can we apply a "hopping" model to energy migration in an aromatic polymer, or is the migration propagated by an excitonic wave, as in aromatic crystals?

The next part of this chapter includes a comprehensive survey of the fundamentals of the EET phenomenon from the perspective of radiationless transition theory. A detailed connection between EET and radiationless transition theory was first given by Lin, et al.[5-7] The work also rigorously elucidates the limits and situations within which the usual Fermi Golden Rule rate expression may be used, as well as providing the physical basis for the strong and weak coupling delimitations. This provides the foundation for examining the relayed EET processes which are investigated in the final sections.

The expressions of Förster[1-3] and of Dexter,[8] define a *rate* of EET, implying that excitation must initially be *localized* on one chromophore. It is important to understand the limits of applicability of such rate expressions to resonance EET under various experimental conditions. For example, in the apparently simple case of a homo-bichromophore, is it, in principle, possible to observe a *rate* of excitation transfer from an initially excited chromophore to the other, identical, chromophore? Or does an excitonic resonance predominate - that is, should we consider only stationary states representing a delocalization of the excitation over both chromophores?

In the following sections a number of questions are explored and answered. The first question is: *'is it possible to apply a Golden Rule-type rate expression to resonance excitation transfer?'* Stemming from this are the ancillary questions: 'can we prepare an appropriate initial state?' and 'within what limits is the rate expression valid?' Green's function methods, derived from scattering theory of resonance phenomena, and utilized in some treatments of radiationless transition theory, are applied to the problem in order to incorporate the vibrational component of the wavefunction. In this way the theory is not confined to dealing with interactions between discrete Born-Oppenheimer levels, but considers "real", broadened vibronic states. Within this framework the cases of weak and strong coupling are formalized prior to deriving a general equation relating to any type of EET process.

The second question to be addressed is: *'Can we describe both superexchange-mediated EET and energy migration phenomena using a general Golden Rule formalism?'* Typically either "hopping" or "diffusion" models[9,10] founded on the Förster model form the basis for analysis of energy migration. The high efficiency of trap sensitization, observed experimentally, is rationalized here within a quantum mechanical framework; it is seen that a key component in trap sensitization can be a coherent process, related to superexchange.

The format of the Chapter is summarized as follows:

§ II. A general discussion on preparation of an initial state, such that excitation is localized on just one chromophore.

§ III. Basis of the general model of excitation transfer based upon radiationless transition theory.

§ IV. A formulation of the general "Golden Rule" rate expression, emphasizing the approximations inherent in its development.

§ V. Discussion of the limits of weak and strong coupling and their relation to the theory of Section IV.

§ VI. Defining a rate expression for EET in molecular systems.

§ VII. Overview of energy migration and superexchange-mediated EET.

§ VIII. Expansion and evaluation of the general Golden Rule, and interpretation of superexchange-mediated EET and energy migration.

§ IX. Consideration of vibronic mixing effects.

§ X. Discussion of the theory and its consequences

§ XI. Conclusions.

II. Preparation and Observation of the System

Whether one can talk of a stationary state or of the dynamics of EET depends on the nature of the initially prepared state and the observation conditions; both of which may be controlled, to an extent, by the experimenter, as well as, importantly, by factors affecting the probability of *recurrence* (the return of excitation to the initially excited chromophore). These ideas are linked to the concepts of delocalization and localization of electronic excitation, which were examined by Simpson and Peterson,[11] Witkowski and Moffit,[12] Fulton and Gouterman,[13] and Förster[3]. Hence the ideas of strong, weak, and later, very weak coupling were introduced. Förster concluded that these three apparently different cases are intimately related: their different characteristics resulting from relative magnitudes of the electronic coupling, electronic band width, and individual vibronic band width.

One can see by inspection of the numerical results of Langhoff and Robinson[14] that the concepts of stationary states and dynamics are not entirely complementary. For instance, it can be seen that the coupling of a primary state (a state carrying all the oscillator strength) to a sparse "ladder" of secondary states (which couple to the primary state and are usually assumed to contribute no oscillator strength to the transition), is somewhat analogous to the spectral perturbations in a dimer (where there may be just one secondary state). Given that it is possible to prepare the primary state initially (obviously not a problem if it really did carry all the oscillator strength) then with sufficient time resolution it is, in principle, possible to observe *any* time-development of the system, including EET to the secondary state. It is equally possible, in another experiment, to observe the perturbed steady-state electronic absorption spectrum of the dimer. An interesting example of application of this theory is in the untangling of the higher vibrational levels of the naphthalene first singlet transition from the origin of the second singlet transition.[15,16] In systems relevant to the present exposition both states carry oscillator strength. This doesn't change the theory, but means we should first consider how to prepare the initial state.

The correct excited *electronic* states of two coupled identical chromophores are the stationary state wavefunctions $(|A^*B\rangle \pm |AB^*\rangle)/\sqrt{2}$, which represent a splitting of the single-molecule excited state potential energy surface by twice the electronic coupling,[17-19] $V_{AB} = \langle A^*B |H| AB^*\rangle$. Therefore, with sufficient spectral resolution, a doublet will be observed in the dimer absorption spectrum - the splitting of which gauges the magnitude of the electronic coupling. If such a stationary state (i.e., one line of the doublet) is electronically excited, then excitation is instantly *delocalized* over both chromophores; at no stage is excitation localized on any *one* of the chromophores (and therefore a rate of EET cannot be defined). The somewhat abstract concept of a period of excitation resonance, $\tau^{res} = \pi\hbar/|V_{AB}|$, is however, often introduced,[3,20] and has meaning through the concept of decoherence: an interpretation of quantum mechanics whereby an initial quantum superposition of two states becomes a simple statistical alternative for the two disjoint events via dynamics.[21]

It is apparent that *purely electronic* stationary states are observable only where the natural linewidth of the state so formed is much less than $|V_{AB}|$, which provides a partition between the small separation limit and the intermediate separation regime discussed by Craig and Thirunamchandran.[22] If the bandwidth of the exciting light is less than the spacing between these stationary Born-Oppenheimer (BO) states then excitation transfer transitions cannot be associated with their excitation. Only spontaneous emission from the "supermolecular" BO state may ensue.

In order to observe excitation transfer then, the system must initially be prepared in a non-stationary state. Provided that the electronic coupling between two non-stationary states is weak, this may be achieved as discussed by Robinson and Frosch[23] (see especially footnote 10 of this reference). Essentially, an experimenter can choose which non-stationary states to initially excite, and given a short enough exciting light pulse, may monitor the subsequent dynamics. This is achieved by exciting the whole envelope of lines.[24-26] Then, immediately following excitation, the entire ensemble of molecules will exhibit the properties of a weighted average over the sharp states in the envelope. Thus the nonstationary BO state $|A^*B\rangle$ (expressed as a linear combination of the complete set of compound states) could be prepared, and a rate of EET to $|AB^*\rangle$ may often be defined. There will, of course, be a finite recurrence time to $|A^*B\rangle$ so that this rate will only be strictly defined for short times after preparation of the initial state. This will be made clear in the following sections.

Experimental observation of the excitation transfer process is often made using time-resolved fluorescence anisotropy methods.[27,28] Such experiments demonstrate the idea of preparing $|A^*B\rangle$ and observing formation of the stationary state $(|A^*B\rangle + |AB^*\rangle)/\sqrt{2}$. A useful (but simplistic) analogy is a pair of coupled oscillators.[29-31]

A further point, the implications of which will become evident, is that in general we are not dealing with sharp interacting levels. Spectral lines will be broadened by normal non-radiative energy transfer to lower electronic states. This means that in actuality each level of the broadened $|A^*B\rangle$ *compound* (non-stationary) state interacts with the entire $|AB^*\rangle$ *compound* state. Therefore a rate model dealing with the interaction of two discrete states via a (quasi-) continuum should be appropriate under suitable coupling conditions.

III. Radiationless Transitions and Compound States

The basic model used in radiationless transition theory has remained unchanged since the treatment of Robinson and Frosch.[23,32,33] It usually involves single discrete states (*primary levels*), which carry all the oscillator strength, interacting with a quasi-continuum of states (the *secondary levels*, which may be for example, high vibrational levels of a lower electronic excited state) to form *compound states*. Such a model has been used previously to compute spectral line shapes and rates of radiationless transitions.[14,24,34-40]

Considerable insight into the physics of a discrete state interacting with a continuum can be obtained from some additional papers[41-44] and books dealing with such phenomena,[45,46] which essentially provide the basis for the mathematical elaboration of the Robinson and Frosch model by Lin,[47] Bixon and Jortner,[37,38] Chock, Jortner and Rice,[36] and Robinson and Langhoff.[35] These latter models attributed the perturbation acting to induce the non-radiative transition to deviations from the Born-Oppenheimer approximation, bringing about a coupling between vibrational and electronic motion.

In the present work the electronic coupling between an excited state donor and ground state acceptor is of interest. If all levels were discrete then perturbation theory for degenerate levels would need to be applied.[46,48-51] However, given weak coupling conditions (see Section V) and that the levels are actually broadened (even in the dilute gas phase), we may employ the Green's function methods of radiationless transition theory to gain considerable physical insight into the process. A powerful mathematical tool for treating all facets of non-radiative processes is the Green's operator model.[14,34,35,39] A key advantage is that it closely approximates a real system; that is, one where spectral lines are not discrete, but reflect line broadening and levels shifts resulting from the near infinite number of terms in the system Hamiltonian. Similar methods have been employed by Mukamel and Jortner to describe molecular dissociation, [52,53].

The model central to this discussion (Figure 2.1) stems from the intuitive reasoning of Robinson and Frosch.[23,32,33,54] It relates to the interaction between two discrete states and a quasicontinuum; *where the discrete states are non-stationary*. As stated by Robinson and Frosch:[23] the matrix elements for the electronic relaxation problem in the limit of weak coupling are determined by the presence of perturbations between electronic states which are almost, but not quite, stationary. This is because a stationary state may undergo transitions to other stationary states only by coupling with the radiation field, so that all time-dependent transitions between stationary states are radiative in nature. However, if the system is prepared in a non-stationary state of the total Hamiltonian, nonradiative "transitions" will occur.[39]

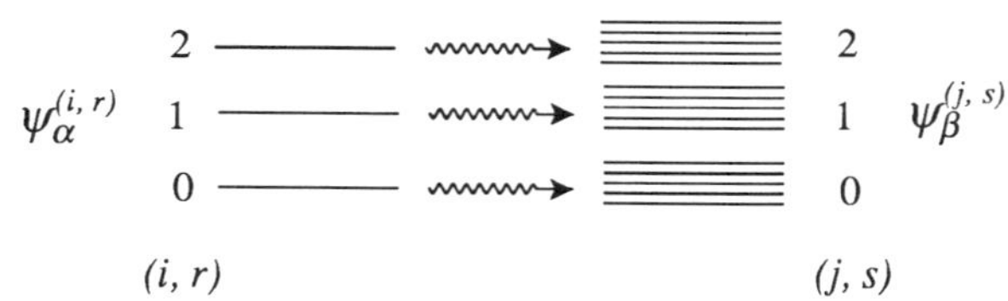

Figure 2.1. A schematic view of a physical model for EET, consisting of *discrete* initial and final states interacting via a continuum (see, e.g., Ref. 36). It is equivalently represented as a sum over the discrete initial states $\psi_\alpha^{(i,r)}$ (= $\psi_a^{(i)}\psi_{b_0}^{(r)}$) which undergo a transition to the quasi-continuous band of final states $\Psi_\beta^{(j,s)}$ (the compound state [cf. Eq. (3.3)] of $\psi_{a_0}^{(j)}\psi_b^{(s)}$).

The basis of theories for non-radiative processes is that an intramolecular nonradiative transition ensues after preparation of a non-stationary initial state. Similarly, the intermolecular process discussed in the present work proceeds from an initial (non-stationary) state where excitation is localized on the donor chromophore, and is induced by electronic coupling terms in the total Hamiltonian. The nature of this electronic coupling is discussed in Chapter 3.

The mathematical development which follows is similar to that given by Lin[47] and Jortner, et al,[36-39] but may be generalized to include any number of intermediate "relay" states. In this manner the theory may be applied to elucidate rates for processes such as energy migration and light harvesting.

Only the discrete states are assumed to have oscillator strength and the quasicontinuum generally represents closely spaced, high vibrational levels of the ground state. Each electronic state, a, is characterized by the zero-order Born-Oppenheimer (BO) levels (primary levels),

$$\psi_a^{(i)}(q,Q^{(a)}) = \phi_a(q,Q^{(a)})\,\chi_a^{(i)}(Q^{(a)}) \tag{3.1}$$

which are coarsely spaced and coupled to the dense quasicontinuum of vibronic (secondary) levels:

$$\psi_l^{(m)}(q,Q^{(l)}) = \phi_l(q,Q^{(l)})\,\chi_l^{(m)}(Q^{(l)}) \tag{3.2}$$

where ϕ and χ represent electronic and vibrational wavefunctions respectively, q are the electronic coordinates and $Q^{(a)}$ are the normal coordinates in the electronic state a.

The adiabatic approximation is applicable provided that the energy difference between vibronic states is large relative to the nuclear matrix elements connecting these states. The BO approximation therefore fails when a given BO level (e.g. $\psi_s^{(i)}$) is degenerate or quasidegenerate with one or more other BO levels (e.g. $\psi_l^{(m)}$). In such a situation the BO states must be mixed in order to obtain an appropriate wavefunction. So, for the case depicted in Figure 2.1, we may write

$$\Psi_a = \sum_i c_n^{(i)} \psi_a^{(i)} + \sum_m d_n^{(m)} \psi_l^{(m)} \tag{3.3}$$

which expresses the actual (*compound*) state as a linear combination of the discrete primary levels and the quasicontinuum of secondary levels.

Thus we have, for the complete system:

$$\mathrm{H}(q,Q)\ \Psi_a(q,Q) = E_\lambda\ \Psi_a(q,Q) \tag{3.4}$$

and for the BO states:

$$\mathrm{H}_0(q,Q)\ \psi_a(q,Q) = E_a\ \psi_a(q,Q) \tag{3.5}$$

Where the total Hamiltonian, a function of both electronic and nuclear coordinates, is partitioned into a component H_0 as well as a small perturbation V which mixes the zero-order states,

$$\mathrm{H} = \mathrm{H}_0 + \mathrm{V} \tag{3.6}$$

with

$$\mathrm{H}_0 = \mathrm{H}_{\mathrm{mol}} + \mathrm{H}_{\mathrm{rad}} \tag{3.7}$$

$$\mathrm{V} = \mathrm{H}_{\text{int-elec}} + \mathrm{H}_{\text{int-rad}} \tag{3.8}$$

and H_{mol} is the unperturbed molecular BO Hamiltonian, H_{rad} is the Hamiltonian for the unperturbed radiation field, $H_{int\text{-}elec}$ is an interaction Hamiltonian which mixes electronic levels, and $H_{int\text{-}rad}$ describes an interaction between an electronic level and the radiation field. Furthermore, it is assumed that the BO states form a complete orthonormal set, as do the true eigenstates of the total Hamiltonian.[36,44]

We begin by deriving a stationary state expansion of the transition matrix element. Our model, depicted in Figure 2.1, is that of an initially prepared discrete BO level $|\psi_\alpha\rangle$ interacting with a quasi-continuum of levels in the final compound state $|\Psi_\beta\rangle$. Such a final state quasi-continuum arises due to broadening of the discrete acceptor primary state by non-radiative effects. As will be clarified later, the implication of this model is that intrachromophore non-radiative processes are assumed to occur faster than EET.

This essential physical model, which relates to the interaction between two discrete states and a quasicontinuum,[36] may be generalized to include any number of intermediate "relay" states (Figure 2.2). In this manner it may be applied to determine rates for processes such as superexchange, energy migration and light harvesting.

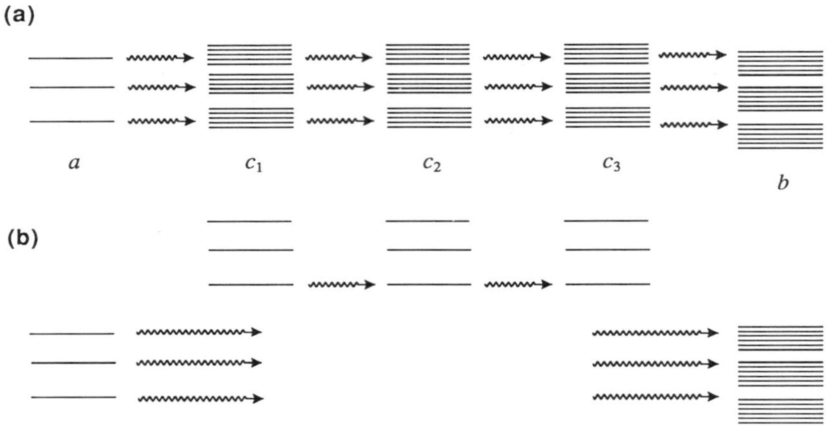

Figure 2.2. Schematic representations (cf. Figure 2.1) of (a) coherent energy migration and (b) superexchange energy transfer. a and b are the donor and acceptor respectively, while c_i represent intermediate (relay) moieties (which in energy migration are donor-like). All molecular states are broadened by non-radiative processes; but note that for case (b), in the limit of large energy gaps, the intermediate states should be considered to be discrete.

IV. Form and Properties of the Rate Expression

We proceed by obtaining a formal rate expression for EET from donor to acceptor, which may account for the possibility of intermediate states being close in energy to the initial (donor) state (i.e., the situation where normal perturbation theory breaks down). Such a rate expression may be used, for example, to determine the expected rise time of trap fluorescence in energy migration and trapping processes.[4,55]

As is usual in scattering problems, we partition the total Hamiltonian governing the system, H, into a part H_0, which describes the components of the system in the absence of interaction (the quasi-continuous states Ψ_s), and a small component V, which effects the interaction. The form of this interaction Hamiltonian operator is given by Eq. (3.8), but it should be emphasized that it is a function of both electronic and nuclear coordinates. The implications of nuclear motion are examined in Section VI.

The question to be addressed in this section is: What is the rate of transition from one noninteracting (initially-prepared) state $|a\rangle$, the donor, to another, $|b\rangle$ the acceptor or trap? In general the trap will be assumed to be different (spectroscopically) from the donor and any intervening states, so that we will discuss the grow-in of the trap state $|b\rangle$, which is an experimental observable. The theory is established along the lines presented by Gell-Mann and Goldberger,[56] and is compared to Dirac's variation of constants method (given in Appendix A).

The time-development of the system, given that we have initially prepared the state ψ_a, is given by

$$|\Psi(t)\rangle = \exp(-\mathrm{i}\mathrm{H}\,t\,/\,\hbar)\,|\psi_a\rangle \tag{4.1}$$

which may be written in the interaction picture[57,58] as

$$|\Psi(t)\rangle \;=\; \mathrm{U}(t,t_0)|\psi_a\rangle \tag{4.2}$$

where the evolution operator is written as

$$\mathrm{U}(t,t_0) \;=\; \mathrm{U}_0(t,t_0) \;+\; \frac{-\mathrm{i}}{\hbar}\int_{t_0}^{t} d\tau\, \mathrm{U}_0(t,\tau)\mathrm{V}(\tau)\mathrm{U}(\tau,t_0) \tag{4.3}$$

with

$$\mathrm{U}_0(t,t_0) \;=\; \exp\!\left(\frac{-\mathrm{i}}{\hbar}\,\mathrm{H}_0(t - t_0)\right) \tag{4.4}$$

which may be iterated to any order.

It is convenient to re-write Eq. (4.2) and insert a convergence factor, η:

$$|\Psi(t)\rangle = \lim_{\eta\to 0+}\left\{U_0(t,t_0) + \frac{-i}{\hbar}\int_{t_0}^{t} d\tau\, U_0^{\dagger}(\tau,t)V(\tau)e^{-\eta|\tau|}U_0(\tau,t_0)\right\}|\Psi(0)\rangle \tag{4.5}$$

where, ensuring that we implicitly maintain the explicit time-ordering of operators from Eq. (4.3) and assuming that V is time-independent, the initial state vector for the system may be written as an expansion of the "initial state", (an eigenvector of the zeroth-order Hamiltonian, H_0),

$$|\Psi(0)\rangle = |\psi_a\rangle + G^{+}(E_a)V|\Psi_a\rangle \tag{4.6a}$$

$$= |\psi_a\rangle + G_0^{+}(E_a)V|\Psi(0)\rangle \tag{4.6b}$$

after algebraic manipulation. $G^{+}(E_a)$ is the retarded Green's function for the entire system (i.e., incorporating the complete Hamiltonian, H), whilst $G_0^{+}(E_a)$ is the retarded Greens function (and $G_0^{-}(E_a)$ is the advanced Green's function) for the zeroth-order Hamiltonian, H_0, and is given by:

$$G_0^{\pm}(E_a) = \lim_{\eta\to 0+}\left(\sum_n \frac{1}{E_a \pm i\eta - H_0}|\psi_n\rangle\langle\psi_n| + \int dn\, \frac{1}{E_a \pm i\eta - H_0}|\psi_n\rangle\langle\psi_n|\right)$$

$$= \lim_{\eta\to 0+}\left(\sum_n \frac{|\psi_n\rangle\langle\psi_n|}{E_a \pm i\eta - E_n} + \int dn\, \frac{|\psi_n\rangle\langle\psi_n|}{E_a \pm i\eta - E_n}\right) \tag{4.7}$$

where $\sum_n |\psi_n\rangle\langle\psi_n| = 1$ spans a complete set of states.

The probability that we have a transition to the state ψ_b is, taking the limits $t_0 \rightarrow -\infty$ and $t \rightarrow +\infty$:

$$P_{ba}(t) = |\langle \psi_b | \Psi(t) \rangle|^2 \tag{4.8}$$

$$P_{ba}(t) = \lim_{t \rightarrow +\infty} \lim_{t_0 \rightarrow -\infty} \left| \lim_{\eta \rightarrow 0+} \frac{-i}{\hbar} \int_{t_0}^{t} d\tau \exp\left\{ (i\tau[E_b - E_a]/\hbar) - \eta|\tau| \right\} \langle \psi_b | \Psi(0) \rangle \right|^2 \tag{4.9}$$

When the eigenvalues of the total Hamiltonian are discrete and coarsely-spaced (e.g. when ψ_b is a discrete state), a true transition out of the initial state (in the sense that the probability of the system being in the initial state $|\psi_a\rangle$ approaches zero as t tends to infinity) does not occur.[45] The system appears to oscillate between closely-degenerate initial and final (non-stationary) states [cf. Equation (A.6)]. However, when the energies of a large number of final states are closely spaced the dynamics of the system may closely approximate a transition. In this case we have [cf. Equation (A.10)]

$$P_{ba}(t) = \frac{2\pi}{\hbar} |\langle \Psi_b | T^+(E_a) | \psi_a \rangle|^2 \, \delta(E_b - E_a)\, t \tag{4.10}$$

where contours of integration in the complex plane have been taken around the poles, enabling us to use the relation

$$\lim_{\eta \rightarrow 0+} \frac{1}{x \pm i\eta} = \mathcal{P}\frac{1}{x} \mp i\pi\,\delta(x) \tag{4.11}$$

where $\mathcal{P}(f)$ denotes the principal value of f, and the transition operator is given by the Lippmann-Schwinger Equation [59] Eq. (4.12),

$$T^+(E_a) = V + VG_0^+(E_a)T^+(E_a) \tag{4.12}$$

Thus the transition probability per unit time is given by:

$$w_{ba} = \frac{2\pi}{\hbar} |\langle \psi_b | T^+(E_a) | \psi_a \rangle|^2 \, \delta(E_b - E_a) \tag{4.13}$$

This is the Fermi Golden Rule rate expression. The transition is from ψ_a into a density of final states approximately degenerate with the initial state; which is ensured by the Dirac delta function, $\delta(E_b - E_a)$. A similar formulation (truncated at second order) has been recently applied to superexchange-mediated EET.[5] The importance and utility of this general expression involving the transition operator will be especially useful in Section VIII.

Thus we see that, as long as there are a large number of closely spaced final states, we may define a Fermi Golden Rule rate for the transition. Such an expression is valid if the probability of finding the system in the initial state deviates only slightly from unity (see Appendix A), thus t must be short enough that $w_{ba} << t^{-1}$, but larger than atomic or molecular periods. That is, the experimental time-scale across which one measures the rate of an EET process should be shorter then the inverse of the measured rate in order that the probability of energy transfer be *strictly* linear in time. Such a limit implicitly ensures the condition of weak electronic coupling.

V. Weak and Strong Coupling Limits

On the basis of both stationary-state and time-dependent considerations Simpson and Peterson[11] showed that in a system of identical chromophores there is a relationship between the relative magnitudes of the interchromophore interactions and an energy parameter characterizing the electronic absorption spectrum of the system. This relationship determines whether the spectrum reflects absorption by the system as an entirety, or as a collection of independent chromophores. In this way the cases of strong and weak electronic coupling could be related to the ideas of delocalization and localization of excitation respectively.

Later, Förster related these ideas to EET by consideration of potential energy surfaces.[3] He showed that the weak coupling case (i.e. when the magnitude of the electronic coupling is smaller than the vibronic band width) corresponds to essential localization of excitation, and that the cases of strong and weak coupling are intimately related. The treatment of Engleman and Jortner[60] is used here to show they are in fact just different limits of the radiationless energy transfer process. The work of Freed and Jortner[61] is also of relevance.

We apply the harmonic approximation and denote normal coordinates Q_j (j = 1,...,N), with effective masses M_j and frequencies ω_j. The equilibrium configuration of state ψ_α is characterized by $Q_j^{0(\alpha)}$ (j = 1,...,N) and $\Delta Q_j^0 = Q_j^{0(\beta)} - Q_j^{0(\alpha)}$ corresponds to the displacement of the normal coordinate Q_j in the equilibrium configuration of state ψ_β relative to ψ_α. The initially-prepared state ψ_α embodies an excited state chromophore A* and a ground state B. ψ_β, the target final state, may only be observable in the weak coupling case.

The dimensionless coordinates and displacements are defined by

$$q_j = \left(\frac{M_j\omega_j}{\hbar}\right)^{1/2}\left(Q_j - \tfrac{1}{2}Q_j^{0(\beta)}\right) \tag{5.1}$$

$$\Delta_j = \left(\frac{M_j\omega_j}{\hbar}\right)^{1/2}\Delta Q_j^0 \tag{5.2}$$

The adiabatic potentials E_α and E_β for the two electronic states are given as

$$E_\alpha = \tfrac{1}{2}\sum_j \hbar\omega_j q_j^2 \tag{5.3}$$

$$E_\beta = E_\alpha - \sum_j \hbar\omega_j\Delta_j q_j - \Delta E + E_m \tag{5.4}$$

where

$$E_m = \tfrac{1}{2}\sum_j \hbar\omega_j\Delta_j^2 \tag{5.5}$$

represents the molecular nuclear relaxation energy due to Δ_j.

The conditions for weak coupling are that $\Delta_j = q^{0(\beta)} - q^{0(\alpha)} \approx 0$ and $E_m \approx 0$, that is, the relative displacement of the potential energy surfaces is small; corresponding to the Robinson-Frosch-McCoy-Ross[23,33,62-64] picture for radiationless processes. Such a situation represents a case where the electronic coupling between initial and final configurations is small compared with vibrational level spacings.[13] It is seen in Figure (2.3a) that these conditions correspond to approximate localization of the excitation, consistent with other approaches (see, e.g, Figure 1 of Ref. 13). Thus a rate of excitation transfer may be defined for this situation.

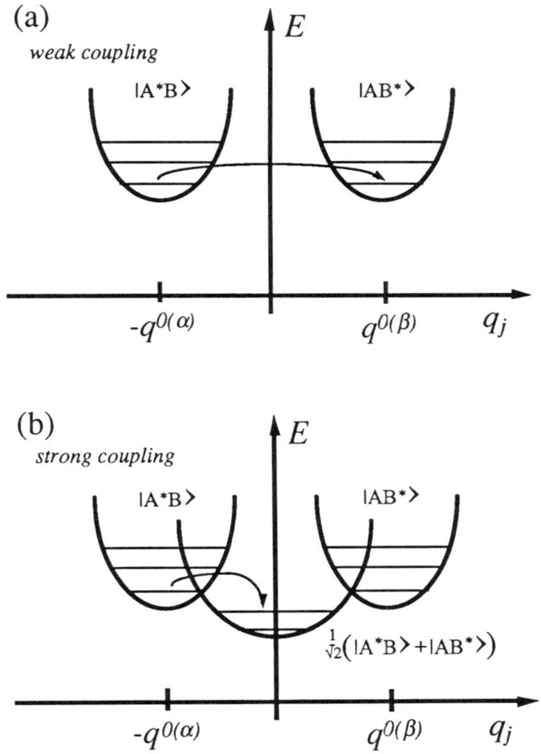

Figure 2.3. Potential energy surfaces as a function of the dimensionless coordinates q_j (cf. Eq. 5.1), where the equilibrium coordinates of each state are denoted by the 0 superscript. The weak coupling case (a) depicts approximate localization of the excitation, whilst the strong coupling case (b) demonstrates the rapid dephasing of the initial wavefunction into the correct stationary vibronic states of the supermolecule.

In the strong coupling limit the relative displacement of the potential energy surfaces is expected to be large, with the surfaces crossing close to the minimum of the higher electronic state. Such a case is depicted in Figure (2.3b), where, even if |A*B> is prepared initially, it will rapidly dephase into the correct stationary state of the system (|A*B> + |AB*>)/√2, and intermolecular vibrational modes assume importance (see, e.g., Figure 3 of Ref. 13). The rate of its formation will largely depend upon the rate of nuclear rearrangement. Now E_m is large, greater than the spectral band-width. This "strong coupling" stationary state will be observed experimentally in the *supermolecule* emission spectrum - that is, as excimer or exciplex emission. Hence such phenomena may be considered as manifestations of the strong coupling limit of EET, as was alluded to recently.[65]

The impetus underpinning such a categorization into delimitations of strong and weak coupling is that the applicability of the Fermi Golden Rule rate expression (Eq. 4.13) requires that non-radiative line broadening must occur faster than EET. Such considerations of a partitioning between strong and weak coupling limits are also of relevance to other theoretical explications of EET. For instance, theories are typically formulated to apply outside of a characteristic coherent transfer distance such that an overall multistep energy migration process is propagated via uncorrelated acts of incoherent energy transfer[66] (the weak coupling regime). On the other hand, coherence is associated with an oscillatory resonance of excited state population (see, for example, the apparent analogy with the coupled, displaced anharmonic oscillator system[31]).

Another perspective on the strong and weak coupling regimes is afforded by analogy with the "small" and "large" molecule limits of electronic relaxation respectively.[54] In the *small molecule limit* the spacing between final states for the radiationless transition is much greater than the lifetime broadening and the electronic coupling which connects initial and final states. In this limit, levels in the "initial" electronic state are only occasionally degenerate with levels with which they can mix in the "final" electronic state. Thus when the perturbation mixing such levels is "turned on," the levels merely split apart, the two resulting levels being mixtures of the "initial" and "final" electronic states. In the *large molecule limit* there is a virtually infinite number of "final" states, providing a multitude of nearly degenerate states accessible to the initial state. One says that an electronic relaxation process takes place from a discrete state into an adjoining region of the quasicontinuum.

The analogy with the strong and weak coupling cases of excitation transfer is obvious, and such ideas will be pursued in the discussion. These classifications of non-radiative processes correspond to the resonance and statistical limits later surveyed by Bixon and Jortner.[67]

VI. Defining a Rate of Excitation Transfer

We proceed by writing a formal rate expression for EET from donor to acceptor. It is presented in a very general form which may deal with the possibility of intermediate states being close in energy to the initial (donor) state[68] (i.e. the situation where normal perturbation theory breaks down). We then inspect the separability of Franck-Condon vibrational overlap factors from electronic factors (as applied to the T-matrix formalism) and finally derive a rate expression in a form related to that presented by Förster,[2,3].

Quantum mechanical theories of scattering[45,69] and multiphonon processes[70-73] provide the foundation for much of the theory presented in this Section. The initial transition probability per unit time for non-radiative decay to a dissipative continuum can be written[47,74] [c.f. Equation (4.18)]:

$$w = \frac{2\pi}{\hbar} \sum_{i} \sum_{j} p(a^*{}_i b_r) \left| \langle \psi_\beta^{(j,s)} | \mathrm{T}^+(E_{a^*{}_i b_r}) | \psi_\alpha^{(i,r)} \rangle \right|^2 \delta(E_{a_j b^*{}_s} - E_{a^*{}_i b_r}), \qquad (6.1)$$

where $\mathrm{T}^+(E_{a^*{}_i b_r})$ is defined in Equation (4.12) and $p(a^*{}_i b_r)$ is the probability that the system is initially in the zero-order state $\psi_\alpha^{(i,r)}$, representing vibrational levels i and r of electronic states a^* (of molecule A) and b (of molecule B) respectively.

This equation is valid only when interference effects between resonances are negligible (i.e. the weak coupling case). The labels of vibronic states i and s refer to primary levels. By referring to a summation over such levels it is implicit that the summation is carried out over all levels of the corresponding compound state.

Figure 2.4, adapted from Ref. 3, depicts the "coupled" isoenergetic transitions which provide synchronous de-excitation and excitation events in the donor and acceptor chromophores respectively. It highlights the importance of electronic as well as vibrational levels. This diagram is cognate to that in Figure 2.1, where the vibrational levels (i,r) of BO state α actually designate vibronic transitions between a^* and the ground state.

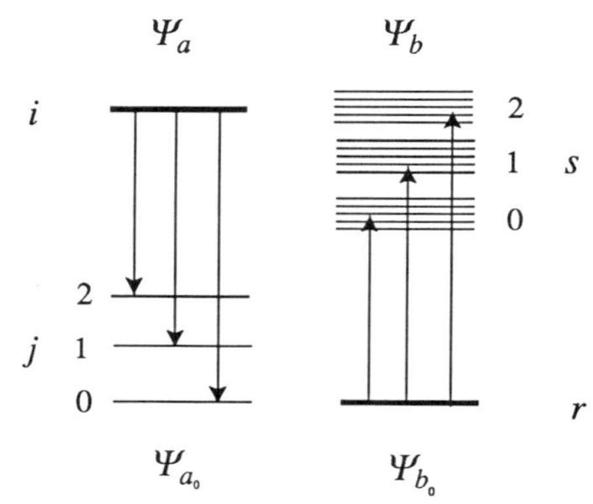

Figure 2.4. An alternative representation of Figure 2.1, which highlights the coupled isoenergetic de-excitation - excitation transitions of the donor and acceptor chromophore respectively. This clearly illustrates the relationship of the process with the absorption and emission spectra.

The interaction Hamiltonian which couples the initial and final vibronic states may be given by a Taylor expansion about the equilibrium geometry, Q_0:[37,75]

$$\mathrm{V}(q,Q) = \mathrm{V}(q,Q_0) + \sum_{\iota} \left(\frac{\partial \mathrm{V}(q,Q)}{\partial Q_\iota}\right)_0 Q_\iota + \frac{1}{2}\sum_{\iota,\kappa} \left(\frac{\partial^2 \mathrm{V}(q,Q)}{\partial Q_\iota \partial Q_\kappa}\right)_0 Q_\iota Q_\kappa + \ldots \quad (6.2)$$

where q represents electronic coordinates, and Q nuclear coordinates.

Consider the matrix element for a transition from $\psi_\alpha^{(i,r)}$ to $\psi_\beta^{(j,s)}$ where, from the BO approximation,

$$\psi_\alpha^{(i,r)} \approx \phi_\alpha(q,Q)\, \chi_{i,r}^{(\alpha)}(Q) \quad (6.3)$$

$$\approx \phi_{a^*}(q,Q)\, \phi_{b_0}(q,Q)\, \chi_i^{(a^*)}(Q)\, \chi_r^{(b)}(Q) \quad (6.4)$$

where the superscript * indicates an excited electronic state. So the vibronic matrix element $V_{\alpha\beta}$ is given as Eq. (6.5), where the Hamiltonian operator $\mathrm{V}(q,Q)$ is partitioned into an electronic potential energy operator, $\mathrm{U}(q,Q)$, and an electronic kinetic energy operator, $\mathrm{T}(q)$.

$$V_{\alpha\beta} \approx \langle \phi_\beta(q,Q) \mid \phi_\alpha(q,Q)\rangle \langle \chi_{j,s}^{(\beta)}(Q) \mid \mathrm{U}(q,Q) \mid \chi_{i,r}^{(\alpha)}(Q)\rangle \quad (6.5)$$

$$+ \langle \chi_{j,s}^{(\beta)}(Q) \mid \langle \phi_\beta(q,Q) \mid \mathrm{U}(q,Q) + \mathrm{T}(q) \mid \phi_\alpha(q,Q)\rangle \mid \chi_{i,r}^{(\alpha)}(Q)\rangle$$

The first term on the right hand side of Equation (6.5) is zero unless $|\phi_\alpha\rangle \equiv |\phi_\beta\rangle$. Now, employing the Herzberg-Teller expansion (Equation 6.2), we obtain Eq. (6.6) (e.g. see Ref. 76).

$$\langle\chi_{j,s}^{(\beta)}(Q)\,|\,\langle\phi_\beta(q,Q)\,|\,\mathrm{U}(q,Q)+\mathrm{T}(q)\,|\,\phi_\alpha(q,Q)\rangle\,|\,\chi_{i,r}^{(\alpha)}(Q)\rangle \tag{6.6}$$

$$= \langle\phi_\beta(q,Q_0)\,|\,\mathrm{V}(q,Q_0)\,|\,\phi_\alpha(q,Q_0)\rangle\,\langle\chi_{j,s}^{(\beta)}(Q)\,|\,\chi_{i,r}^{(\alpha)}(Q)\rangle$$

$$+\sum_\iota\left(\frac{\partial\langle\phi_\beta(q,Q)\,|\,\mathrm{U}(q,Q)\,|\,\phi_\alpha(q,Q)\rangle}{\partial Q_\iota}\right)_{Q_0}\langle\chi_{j,s}^{(\beta)}(Q)\,|\,Q_\iota\,|\,\chi_{i,r}^{(\alpha)}(Q)\rangle\;+\ldots$$

The first term on the right hand side represents the Condon approximation result; and for the second term, if the coupling is approximately constant with small changes in nuclear coordinates, we have

$$\sum_\iota\left(\frac{\partial\langle\phi_\beta(q,Q)\,|\,\mathrm{U}(q,Q)\,|\,\phi_\alpha(q,Q)\rangle}{\partial Q_\iota}\right)_{Q_0}$$

$$=\sum_\iota\sum_\gamma\Bigg(\langle\phi_\alpha(q,Q)|\,\mathrm{U}(q,Q)\,|\phi_\gamma(q,Q)\rangle\langle\phi_\gamma(q,Q)|\,\frac{\partial}{\partial Q_\iota}\,|\phi_\beta(q,Q)\rangle$$

$$-\ \langle\phi_\alpha(q,Q)|\,\frac{\partial}{\partial Q_\iota}\,|\phi_\gamma(q,Q)\rangle\langle\phi_\gamma(q,Q)|\,\mathrm{U}(q,Q)\,|\phi_\beta(q,Q)\rangle\Bigg)_{Q_0} \tag{6.7}$$

with

$$\sum_\gamma\langle\phi_\beta(q,Q)|\,\frac{\partial}{\partial Q_\iota}\,|\phi_\gamma(q,Q)\rangle = \sum_\gamma\frac{\langle\phi_\beta(q,Q)|\,\frac{\partial\mathrm{U}}{\partial Q_\iota}\,|\phi_\gamma(q,Q)\rangle}{E_\beta(Q)-E_\gamma(Q)} \tag{6.8}$$

Thus, Equation (6.7) accounts for excitation transfer via, or to, vibronically-induced intensity.

When the molecule is in an inert medium which acts as a heat bath (e.g. solvent), thermal equilibrium amongst these initial levels can be assumed. Provided that the vibrational relaxation (and excitation) rates considerably exceed the non-radiative decay time, the initial level populations, $p(a^*_i)$ and $p(b_r)$ should be given by a Boltzmann average over all thermally-available levels.

The Condon approximation result will now be formalized and discussed; the generalization to include the higher-order terms of Eq. (6.6) is straightforward. As first demonstrated by Lin,[6] the rate of EET, w, Eq. (6.1), may be written as a product of an electronic part, Γ^{elec}, Eq. (6.9a) (which controls the mechanisms of EET, c.f. Chapter 3) and a nuclear part, Γ^{nucl}, Eq. (6.9b). This latter term imparts information about temperature, isotope and energy gap effects in EET along the lines of radiationless transition theory.

$$\Gamma^{\text{elec}} = \frac{2\pi}{\hbar} \left| \langle \varphi_{ab^*}(q,Q_0) | \mathrm{T}^+(E_{a^*_i b_r}) | \varphi_{a^*b}(q,Q_0) \rangle \right|^2 \tag{6.9a}$$

$$\Gamma^{\text{nucl}} = \sum_{i,r} \sum_{j,s} p(a^*_i) p(b_r) \left| \langle \chi_j^{(a)} \chi_s^{(b^*)} | \chi_i^{(a^*)} \chi_r^{(b)} \rangle \right|^2 \delta(E_{a_j b^*_s} - E_{a^*_i b_r}) \tag{6.9b}$$

where it is assumed that the electronic coupling is approximately constant over nuclear coordinates† and that the nuclear wavefunction is momentarily unchanged after the excitation transfer.

For singlet-singlet EET Eq. (6.9b) may be further simplified using Eq. (6.10a)

$$\left| \langle \chi_j^{(a)} \chi_s^{(b^*)} | \chi_i^{(a^*)} \chi_r^{(b)} \rangle \right|^2 \delta(E_{a_j b^*_s} - E_{a^*_i b_r})$$
$$\approx \sum_{i,r} \sum_{j,s} |\langle \chi_j^{(a)} | \chi_i^{(a^*)} \rangle|^2 \, |\langle \chi_s^{(b^*)} | \chi_r^{(b)} \rangle|^2 \, \delta(\Delta E_{a^*a}^{(i,j)} - \Delta E_{bb^*}^{(r,s)}) \tag{6.10a}$$

---

† Q_0 refers to equilibrium excited state geometry of the electronically excited donor a^* and the equilibrium ground state geometry of the acceptor, b.

where $\Delta E_{a*a}^{(i,j)}$ and $\Delta E_{bb*}^{(r,s)}$ represent emission and excitation energies between the appropriate vibronic levels of isolated donor and acceptor respectively, which therefore determine the energy conservation condition dictated by the Dirac delta term. If however, there is a contribution to the electronic coupling from short-range interactions (i.e., interactions engendered by interchromophore orbital overlap[65,77,78]) or we are dealing with triplet-triplet EET, then the associated contribution to Eq. (6.9b) is given by Eq. (6.10b):

$$\left|\langle \chi_j^{(a)} \chi_s^{(b*)} | \chi_i^{(a*)} \chi_r^{(b)} \rangle\right|^2 \delta(E_{a_j b*_s} - E_{a*_i b_r})$$

$$\approx \sum_{i,r} \sum_{j,s} |\langle \chi_j^{(a)} | \chi_r^{(b)} \rangle|^2 \, |\langle \chi_s^{(b*)} | \chi_i^{(a*)} \rangle|^2 \, \delta(\Delta E_{a*a}^{(i,j)} - \Delta E_{bb*}^{(r,s)}) \qquad (6.10b)$$

Such Franck-Condon factors are similar to those important in electron-transfer, highlighting the correspondence between the phenomena of electron-transfer and short-range EET. Quantum mechanical theories of electron-transfer may be applied to the rate expression given here (see, e.g., Ulstrup and Jortner[79] and Bixon and Jortner[80]). In fact non-radiative rate theory is intimately related to the classical view afforded by Marcus-Hush theory of electron transfer processes.[5-7]

Following Förster,[3] we may re-write expression (6.9b) with Eq. (6.10a) in terms of an overlap integral between donor emission and acceptor absorption spectra:

$$\Gamma_{\text{Förster}}^{\text{nucl}} = \frac{\hbar}{2\pi} \frac{9(\ln 10)10^{-9}}{128\pi^5 L} \left(\frac{J\Phi_A}{\tau_A}\right) \left(\frac{4\pi\varepsilon_0}{|\mu_A||\mu_B|}\right)^2 \qquad (6.11)$$

which also accounts for the influence of vibronically-induced intensity via Eqs. (6.5 - 6.8) since just the direct electronic coupling has been removed via the last term. L is Avagadro's constant, Φ_A and τ_A are the donor fluorescence quantum yield and lifetime, respectively, and $|\mu_A|$ and $|\mu_B|$ are the donor and acceptor dipole transition moments (expressed in S.I. units). The spectral overlap integral J ($M^{-1}cm^{3}$) contains all the relevant details of population distribution functions, Franck-Condon factors and density of final states (c.f Eq. 6.10a). It is defined from the donor emission spectrum (normalized to unit area on a wavenumber scale), $F_A(\nu)$, and acceptor absorption spectra (intensity in units of extinction coefficient, $M^{-1}cm^{-1}$), $\varepsilon_B(\nu)$, as

$$J = \int_0^\infty F_A(\nu)\varepsilon_B(\nu)\frac{\mathrm{d}\nu}{\nu^4} \tag{6.12}$$

For direct Coulombic EET the rate Equation (6.1) may therefore be written

$$w = \frac{2\pi}{\hbar}|V_{\alpha\beta}|^2\, \Gamma^{\mathrm{nucl}}_{\mathrm{Förster}} \tag{6.13}$$

and we may estimate the electronic coupling matrix elements (i.e. $V_{\alpha\beta}$) using the theory presented in Chapter 3. When the electronic coupling matrix element is given in the limit of a dipole-dipole Coulombic interaction, Equation (6.13) reduces to the usual Förster Equation.[1-3] This has been discussed in detail by Lin.[5-7]

VII. Energy Migration and Superexchange Phenomena

In the first part of this Chapter the cases of weak and strong coupling were formalized prior to deriving a general equation relating to any type of EET process. The theory presented so far is not confined to dealing with interactions between discrete Born-Oppenheimer levels, but considered "real", broadened, vibronic states. The nature of the problem was discussed with especial reference to resonance transfer; with the conclusion that an energy transfer rate may be defined, for practical time-scales, in the weak coupling limit. In this case a "hopping" model for energy migration (as opposed to a completely coherent mechanism) *may* be realistic. In the next sections the important cases of superexchange-mediated EET and energy migration (possibly to a trap) are examined and compared. This is the first time that such a formal quantum mechanical treatment of the microscopic mechanisms contributing to energy migration has been attempted, mainly due to the difficulties imposed by intermediate electronic states being (quasi-) degenerate with the initial state. A thorough discussion of the results, in context with other work, is reserved for the discussion section of this chapter (Section X).

Current models of energy migration processes, whether they relate to natural photosynthetic or synthetic (e.g. polymer) systems, consider the migration to involve a random walk (i.e. incoherent, random excitation hops between chromophores, which may possibly lead to trap sensitization).[81,82] Each donor chromophore may therefore either fluoresce or transfer its excitation energy. The rate of trap sensitization is thus determined by this series of excitation "hops".[9,83-85] Such hopping models have been employed to treat the incoherent motion of triplet excitation in aromatic molecular crystals.[86-88]

The quantum mechanical picture presented in this work is complementary to the random walk picture in that it attempts to ascertain the nature of each donor-donor or donor-trap excitation transfer. Each step is coherent in the sense that (as in the double-slit experiment) all possible pathways connecting donor and acceptor are traversed simultaneously. Given the speed of light compared to typical donor-trap separations such a model would appear essential, that is, in the quantum mechanical picture all interactions must be accounted for when determining the transition probability from donor to acceptor. The extent to which an excitation transfer step maintains coherence will largely depend upon temperature, as is discussed in Section X: the coherence will be more rapidly dephased at higher temperatures. The inference of this study is that each excitation transfer step may be over a considerably larger distance than previously recognized; principally due to such a coherent relay mechanism.

The theory is developed by first obtaining a rate expression for energy transfer between a specified donor and acceptor (trap) in any system. The part involving the donor-acceptor interaction is then expanded in terms of all possible interactions involving intervening (possibly donor-like) moieties. Such interaction-terms are often interpreted as pathways.[89] An implicit time-ordering of operators is maintained (as in scattering problems) so that in the limit the of weak interactions (or large separations) the completely classical model is recovered. In the quantum mechanical model the energy transport is highly efficient due to the superposition of pathways and the vectorial (directed) nature of EET to a trap.

(i)

(ii)

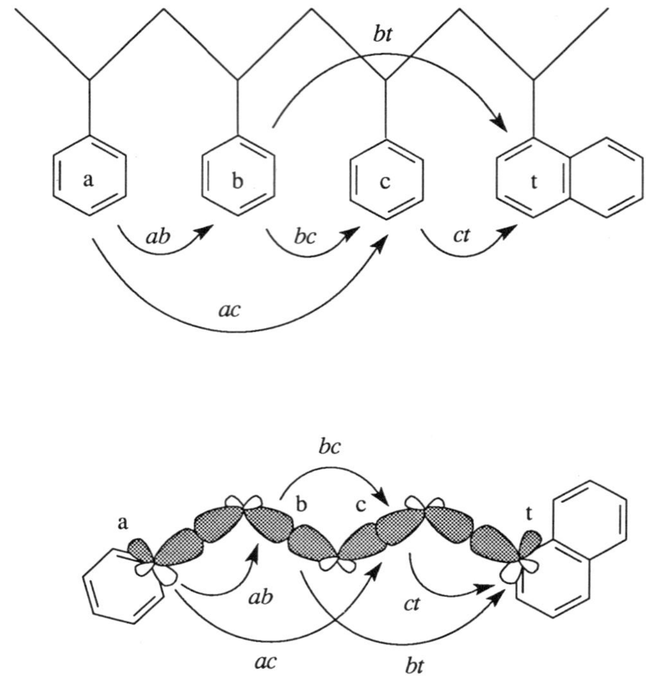

Figure 2.5. (i) Schematic depiction of some matrix elements leading to various pathways for energy migration from donor-like chromophores (a,b and c) to a trap, t, in an aromatic polymer.; and (ii) Dissection of the matrix element, ct, of (i) into contributions from superexchange pathways mediated via σ-bonds of the polymer backbone.

In such a model, energy migration may be considered a special case of donor-acceptor superexchange-mediated energy transfer, in which long-range interactions are, for example, mediated via intervening bonds or solvent (Figure 2.5). The first evidence that a superexchange mechanism could significantly enhance rates of energy transfer was reported by El-Sayed, et al.[90] for dilute mixed aromatic crystals; their observations being later explained by Sternlicht, et al.[91] These authors considered a mixed crystal consisting of host doped with a small concentration of isotopic guest molecules. They postulated that the most important mechanism for promoting long-range triplet-triplet energy migration amongst the traps (isotopic guests) at liquid helium temperature involved indirect trap-to-trap interactions through the virtual states of the host. This work was later extended by Colson and Robinson who employed two different isotopic guests, producing traps of different depths (a trap and a supertrap).[92]

The involvement of σ-bond "spacer" groups in promoting intramolecular triplet-triplet EET was first demonstrated by Closs, et al.[93,94] Subsequent studies have uncovered further evidence to support this observation, as well as extending the investigation to singlet-singlet EET.[77,95-98] It is well-established that such indirect electronic couplings also effect long-range intramolecular electron transfer.[99-103] A particularly lucid introduction is given by Kuki.[104]

A theoretical treatment of the phenomena of electron and energy transfer through bridged systems has been presented by Reimers and Hush.[105,106] In their first paper[105] a time-dependent formalism, modelled about the Robinson and Frosch theory of energy transfer within two-level systems,[23,33] was developed for the coherent transfer of energy between donor and acceptor centres via a bridge. A three-level system was examined, including the case of a resonant bridge. The recent work of Juzeliūnas and Andrews[66] investigates the quantum electrodynamics of resonant energy transfer in condensed matter. The approach employs the concept of bath polaritons mediating the energy transfer (and destroying coherence), and may be considered as a detailed extension of the Löwdin partitioning methods[107] used by for example Evenson and Karplus[108] for purely electronic interactions. A discussion of the key studies of "quasi-incoherent" energy migration in molecular crystals, stemming from the work of Haken and Stobl[109] is provided in Section X.

Quite a number and diversity of experimental investigations into this topic have since ensued,[4] particularly with a view to understanding the dynamics of the energy migration and trapping process. Typically, either "hopping" or "diffusion" models form the basis of such analyses.[9,10] The present work addresses microscopic contributions to this energy migration to a trap process, but from a quantum mechanical viewpoint. It is seen clearly that trap sensitization can be achieved via a coherent process related to superexchange. This may contribute to the high efficiency of trap sensitization which has been observed experimentally.[4]

VIII. Expansion and Evaluation of the General Golden Rule

Projection Operators. The diagrammatic expansion of T (Figure 2.6) essentially illustrates the objective of this Section: to obtain an expression for the transition probability for the system to pass from state $|a\rangle$ to $|b\rangle$, probably via a number of intermediate states, some possibly degenerate with $|a\rangle$ (i.e. we wish to evaluate higher order contributions to Eq. (4.13)). Rather than tackling the problem by attempting to diagonalize an $M \times M$ matrix (where M is the number of states in the sequence), Mower[110] has introduced an expansion of the level shift operator based on a sequence of partial projections. In such a way interpretation of each term in the expansion may be interpreted as a pathway, as when related methods are applied to superexchange-mediated electron-transfer.[89,104,111-113] The most important pathways* give the greatest contribution to the rate. These techniques are, in a sense, an interpretive extension of the Löwdin partitioning method.[107,108]

---

* Where the "importance" of a pathway is determined by a combination of the *magnitude* of the matrix element contributed by it and the *number* of similar pathways of this type.

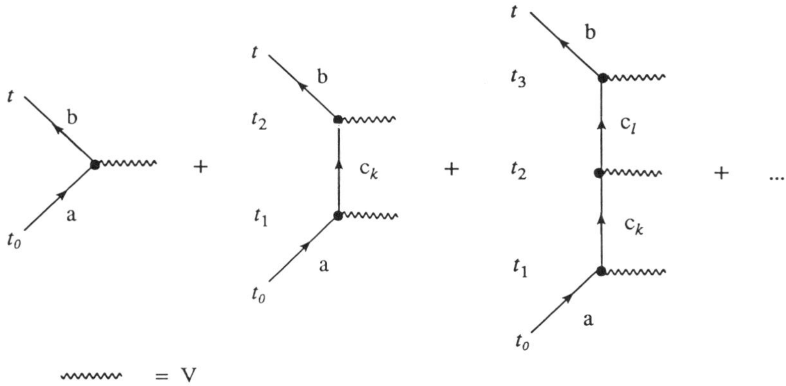

Figure 2.6. Diagrammatic depiction of the first three terms in the expansion of the T-matrix, where time proceeds upwards. Vibronic states, represented by a full line, interact via V at vertices (highlighted by a filled circle).

Consider the situation depicted in Figure 2.6. We define projection operators onto each intervening chromophore (or group) between a and b as P_i, corresponding to a projection onto c_i, and the projection operator Eq. (8.1),

$$P_0 = |a\rangle\langle a| + |b\rangle\langle b| + |c_{-1}\rangle\langle c_{-1}| + |c_{-2}\rangle\langle c_{-2}| + \ldots \tag{8.1}$$

where by c_{-i} we mean relay units or chromophores spatially located away from the donor-trap pathway "vector". Then by defining the complementary projection operators of Eq. (8.2),

$$
\begin{aligned}
Q_0 &= 1 - P_0 \\
Q_1 &= 1 - P_0 - P_1 \\
&\vdots \\
Q_i &= 1 - P_0 - P_1 - P_2 - \ldots - P_i
\end{aligned} \tag{8.2}
$$

we can write a T-matrix describing the vectorial transport of excitation which excludes the possibility of back transfers:

$$
T^{+}(E_a) = V + \sum_{i} VQ_i G_0^{+}(E_a) Q_i V + \sum_{i'<j'} VQ_{i'} G_0^{+}(E_a) Q_{i'} VQ_{j'} G_0^{+}(E_a) Q_{j'} + \ldots \tag{8.3}
$$

Alternatively, an expression which allows all possible interactions could be written, as above, but with complementary projection operators defined in Eq. (8.4) and sums in Eq. (8.3) over $i' \neq j'$, etc.,

$$
\begin{aligned}
Q_0 &= 1 - P_a - P_b \\
Q_i &= 1 - P_i - P_b \\
P_b &= |b\rangle\langle b|
\end{aligned} \tag{8.4}
$$

Note that P_b and P_a are always excluded from the set of intermediate states because b is an energy trap and a is formally omitted to prevent coherence leading to an oscillatory resonance (which would violate the irreversibility criteria of the rate expression).

Expansion of the Coupling Term. We now look at the rate expression Eq. (4.13) more closely. The squared modulus of the transition matrix element in this equation can be evaluated if we note that

$$(\langle b|\mathrm{T}^{+}(E_a)|a\rangle)^{\dagger} = \langle a|\mathrm{T}^{-}(E_a)|b\rangle \tag{8.5}$$

where T^{+} and T^{-} are assumed to be Hermitean operators† and a dagger means Hermitean conjugate, and where

$$\mathrm{T}^{-}(E_a) \equiv (\mathrm{T}^{+}(E_a))^{\dagger} = \mathrm{V}^{*} + \mathrm{V}^{*}\mathrm{G}_0^{-}(E_a)\mathrm{T}^{-}(E_a) \tag{8.6}$$

It is evident that the squared modulus of the T-matrix may be expanded into a sum of "squared" terms and "cross" (or "interference") terms. Employing the relations Eq. (4.12, 4.13, 8.5 & 8.6), the T-matrix will now be expanded.

(i) For the *squared terms* we obtain:

$$|V_{\mathrm{ab}}|^2 + \sum_{\mathrm{c,c'}} V_{\mathrm{ac}}V^{*}_{\mathrm{ac'}}\, V_{\mathrm{cb}}V^{*}_{\mathrm{c'b}} \left\{ \mathcal{P}\frac{1}{E_{\mathrm{a}} - E_{\mathrm{c}}}\mathcal{P}\frac{1}{E_{\mathrm{a}} - E_{\mathrm{c'}}} + i\pi\mathcal{P}\frac{1}{E_{\mathrm{a}} - E_{\mathrm{c}}}\delta(E_{\mathrm{a}} - E_{\mathrm{c'}}) \right. \tag{8.7}$$

$$\left. - i\pi\mathcal{P}\frac{1}{E_{\mathrm{a}} - E_{\mathrm{c'}}}\delta(E_{\mathrm{a}} - E_{\mathrm{c}}) + \pi^2\delta(E_{\mathrm{a}} - E_{\mathrm{c}})\delta(E_{\mathrm{a}} - E_{\mathrm{c'}}) + \ldots \right\}$$

Now, if $E_{\mathrm{a}} \approx E_{\mathrm{c}}$, then the Dirac delta functions dominate and principal values vanish at the poles, so we obtain Eq. (8.8).

---

† This may depend upon choice of system Hamiltonian; particularly in complex molecular systems. See, e.g. p. 863 of Messiah.[46]

$$|V_{ab}|^2 + \pi^2 \sum_c |V_{ac}|^2 |V_{cb}|^2 \, \delta(E_a - E_c)\delta(E_a - E_c) + \ldots \tag{8.8}$$

Conversely, if $E_a \neq E_c$, then the Dirac delta functions vanish and we obtain Eq. (8.9).

$$|V_{ab}|^2 + \sum_{c,c'} V_{ac} V^*_{ac'} V_{cb} V^*_{c'b} \frac{1}{(E_a - E_c)(E_a - E_{c'})} + \ldots \tag{8.9}$$

(ii) The *cross terms* corresponding to Eq. (8.7) are

$$V_{ab} \sum_{c'} V^*_{ac'} V^*_{c'b} \left\{ \mathcal{P} \frac{1}{E_a - E_{c'}} + i\pi\delta(E_a - E_{c'}) \right\} + V^*_{ab} \sum_c V_{ac} V_{cb} \left\{ \mathcal{P} \frac{1}{E_a - E_c} - i\pi\delta(E_a - E_c) \right\} + \ldots \tag{8.10}$$

So, when $E_a \approx E_c$ Eq. (8.10) reduces to Eq. (8.11). Cross terms involving an odd number of electronic matrix elements (e.g. terms one and two in Eq. 8.11) will cancel if the matrix elements are real.

$$\begin{aligned} &i\pi V_{ab} \sum_{c'} V^*_{ac'} V^*_{c'b} \delta(E_a - E_{c'}) - i\pi V^*_{ab} \sum_c V_{ac} V_{cb} \delta(E_a - E_c) \\ &\quad - \pi^2 V_{ab} \sum_{c',d'} V^*_{ac'} V^*_{c'd'} V^*_{d'b} \delta(E_a - E_{c'}) \delta(E_a - E_{d'}) \\ &\quad - \pi^2 V^*_{ab} \sum_{c,d} V_{ac} V_{cd} V_{db} \delta(E_a - E_c) \delta(E_a - E_d) + \ldots \end{aligned} \tag{8.11}$$

In contrast, when $E_a \neq E_c$ we obtain Eq. (8.12a), which is non-zero, and will equal Eq. (8.12b) if the matrix elements are real.

$$V_{ab} \sum_{c'} V^*_{ac'} V^*_{c'b} \frac{1}{E_a - E_{c'}} + V^*_{ab} \sum_c V_{ac} V_{cb} \frac{1}{E_a - E_c} + \ldots \tag{8.12a}$$

$$2V_{ab}\sum_{c} V_{ac}V_{cb}\frac{1}{E_a - E_c} + 2V_{ab}\sum_{c,d} V_{ac}V_{cd}V_{db}\frac{1}{E_a - E_d}\frac{1}{E_a - E_c} + \ldots \qquad (8.12b)$$

Summary and Interpretation. In summary, if E_a is very different from E_c and the matrix elements are real then the expansion of the T-matrix in the general Golden Rule expression (Eq. 4.13) may be written as:

$$|\langle b|T^+(E_a)|a\rangle|^2 = \text{Eq. (8.9)} + \text{Eq. (8.12b)} \qquad (8.13a)$$

$$= V_{ab}^2 + 2\sum_{c}\frac{V_{ab}V_{ac}V_{cb}}{E_a - E_c} + \sum_{c}\left(\frac{V_{ac}V_{cb}}{E_a - E_c}\right)^2 + 2\sum_{c,d}\frac{V_{ab}V_{ac}V_{cd}V_{db}}{(E_a - E_c)(E_a - E_d)}$$

$$+ \sum_{c,d}\left(\frac{V_{ac}V_{cd}V_{db}}{(E_a - E_c)(E_a - E_d)}\right)^2 + \ldots \qquad (8.13b)$$

Whilst, if E_a is of a similar order to E_c (note, that when we refer to "E_a" or "E_c" we mean to denote, not a discrete eigenvalue, but a quasi-continuous distribution of eigenvalues centred about the primary level, which *does* have eigenvalue E_a (or E_c) (cf. Figure 1), and the matrix elements are assumed to be real, then we should write

$$|\langle b|T^+(E_a)|a\rangle|^2 = \text{Eq. (8.8)} + \text{Eq. (8.11)} \qquad (8.14a)$$

$$= V_{ab}^2 + \pi^2\sum_{c}(V_{ac}V_{cb})^2\,\delta(E_a - E_c)\delta(E_a - E_{c'})$$

$$- 2\pi^2\sum_{c,d}V_{ab}V_{ac}V_{cd}V_{db}\,\delta(E_a - E_c)\delta(E_a - E_d)$$

$$+ \pi^4\sum_{c,d}(V_{ac}V_{cd}V_{db})^2\,\delta(E_a - E_c)\delta(E_a - E_{c'})\delta(E_a - E_d)\delta(E_a - E_{d'})$$

$$- 2\pi^4\sum_{c,c',d,e}V_{ac'}V_{c'b}V_{ac}V_{cd}V_{de}V_{eb}\,\delta(E_a - E_{c'})\delta(E_a - E_c)\delta(E_a - E_d)\delta(E_a - E_e) + \ldots \qquad (8.14b)$$

where only terms involving an even number of matrix elements remain uncancelled.

Thus the complementarity of the principal value and Dirac delta functions is evident. Note the similar forms of progressions in the even terms of Eq. (8.13) and Eq. (8.14): squared term - cross terms, ... As the expansion is taken to higher order more and more cross terms are recovered.

The case where $E_a < E_c$ has been described previously for long-range triplet-triplet EET,[91-94,114-116] singlet-singlet energy transfer,[77,95-98,105,106] and particularly for electron transfer.[21,99-104,111,113,117] It is evident from Eq. (8.13), and experiment,[91-101] that intervening relay units which connect the donor and acceptor chromophores, and have higher excitation energies than the donor (e.g. σ-bond "spacer" groups, Figure 2.5b), may significantly alter the rate of excitation transfer. This is facilitated by an indirect coupling between donor and acceptor via virtual excitation transfers involving the relay moieties (electronic tunnelling).

Craig and Thirunamachandran, [118], have examined the third-body mediation of dipolar electronic coupling, in particular where the third body is a solvent medium. They view the macroscopic effect of the medium as the summation of microscopic many-body terms connecting donor and acceptor via the molecules of the medium. In the limit of a large number of intermediate solvent molecules, the direct interaction (and most likely low order relayed interactions) are swamped by higher-order indirect interactions. The resultant macroscopic picture is that of the donor-acceptor pair immersed in a dielectic continuum. This is just one example of a superexchange-mediated interaction. Juzeliūnas and Andrews,[66], have also examined this phenomenon. They employ the concept of bath polaritons mediating the EET, and (as in this work) can treat each molecule in the medium as having an arbitrary number of energy levels. The case where the bath is a dissipative subsystem is also addressed. The treatment, as Lin's does,[5] employs an effective two-state model, and may be used where the bath molecules are identical to the donor and/or acceptor molecules. This, and other examples are treated in detail in Chapter 3.

When $E_a \approx E_c$ and the interaction is within the limits of weak coupling, the rate expression assumes a somewhat different form. The most significant observation is that half of the interference terms, important in modifying the rate of excitation transfer when $E_a \neq E_c$ (see e.g. Ref. 119), vanish in the rate expression for energy migration. A detailed discussion of this case is given in the discussion, Section X.

Relationship to Experiment. Lin has described superexchange-mediated electron-transfer involving spacer units with an energy gap (between donor and spacer) of the order of vibrational energies.[120] Evaluation of the pure second-order contribution to the rate by Lin employed a three-time correlation function, thus maintaining an explicit chronological ordering of events (such an ordering being implicit in Eq. (4.12)). Lin's treatment is related to that for resonance raman scattering[121] and an explicit expression for Franck-Condon factors and the density of states is determined.

Using such an approach, [5-7,47,120,122], the general rate expression (Eq. 6.1) will now be partitioned into products of electronic and nuclear factors with the aim of obtaining an expression closely, related to experiment, which may be evaluated numerically. We proceed by using the integral representation of the delta function:

$$\delta(E_{a_j b^*_s} - E_{a^*_i b_r}) = \frac{1}{2\pi\hbar}\int_{-\infty}^{\infty} \mathrm{d}t \exp\left[\frac{it}{\hbar}\left(E^{(e)}_{a^*b} - E^{(e)}_{ab^*}\right)\right]\exp\left[\frac{it}{\hbar}\left(E^{(v)}_{a_j} + E^{(v)}_{b^*_s} - E^{(v)}_{a^*_i} - E^{(v)}_{b_r}\right)\right] \tag{8.15}$$

where superscripts e and v denote electronic and vibrational energies respectively. Then

$$w = \Gamma \Gamma'^{\text{nucl}}_{(1)} \tag{8.16}$$

where

$$\Gamma = \frac{2\pi}{\hbar} \left| \langle \psi_{a_j} \psi_{b*_s} | \mathrm{T}^{+}(E_{a*_i b_r}) | \psi_{a*_i} \psi_{b_r} \rangle \right|^2 \tag{8.17}$$

with $\psi_{a_j} = \phi_a(q,Q)\, \chi_j^{(a)}(Q)$, etc. and

$$\Gamma'^{\text{nucl}}_{(1)} = \frac{1}{2\pi\hbar} \int_{-\infty}^{\infty} \mathrm{d}t \exp[(\mathrm{i}t \Delta\omega_{ab}^{(e)}\, G'(t)] \tag{8.18}$$

with the electronic energy gap represented by the frequency $\Delta\omega_{ab}^{(e)} = (E_{a*b}^{(e)} - E_{ab*}^{(e)})/\hbar$ and

$$G'(t) = \sum_{i,r} \sum_{j,s} p(a*_i) p(b_r) \exp[\mathrm{i}t \Delta\omega_{\alpha\beta}^{(v)}] \tag{8.19}$$

where $\omega_{\alpha\beta}^{(v)} = (\omega_{a_j}^{(v)} + \omega_{b*_s}^{(v)} - \omega_{a*_i}^{(v)} - \omega_{b_r}^{(v)})$ and $\omega_{g_k}^{(v)}$ is the kth vibrational frequency of the electronic state g. The microcanonical distribution factors are given by, for example,

$$p(b_r) = Z^{-1} \exp\left(-E_{b_r}^{(v)}/kT\right); \quad Z = \sum_i \exp\left(-E_{b_r}^{(v)}/kT\right) \tag{8.20}$$

$G'(t)$, and therefore $\Gamma'^{\text{nucl}}_{(1)}$, may be evaluated by assuming the molecular vibrations to be harmonic.[5,6,47,61] The partitioning of the rate expression into electronic and nuclear factors is straightforward,

$$\Gamma = \Gamma^{\text{elec}} \left| \langle \chi_j^{(a)} \chi_s^{(b*)} | \chi_i^{(a*)} \chi_r^{(b)} \rangle \right|^2 \tag{8.21}$$

with

$$\Gamma^{\text{elec}} = \frac{2\pi}{\hbar} \left| \langle \varphi_{a_j} \varphi_{b*_s} | \mathrm{T}^{+}(E_{a*_i b_r}) | \varphi_{a*_i} \varphi_{b_r} \rangle \right|^2 \tag{8.22}$$

which gives the electronic interaction matrix element for direct or superexchange-mediated EET (cf. Eq. 8.13). Then with Eq. (8.23),

$$G(t) = G'(t) \left| \langle \chi_j^{(a)} \chi_s^{(b*)} | \chi_i^{(a*)} \chi_r^{(b)} \rangle \right|^2 \tag{8.23}$$

Eq. (8.18) becomes Eq. (8.24), which encompasses the nuclear factors for donor and acceptor chromophores,

$$\Gamma_{(1)}^{\text{nucl}} = \frac{1}{2\pi\hbar} \int_{-\infty}^{\infty} dt \exp[(it\Delta\omega_{ab}^{(e)}] \, G(t) \tag{8.24}$$

So that

$$w^{(1)} = \Gamma^{\text{elec}} \Gamma_{(1)}^{\text{nucl}} \tag{8.25}$$

However, if the energy difference between the intermediate states and the initial state are of the order of a vibrational spacing then the dissection of Γ should be considered more carefully. Electronic coupling matrix elements V_{nm} will be assumed to be constant and time-independent. Corresponding Franck-Condon factors will be separated (in the Condon approximation), given a time ordering label and defined as, for example,

$$F_{\beta\alpha}(t) = \langle \chi_j^{(a)} \chi_s^{(b*)} | \chi_i^{(a*)} \chi_r^{(b)} \rangle_t \tag{8.26}$$

Using the relation

$$\pi\delta(x) = \frac{1}{\hbar} \int_0^{\infty} dt \exp\left(\frac{it}{\hbar} x\right) \tag{8.27}$$

and Eq. (4.11, 4.12 and 6.1) we can write

$$\Gamma = \frac{2\pi}{\hbar}\left| V_{\rm ba}F_{\beta\alpha}(t) + \sum_{\rm c} V_{\rm bc}V_{\rm ca}F_{\beta\gamma}(t)F_{\gamma\alpha}(t_1)\frac{\rm i}{\hbar}\int_0^\infty {\rm d}t_1 \exp\left({\rm i}t_1\Delta\omega_{ac}^{(e)}\right)\right.$$

$$\times \sum_{\iota,u}\sum_{j,\nu} \exp\left({\rm i}t_1\Delta\omega_{\alpha\gamma}^{(\nu)}\right) + \sum_{\rm c\neq d} V_{\rm bd}V_{\rm dc}V_{\rm ca}F_{\beta\delta}(t)F_{\delta\gamma}(t_2)F_{\gamma\alpha}(t_1)$$

$$\times\left(\frac{\rm i}{\hbar}\right)^2 \int_0^\infty {\rm d}t_2 \int_0^\infty {\rm d}t_1 \exp\left({\rm i}t_2\Delta\omega_{ad}^{(e)} + {\rm i}t_1\Delta\omega_{ac}^{(e)}\right) \sum_{\iota,u}\sum_{j,\nu}\sum_{u',\nu'} \exp\left({\rm i}t_2\Delta\omega_{\alpha\delta}^{(\nu)} + {\rm i}t_1\Delta\omega_{\alpha\gamma}^{(\nu)}\right)$$

$$\left. + \ \ldots \right|^2 \tag{8.28}$$

No averaging of energy distributions of the intermediate states has been carried out. As noted by Kuznetsov and Ulstrup,[123] the system is then expected to reside in the intermediate states only for a period comparable to the relaxation time of nuclear modes. It is thus expected that a longer population of an intermediate state will break the coherency of the energy migration. This is noted also by Juzeliūnas and Andrews, who treat the EET to be mediated by bath polaritons.[66] In their model, at large separations where the virtual polariton mediating the interaction exists for a time appreciably exceeding the duration of the optical cycle, the polariton acquires real character.

Thus we can write

$$w = \sum_n \Gamma_{(n)}^{\rm elec}\Gamma_{(n)}^{\rm nucl} \tag{8.29}$$

where $\Gamma_{(n)}^{\rm elec}$ are the electronic terms from Eq. (8.22) and $\Gamma_{(n)}^{\rm nucl}$ encompasses nuclear information, and to second order in electronic coupling are given by Eq. (8.24, 8.30-8.33), etc. (cf. Eq. 8.14).

$$\Gamma_{(2)}^{\rm nucl} = \frac{1}{2\pi\hbar}\frac{1}{\hbar^2}\int_{-\infty}^\infty {\rm d}t \int_0^\infty {\rm d}t_1 \int_0^\infty {\rm d}t'_1 \sum_{\rm c} \exp\left[{\rm i}t\Delta\omega_{ab}^{(e)} + {\rm i}(t_1 - t'_1)\Delta\omega_{ac}^{(e)}\right]G(t, t_1, t'_1) \tag{8.30}$$

$$G(t, t_1, t'_1) = G'(t) \sum_{u,v} \exp\left[\mathrm{i}(t_1 - t'_1)\Delta\omega_{\alpha\gamma}^{(v)}\right] |F_{\beta\gamma}|^2 |F_{\gamma\alpha}|^2 \tag{8.31}$$

$$\Gamma_{(1\times3)}^{\mathrm{nucl}} = \frac{1}{2\pi\hbar}\frac{-1}{\hbar}\int_{-\infty}^{\infty}\mathrm{d}t\int_{0}^{\infty}\mathrm{d}t_2\int_{0}^{\infty}\mathrm{d}t_1 \sum_{c\neq d}\left\{\exp\left[\mathrm{i}t\Delta\omega_{ab}^{(e)} + \mathrm{i}t_1\Delta\omega_{ac}^{(e)} + \mathrm{i}t_2\Delta\omega_{ad}^{(e)}\right]G(t, t_1, t_2)\right.$$

$$\left. + \exp\left[\mathrm{i}t\Delta\omega_{ab}^{(e)} - \mathrm{i}t_1\Delta\omega_{ac}^{(e)} - \mathrm{i}t_2\Delta\omega_{ad}^{(e)}\right]G^{\dagger}(t, t_1, t_2)\right\} \tag{8.32}$$

$$G(t, t_1, t_2) = G'(t) \sum_{i,j}\sum_{u,v}\sum_{u',v'} \exp\left[\mathrm{i}t_1\Delta\omega_{\alpha\gamma}^{(v)} + \mathrm{i}t_2\Delta\omega_{\alpha\delta}^{(v)}\right] |F_{\beta\alpha}||F_{\beta\delta}||F_{\delta\gamma}|F_{\gamma\alpha}| \tag{8.33}$$

The time correlation functions, $G(t)$ etc may be evaluated,[5,120,121] essentially by using generating function methods.[61,72] These methods are useful to compute the generalized line-shape function (the density of states weighted by the matrix elements of an arbitrary operator - in this case vibrational overlap integrals) and has the advantage that vibrational motion in the electronic states can be expressed in terms of independent oscillations. Importantly, the correlation function is evaluated at finite temperatures and has been done for single, $G(t)$, and three-time, $G(t, t_1, t_2)$, correlation functions by Lin et al.[5,120,121] for displaced harmonic oscillators. Displaced-distorted and displaced-distorted-rotated harmonic oscillators as well as anharmonic effects have also been examined.[122]

IX. Consideration of Vibronic Effects

This Section examines one role played by nuclear vibrations in EET: the involvement of vibronic mixing in superexchange-mediated processes. This is essential given the obvious significance of vibronically-induced intensity in electronic absorption and emission spectra. Such theory may also be useful in the analysis of excitonic interactions involving vibronic transitions, such as those investigated recently for jet-cooled naphthalene clusters.[124]

A discussion of the implications of such "non-Born-Oppenheimer" corrections to the calculation of long-distance photoinduced electron transfer (PET) rates has been given by Freed,[125] where the suggestion of Beraton and Hopfield[89] that relaxation of the BO approximation leads to an enhancement of calculated rates of long-distance electron transfer is investigated. The consequences of electronic coupling with higher excited states on the calculated rate of PET in relation to a "vibronic mixing coefficient" are examined. Such considerations are extended and generalized here, and related to the early work of Robinson[126] on intensity enhancement of forbidden electronic transitions by weak (intermolecular) interactions.

We will consider the simple donor-bridge-acceptor system A-L-B, where the excitation energy of the bridge moiety, L, (which may be, for example, a saturated hydrocarbon bridge linking aromatic donor and acceptor species) is higher than that of the donor, A, or acceptor, B. In the Condon approximation we may write:

$$T^{+}(E_a(Q_0)) = \langle \varphi_{a^\circ}^{(j)} \varphi_{l^\circ}^{(0)} \varphi_{b'}^{(s)} | \mathrm{V}(q,Q_0) | \varphi_{a'}^{(i)} \varphi_{l^\circ}^{(0)} \varphi_{b^\circ}^{(r)} \rangle \quad \text{(I)}$$

$$+ \sum_{\kappa} \lim_{\eta \to 0^+} \frac{\langle \varphi_{a^\circ}^{(j)} \varphi_{l^\circ}^{(0)} \varphi_{b'}^{(s)} | \mathrm{V}(q,Q_0) | \varphi_{a^\circ}^{(j')} \varphi_{l^\kappa}^{(0)} \varphi_{b^\circ}^{(r')} \rangle \langle \varphi_{a^\circ}^{(j')} \varphi_{l^\kappa}^{(0)} \varphi_{b^\circ}^{(r')} | \mathrm{V}(q,Q_0) | \varphi_{a'}^{(i)} \varphi_{l^\circ}^{(0)} \varphi_{b^\circ}^{(r)} \rangle}{(E_{a'}^{(i)} - E_{a^\circ}^{(j')}) - (E_{l^\kappa}^{(0)} - E_{l^\circ}^{(0)}) + i\eta} \quad \text{(II)} \qquad (9.1$$

where the Roman numerals label terms relevant to later discussion. The superscripts label vibrational levels, while the subscripts denote the electronic state of the donor (a), bridge (l), or acceptor (b) as ground (°), first excited (') or any excited state (κ). Term I represents a direct transfer interaction between donor and acceptor, whilst term II denotes an interaction mediated by a third, bridging, moiety, L. II thus refers to a superexchange interaction.

Now, if we consider the Taylor expansion of the interaction Hamiltonian about the equilibrium geometry Q_0 (Eq. 6.2) it is seen that there will be additional contributions to the transition matrix element such that

$$\mathrm{T}^{+}(E_a(Q)) = \mathrm{T}^{+}(E_a(Q_0)) + \sum_{\iota} \left(\frac{\partial \mathrm{T}^{+}(E_a(Q))}{\partial Q_{\iota}}\right)_0 Q_{\iota} + \ldots \tag{9.2}$$

where

$$\sum_{\iota} \left(\frac{\partial T^{+}(E_a(Q))}{\partial Q_{\iota}}\right)_0 Q_{\iota}$$

$$= \sum_{\iota} \sum_{\lambda} \Bigg\{ \underbrace{\langle \varphi_{\text{final}} | \mathrm{V}(q,Q) | \varphi_{a^{\lambda}}^{(0)} \varphi_{l^{\circ}}^{(0)} \varphi_{b^{\circ}}^{(r)} \rangle \langle \varphi_{a^{\lambda}}^{(0)} \varphi_{l^{\circ}}^{(0)} \varphi_{b^{\circ}}^{(r)} | \partial/\partial Q_{\iota} | \varphi_{\text{initial}} \rangle}_{\text{(III)}}$$

$$+ \underbrace{\langle \varphi_{\text{final}} | \mathrm{V}(q,Q) | \varphi_{a^{\circ}}^{(j')} \varphi_{l^{\lambda}}^{(0)} \varphi_{b^{\circ}}^{(r)} \rangle \langle \varphi_{a^{\circ}}^{(j')} \varphi_{l^{\lambda}}^{(0)} \varphi_{b^{\circ}}^{(r)} | \partial/\partial Q_{\iota} | \varphi_{\text{initial}} \rangle}_{\text{(IV)}}$$

$$- \underbrace{\langle \varphi_{\text{final}} | \partial/\partial Q_{\iota} | \varphi_{a^{\circ}}^{(j')} \varphi_{l^{\lambda}}^{(0)} \varphi_{b^{\circ}}^{(r)} \rangle \langle \varphi_{a^{\circ}}^{(j')} \varphi_{l^{\lambda}}^{(0)} \varphi_{b^{\circ}}^{(r)} | \mathrm{V}(q,Q) | \varphi_{\text{initial}} \rangle}_{\text{(V)}}$$

$$+ \underbrace{\sum_{\kappa} \lim_{\eta \to 0^{+}} \frac{\langle \varphi_{\text{final}} | \mathrm{V}(q,Q) | \varphi_{a^{\circ}}^{(j')} \varphi_{l^{\kappa}}^{(0)} \varphi_{b^{\circ}}^{(r)} \rangle \langle \ldots | \mathrm{V}(q,Q) | \varphi_{a^{\lambda}}^{(0)} \varphi_{l^{\circ}}^{(0)} \varphi_{b^{\circ}}^{(r)} \rangle \langle \ldots | \partial/\partial Q_{\iota} | \varphi_{\text{initial}} \rangle}{(E_{a^{\lambda}}^{(0)} - E_{a^{\circ}}^{(j')}) - (E_{l^{\kappa}}^{(0)} - E_{l^{\circ}}^{(0)}) + i\eta}}_{\text{(VI)}}$$

$$+ \ldots \Bigg\} \tag{9.3}$$

where in term VI an ellipsis within a bra represents a ditto of the preceding ket, and the Herzberg-Teller coupling matrix element is defined as in Eq. (6.8). The matrix element in such a form promotes the mixing of some character of a higher excited state wavefunction $|b\rangle$ into a wavefunction $|a\rangle$ via a nuclear vibration of appropriate symmetry.

One may depict schemes for superexchange-mediated EET, PET and a combination of the two[120] as:

	$D^*LA \rightarrow DL^*A \rightarrow DLA^*$	EET
	$D^*LA \rightarrow D^+L^-A \rightarrow D^+LA^-$	PET
	$D^*LA \rightarrow DL^*A \rightarrow D^+LA^-$	EET & PET
$\equiv$	$|\varphi_{\mathrm{initial}}\rangle \rightarrow |\varphi_{\mathrm{I}}\rangle \rightarrow |\varphi_{\mathrm{final}}\rangle$	

All are described by Equations (9.1) and (9.3), as long as the nomenclature for electronic and vibrational levels is clarified. For example, in the final state of an EET process $\varphi_{a^o}^{(j)}$ represents the *j*th vibrational level of the ground electronic state of the donor (A), whereas for PET it would define the *j*th vibrational level of a radical cation ground state of *a* (D^+).

Visualization and evaluation of such matrix elements may be simplified by the use of diagrams.[127,128] Such methods have been previously employed by Robinson in an examination of the intensity enhancement of a forbidden electronic transition by the influence of a "perturber".[126] Only the terms considered to carry the most significance are given, with their relation to the matrix elements in Eq. (9.1) and (9.3) indicated by Roman numerals. They have been drawn such that time proceeds upwards and energy increases to the right. Thus, solid lines, representing an interaction promoted by $V(q,Q_0)$, represent virtual transitions in all orders of perturbation except the zeroth (diagram I) and the dashed lines are used in this context to represent interactions mediated by vibronic mixing, $\partial H/\partial Q_\iota$.

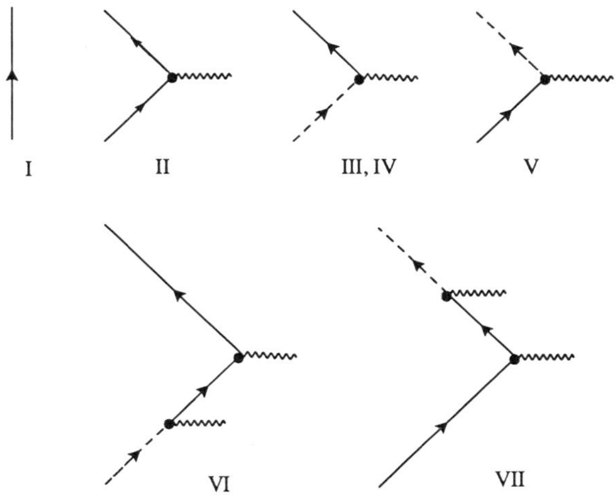

Figure 2.7. Feynman diagrams corresponding to terms in Eq. (9.1) and (9.3). Dashed lines represent interactions mediated by vibronic mixing.

In this picture the direct transfer of energy or an electron [Equation (9.1), term I] is depicted by the corresponding diagram I. If the a' state can couple vibronically with a higher excited state (a^λ) such that the "dashed line" matrix element in diagram III is significant, then some fraction of the total interaction may be mediated via the state a^λ (the "solid line" matrix element in diagram III). Thus the first two terms on the right hand side of Equation (9.3), assigned III and IV, describe a transition from $|\varphi_{\text{initial}}\rangle \rightarrow |\varphi_{\text{final}}\rangle$ via vibronic mixing with a higher excited state of the donor and bridge respectively. In the third term (V) the transition proceeds via a higher excited state of the acceptor; vibronically coupled to the correct final state. Such effects were discussed in Section VI.

The final term, VI, and the next term in Equation (9.3), VII, in comparison with term II of Equation (9.1), demonstrate how borrowed intensity in the donor emission spectrum may contribute to intramolecular EET or PET which is being mediated via a superexchange mechanism. In the case of VI, the superexchange-mediated EET or PET may, in a sense, proceed via a higher electronic excited state of the donor (λ), where such an interaction (even though arising via third-order perturbation theory) may be significant if the vibronic coupling is strong. In this case ($H_{\lambda i}$ / E') is large enough so that $(H_{fl}H_{li} / E) < (H_{fl}H_{l\lambda}H_{\lambda i} / [EE'])$, where the matrix elements and energy gaps have been written symbolically for clarity. Note that there is no energy gap advantage in the denominator, but a significant interaction may be manifested via MO coefficients which provide more favourable overlap (thus improving the efficiency of relayed "short-range" interactions[77]), or larger electronic transition moments (influencing the Coulombic interaction).

An alternative (but equivalent) view is afforded by considering the initial wavefunctions to be an admixture of initially excited and vibronically induced states. Thus diagram III is reduced to diagram I if the initial state $| \varphi_{a'}^{(i)} \varphi_{l\circ}^{(0)} \varphi_{b\circ}^{(r)} \rangle$ + $\sum_{\lambda} \mu_{\lambda} | \varphi_{a\lambda}^{(i)} \varphi_{l\circ}^{(0)} \varphi_{b\circ}^{(r)} \rangle$ is chosen, where the μ_{λ} are small mixing coefficients.

A typical electronic emission or absorption experiment on a molecular system displays a significant degree of borrowed intensity, arising via Equation (9.2). This observation has immediate significance to EET mediated by a direct, through-space dipole-dipole interaction because of the increased effective donor and/or acceptor transition moments. Thus, vibronic interactions should be accounted for in any general theory for energy or electron transfer (they are automatically included in the experimental form of Förster's Equation via the spectral overlap integral). A quite striking observation of increased interchromophore interactions via vibronic coupling is reported by Wessel and Syage[124] for jet-cooled naphthalene dimers.

They observed a larger exciton splitting of vibronically-induced S_1 transitions, relative to the origin system, which may be attributed to the larger Coulombic interaction provided by term III (resulting from the larger transition moment of the naphthalene S_2 transition). One must therefore be careful when matching theory with experiment in such situations.

It is indicated by the experimental results of Oevering et al,[101] that vibronic pathways can be important in some long-distance electron transfer processes. They examined several rigidly-linked dimethoxynaphthalene-dicyanovinyl systems in which, because of the symmetry dictated by the rigid bridge geometry, PET from the lowest-lying naphthalene singlet state (^1L_b) to the dicyanovinyl group is electronically forbidden. The observed rapid through-bond electron transfer was, however, explained by the very strong ^1L_b-^1L_a vibronic coupling in the donor moiety.[129] In such a case the PET proceeds via diagram VI.

In addition, such a pathway may have significance in relayed energy transfer, such as in the naphthalene-bridge-anthracene system recently examined by the author,[98] or in the dimethoxynaphthalene-bridge-ketone systems investigated previously.[95,96] The latter study revealed a distance-dependence for the rate of intramolecular EET which did not agree satisfactorily with a purely superexchange-mediated "Dexter" mechanism, indicating a contribution from direct Coulombic interactions via vibronically-induced intensity is likely. In this case the vibronically-induced emission intensity will contribute not just to the direct Coulombic interaction (formally forbidden by I, however made allowed by, for example, III), but may also affect the through-bond relayed interactions. The contribution via vibronic mixing may be considered a superexchange interaction, which in some instances may provide more favourable pathways for electronic interactions.

X. Discussion and Consequences

In this Chapter the dynamics of electronic excitation transfer (EET) have been described with particular reference to the nuclear component of the interaction. The concepts of preparation (Section II) and evolution (Sections IV and V) of the initial state were discussed, and a general theory for the rate of direct donor to acceptor excitation transfer was derived (Section VI). From this foundation a detailed examination and juxtaposition of relayed interactions was undertaken. In particular, aspects of superexchange-mediated EET and energy migration were investigated (Sections VII to IX).

In order to define a rate of excitation transfer the electronic coupling must be small enough that excitation may be described as being temporarily localized on either donor or acceptor chromophores. It has been shown previously that this corresponds to the case where the magnitude of electronic coupling is less than the vibronic band width.[3,11] These ideas were further elucidated in Section (V) using radiationless transition theory. It is here that the dovetailing of electronic and nuclear aspects of the theory is demonstrated.

When studying *resonance* EET, choice of an appropriate model is not always straightforward. For example, if the system is considered to be well-defined by a two-state model due resonance between $|A^*B\rangle$ and $|AB^*\rangle$ (where an asterisk indicates electronic excitation), with strictly no net energy change or damping (removal) of $|AB^*\rangle$, then a rate of EET cannot be defined.[20,22] This is due to the high probability of recurrence (i.e. return of the excitation to A after the initial transfer to B), and has been pointed out by Robinson and Frosch.[33] However the applicability of an alternative model in "damped" systems has been emphasized by Förster.[3]

Recently Craig and Thirunamachandran[22] have analyzed three ways of dealing with resonance EET within the framework of quantum electrodynamics. Each approach relates to a different regime of electronic coupling or interchromophore separation. The short separation (strong coupling) limit describes initial preparation of one of the $\frac{1}{\sqrt{2}}(|A^*B\rangle \pm |AB^*\rangle)$ stationary states - for which the idea of a rate for EET is inappropriate. The large separation (retarded interaction) limit pertains to uncorrelated events of photon emission by A and absorption by B. This (relativistic) retarded interaction arises due to the finite speed of light and has been discussed by other workers.[130,131] The intermediate separation (weak coupling) regime describes the time-development of the system given initial preparation of the non-stationary state $|A^*B\rangle$, which exhibits the characteristic R^{-6} distance dependence (in the dipole approximation).

This R^{-6} distance dependence reflects use of a Fermi Golden Rule rate expression - essentially describing the decay of a discrete state into a (quasi-) continuum. Such a rate expression is appropriate for this intermediate regime because electronic coupling is less than the natural line width. In this case the final state (quasi-) continuum results from interaction of the molecular electronic excited state with the radiation field.

Kenkre and Knox[132] have been able to expose a clear connection between the strong and weak coupling cases and have employed this in studies of fluorescence depolarization in molecular pairs.[133]

The conditions within which a theoretical expression for the rate may be determined are particularly dependent upon the magnitude of the electronic coupling relative to the width of spectral lines. The emphasis of the present work is on the substantial broadening of spectral lines by *non-radiative decay processes*. The result being that a suitable Fermi Golden Rule rate expression may be employed for much larger electronic couplings than in the absence of such processes.

The basis of the theory presented in this chapter is that the system is not properly represented by a two-level model. In fact, there is a near infinite number of terms in the system Hamiltonian which lead to line broadening and level shifts. Thus, under suitable coupling-strength conditions we may utilize the well-characterized model of a discrete non-stationary state (or quasicontinuum) interacting with a final state quasicontinuum (i.e. a high density of final states, approximately resonant with the initial state). Microscopic dynamics of a transition from one (initial) non-stationary state to another were examined, where the initial and final states represent approximately localized excitation upon donor or acceptor respectively. It was shown that the dynamics could be determined by a Fermi Golden Rule rate expression, Equation (6.1), together with (6.10a) or (6.10b) as appropriate.

In order to unambiguously define a rate of resonance EET the non-stationary state $|A^*B\rangle$ must be prepared initially. As discussed in Section II this may be achieved by exciting the whole envelope of lines (by the superposition principle), or if the Born-Oppenheimer stationary state $\frac{1}{\sqrt{2}}(|A^*B\rangle \pm |AB^*\rangle)$ absorption bands overlap (the weak coupling regime), then excitation into that band will prepare the initial state $c_1|A^*B\rangle + c_2|AB^*\rangle$. Subsequent dynamics of the $|A^*B\rangle \rightarrow |AB^*\rangle$ EET process will then be described by the Fermi Golden Rule expression. It will, in principle, be observable if $\hbar/|V_{AB}| < \Delta t_{\text{prep}}$ (the preparation time, for example the width of the exciting laser pulse). It will exhibit a normal (unretarded) form of distance dependence as long as $\hbar/|V_{AB}| < \tau^{\text{rad}}$ (the radiative lifetime), that is, the excitation may *theoretically* be transferred several times within the lifetime of the excited state.

To describe the decay of the initial state, the total Hamiltonian for the system is separated into two parts: a zero-order part that includes terms corresponding to processes that happen *faster than the time-scale of the radiationless decay,* and a perturbation term that takes the system from one zero-order eigenstate to another *during the experimental time.*[75] Thus, in the *strong coupling case* the interchromophore interaction terms that promote the transfer of electronic excitation would be included in the zero-order part of the Hamiltonian, as is often done in quasi-degenerate perturbation theory.[46] The only non-radiative processes observable then are internal conversion and intersystem crossing from the split supermolecule electronic excited states. Whereas in the *weak coupling case* those terms which promote excitation transfer would contribute to the perturbation term.

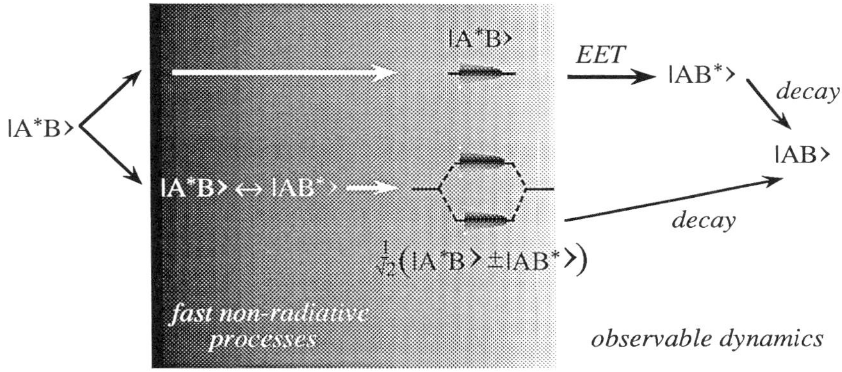

Figure 2.8. A partitioning of photophysical dynamics subsequent to excitation of the initial state (corresponding to excitation localized on chromophore A) into "fast" and "observable" processes. A key difference between the weak and strong coupling cases is that the EET is categorized into the "observable" time regime only for the weak coupling case.

The regimes of strong and weak coupling in electronic excitation transfer may therefore be considered to arise from a *chronological heirachy of events* subsequent to preparation of the initial state (see Figure 2.8). Just as one normally assumes fast vibrational relaxation to an initial Boltzmann distribution, in the strong coupling case there would be a concomitant rapid delocalization of electronic excitation over both (all) chromophores prior to the onset of observable deactivation processes. That is, the excitation transfer process is *not* a rate-determining step in this instance.

In the weak coupling limit, the EET process is slow compared with vibrational relaxation. Thus, once the molecule finds itself in the final state, vibrational relaxation to lower levels causes return to the initial state to be highly unlikely. The problem is therefore related to, but is not quite the same as, that of "quantum mechanical resonance"[30] where the probability for energy transfer oscillates harmonically with time between 0 and 1 [cf. Equation (A.6)] (p.79 of Ref. 32).

It was shown in Section IV that the conditions under which Fermi's Golden Rule applies are that t must be small enough that the transition probability is much less than unity, $w_{ba}t << 1$, with $t >> \hbar(E_\alpha - E_\beta)$ and the density of final states as well as the electronic coupling, $|V_{\alpha\beta}|$, must be approximately constant over nuclear coordinates. For the perturbation method to be a good approximation $|V_{\alpha\beta}|$ must be small. In general the Fermi Golden Rule rate expression is valid as long as $|V_{\alpha\beta}|$ is less than the vibronic band width.

In Sections VII, VIII and IX aspects dictating the rate of EET in supramolecular systems, both natural and synthetic, were investigated by expanding the T-matrix formalism of time-dependent Green's function methodology developed in Section IV. The proposed theory is perturbation-based, but includes the possibility of intermediate chromophores identical to the donor being present in the system (such as is the case in energy migration). This is the first time that such a theory for examining microscopic contributions to the rate of energy migration has been presented.

The goals of these sections were to establish a unified, but lucid, theory for microscopic aspects which determine the rate of energy migration and superexchange-mediated EET processes. Similarities and differences between the mechanisms promoting EET in these two situations were ascertained, particularly enabling us to learn more about the energy migration process. This is pivotal in addressing issues concerning energy migration such as: is it essentially a manifestation of a series of completely uncorrelated, random, incoherent EET steps, or are coherent, superexchange-like pathways available?

We will now focus on these cases; where moieties between donor and acceptor assist the EET process by participating as relay units. After consideration of the theory derived in this Chapter one can envisage three ways by which remote (trap) chromophores may be efficiently sensitized (cf. Figure 2.9):

(i) By a direct donor to trap excitation transfer, as discussed in Section VI.

(ii) The instance where the intermediate moieties have quite different excitation energies to that of the donor. Here, donor-acceptor excitation transfer may be promoted by a superexchange mechanism, as previously discussed.[5,77,91,93-98,114] In this case the T-matrix may be obtained from normal Rayleigh-Schrödinger perturbation theory and thus intermediate excitations are virtual (energy need not be conserved - only between initial and final states).

(iii) When the excitation energies of the relay chromophores are close to those of the donor then an energy migration process may mediate the EET between donor and acceptor. Energy migration is a familiar phenomenon in polymer photophysics, where rates of energy migration to a trap may be measured using time-resolved fluorescence techniques.[4] The relayed excitation transfer in this case appears to involve an electronic tunnelling pathway in addition to a random walk "hopping" mechanism. This is borne out by the similar structures of the rate expressions for (ii) and (iii); whilst the modified form of interference (cross) terms in the expansion of the general Golden Rule rate expression for the case of energy migration is a key difference. In addition, the superexchange mechanism via higher electronic excited states of intermediate chromophores may contribute (cf. the upper potential energy surface of Figure 2.9b).

A further mode of EET may be operating in dispersive excitation transport, where the extent of energy transfer depends on the excitation wavelength.[134] In such an experiment, where the red edge of an inhomogeneously broadened absorption line is excited, energy transport may occur by a coherent (superexchange) mechanism involving chromophores which, due to their local environment, are too high in energy to participate in energy migration.

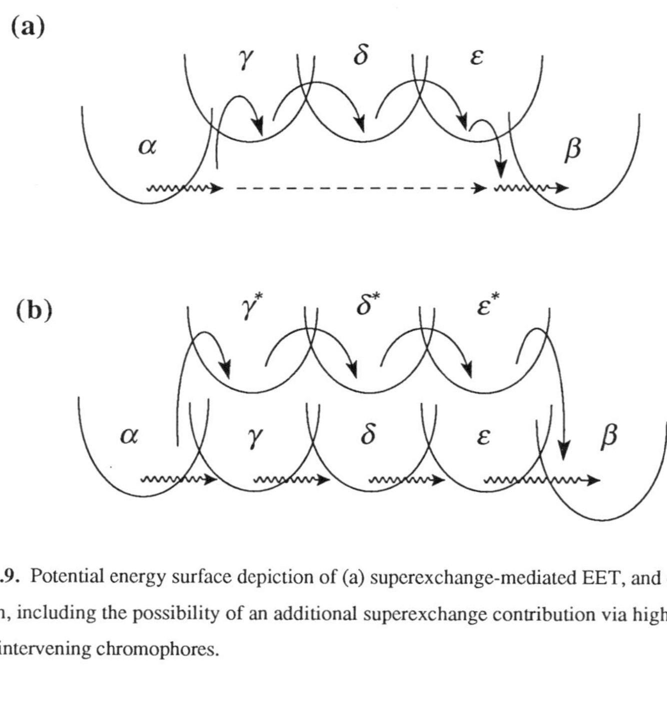

Figure 2.9. Potential energy surface depiction of (a) superexchange-mediated EET, and (b) energy migration, including the possibility of an additional superexchange contribution via higher excited states of intervening chromophores.

Additionally, systems could be synthesized in which the mechanism proposed by Marcus and Almeida[135] is operative. Here, the donor and intermediate chromophore are weakly coupled, but the intermediate chromophore is strongly coupled to the acceptor. The possibility of realizing such a configuration has been recently reported by Kasha, et al,[136] who have investigated an "antenna molecule-fluorescence probe pair" as a means of enhancing the efficiency of probe sensitization.

A theme underlying the present work is the importance of a quantum mechanical viewpoint; so as to account for the interaction *simultaneously* traversing all possible pathways (in Hilbert space) connecting donor and acceptor. This is incongruous with classical principles, which require kinetic models governing the behaviour of the system.

The conceptualization of superexchange pathways is in many ways analogous to Feynman's probability amplitude for a space-time path.[137] In fact, Feynman's review of the quantum mechanical superposition of probability amplitudes given in this article greatly aids one's conceptual understanding of an experiment which probes energy migration. Consider a single photon absorbed by the donor chromophore, a, (which is separated from the trap, b, by just one intermediate donor-like chromophore, i). The probability that the photon is emitted by b is given as the sum of the probability that the excitation is transferred directly from a to b (P_{ab}) and the probability that it is transferred from a via i to b (P_{aib}),

$$P_{ab}^{tot} = P_{ab} + P_{aib} \tag{10.1}$$

Where, using classical principles one would write

$$P_{aib} = P_{ai}P_{ib} \tag{10.2}$$

In quantum mechanics there exist complex numbers φ_{ai}, φ_{ic} and φ_{ab} such that

$$P_{ab} = |\varphi_{ab}|^2,\ P_{ai} = |\varphi_{ai}|^2 \text{ and } P_{ib} = |\varphi_{ib}|^2. \tag{10.3}$$

That is, the P_{mn} are probability amplitudes. Then the classical law (Eq. 10.2) is replaced by

$$\varphi_{aic} = \varphi_{ai}\varphi_{ic} \tag{10.4}$$

A consequence of this quantum mechanical law is that if one makes an attempt to verify it (by observing a photon emitted by i, for example), then the classical law (Eq. 10.2) is recovered. The transposition of such concepts to the energy migration to a trap experiment implies that if one is able to observe an intermediate state (i.e. diagonal matrix element) then the two-step transfer is incoherent (sequential, and defined by a classical probability law). However, if the intermediate state is not an observable, but is virtual in the sense that the overall matrix element (φ_{aic}) involves only off-diagonal terms, then a coherent, or purely quantum mechanical, mechanism (i.e., probability law) is operative. A superexchange (or tunnelling) description is then appropriate.

The interplay between sequential and superexchange mechanisms for electron transfer has been studied by Mukamel et al.,[138,139]. The sequential mechanism was said to proceed via populations of intermediate electronic states (diagonal density matrix elements, which therefore may be regarded as kinetic intermediates) in a succession of incoherent hops, whereas the superexchange proceeds through coherences (off-diagonal matrix elements), that is, an interaction between donor and acceptor is introduced via a coherent mixing of these states with the bridge states (see also Ref. 140). A motivation for such work is discussion concerning the mechanism of electron transfer from photoexcited bacteriochlorophyll dimer to pheophytin in the reaction centres of photosynthetic bacteria.[141] It was postulated that a bacteriochlorophyll monomer may mediate the electron transfer, either as a kinetic intermediate anion, or by participating in a superexchange interaction.

The problem was formulated using the density matrix and was free of limitations imposed by the usual perturbation theory approach. An ancillary conclusion gleaned from the results is that the incoherent and coherent mechanisms do not operate exclusively, but in accord with quantum mechanical ideas, may operate in concert as the energy gap between intermediate and initial states decreases. Thus for some (small) energy gaps it is likely that a complete evaluation of Eq. (8.7) may be required.

In a recent communication by Guo et al.[142] the dynamics of superexchange-mediated electron transfer were investigated by numerically solving the exact time-dependent Schrödinger equation. An interesting feature of this work is the investigation of the (virtual) population on the bridge state. When the bridge state was set higher than *or degenerate with* the donor state, the superexchange mechanism was found to dominate. In such cases the population on the bridge state at any time was extremely small, and decreased with that on the donor state.

Reimers and Hush have developed[105] and investigated[106,143] a time-dependent formalism for reactions in which energy is transferred coherently between centres via a bridge. It is essentially an extension and application of the treatment presented by Robinson and Frosch,[23,33] but with donor and acceptor coupling *only via the bridge state*. An important observation for the case of symmetric, resonant, through-bridge coupling was that the Fermi Golden Rule rate expression was only slightly reduced from that for a symmetric, resonant, two-level system.[105] It was concluded that resonant bridge states would give rise to fast superexchange-mediated reaction rates, even for large bridge lengths.

The second-order and higher terms in the general rate expression (Eq. 4.13) describe coherent, tunnelling, EET mechanisms. This is true whether the virtual states have significantly different energies or similar energies to the initial state, as is evinced by the congruent forms of the energy gap terms (Eq. 8.8, 8.11) and the

Dirac delta terms (Eq. 8.9, 8.12a). Thus, even in energy migration the chromophores intermediate between any donor-acceptor pair (where the acceptor may be either the trap or another donor-like chromophore) can act as virtual relay states to assist EET via a superexchange-like mechanism. Therefore both coherent and incoherent mechanisms operate inseparably. However, because of the small energy gap, the excitation may reside on an intermediate chromophore for a relatively long time without violating the Heisenberg uncertainty principle. If nuclear relaxation occurs during this time (as is quite probable for large kT) then a diagonal (observable) matrix element results – destroying the coherency. That is, incoherencies in the energy transport should become significant at elevated temperatures.

In the Haken-Strobl model[109,144] for the motion of excitons in molecular crystals, the system Hamiltonian is divided into a time-independent part which describes the coherent motion between sites and a time-dependent part which takes into account the influence of vibrations by random fluctuations. It is this second part of the total Hamiltonian which may break the coherency of exciton motion, by introducing interactions with lattice vibrations. Subsequent to this the exciton moves via a hopping process. This model was used to show that at low temperatures the exciton motion is coherent[145] (i.e., whereby an exciton moves without disturbing the thermal equilibrium of the lattice). However, at room temperature the motion of triplet excitons was described as "quasi-incoherent"; that is, the exciton motion may be best described as a hopping process, since phase relations are rapidly destroyed, but on the other hand the hopping rates are essentially determined by the coherent part of the Hamiltonian. The theory was further generalized so as to treat the quasi-incoherent motion of excitons in amorphous, or non-regular, materials.[146]

Munn and Siebrand,[147] using the model of Holstein,[148] also showed that exciton diffusion could be regarded as proceeding by coherent transport at low temperatures and by hopping transport at high temperatures. The meaning of "high" and "low" temperature is, however, difficult to ascertain for large polyatomic molecules. Grover and Silbey[149] found that the exciton migration was be diffusion-like on a long time-scale for all temperatures, as long as exciton-phonon coupling was present.

If the motion of the excitation is best described as purely hopping, or diffusive, then the dynamics of excitation transport in the aggregate are described by an incoherent random walk (i.e., Markoffian dynamics). In this case, each EET step is incoherent and independent of events which occurred previously (i.e., there is a perfect absence of "memory"). A description of the dynamics of energy migration may be developed for these conditions using the Fermi Golden Rule expression of Eqs. (6.1) and (6.9) (where, for this application, the energy gaps between initial and intermediate states must be significant) to obtain Eq. (10.5) – the Pauli master equation[150] (PME).

$$\dot{\rho}_\beta = \sum_\alpha w_{\alpha\beta}\left[\rho_\alpha(t) - \rho_\beta(t)\right] \tag{10.5}$$

Thus, the rate of change of $\rho_\beta(t)$ (the probability density for chromophore β being excited at time t) is the rate of transition to β from all other states in the system, minus the rate of transition out of β (which would be zero for a trap chromophore). $w_{\alpha\beta} = w_{\beta\alpha}$ is the rate constant for EET from α to β and a rate constant for fluorescence decay may also be incorporated into Eq. (10.5). As noted by Kenkre and Knox,[151] the PME approach (e.g. as developed by Förster[1]) makes two assumptions: firstly that after each step the exciton undergoes complete vibrational relaxation (to achieve a thermalized distribution of levels), and secondly the Markoffian (incoherent random walk) approximation, which disregards the possibility of coherent transport mechanisms.

So as to avoid these limitations imposed by the approach described above, Kenkre and Knox[151] have developed a generalized master equation (GME, cf. Ref. 152) to describe excitation transport, Eq. (10.6),

$$\dot{\rho}_\beta = \int_0^t ds \sum_\alpha \left[W_{\beta\alpha}(t-s)\rho_\alpha(s) - W_{\alpha\beta}(t-s)\rho_\beta(s) \right] \tag{10.6}$$

which involves the tetradic $W_{\alpha\beta}(t)$, the memory kernel defined in Ref. 152. The transport dynamics are hence described by a *non-Markoffian* model, with no thermalization assumption, via introduction of this memory function rather than the rate constant $w_{\alpha\beta}$. Kenkre and Knox used a coarse-grained approach so as to include states of the surrounding medium, which may act as a heat bath. They also note the connection of their theory with experimentally-obtained optical spectra.

The generalization of the (Markoffian) PME to the (non-Markoffian) GME has been demonstrated by van Hove,[153] by extending the EET rate constants $w_{\alpha\beta}$ to higher order via Eq. (4.3), thus obtaining $W_{\alpha\beta}(t)$. Under the weak coupling conditions imposed in this work, $W_{\alpha\beta}(t)$ corresponds to the rate expression Eq. (8.29), where the nuclear component includes information regarding the temperature and energy dependencies of damping of the intermediate states; which may lead to destruction of coherency. Such a model is consistent with the "quasi-incoherent" Haken-Strobl model.[144]

Electronic aspects of a coherent energy migration pathway may be illuminated from a nonperturbative variational perspective. One approach is to weakly couple each unperturbed donor (ψ_a) and intermediate (ψ_1, ψ_2, ψ_3) wavefunction with the succeeding chromophore wavefunction:

$$\zeta_3 = \psi_3$$

$$\zeta_2 = \psi_2 + \rho\psi_3 \tag{10.7}$$

$$\zeta_1 = \psi_1 + \sigma\psi_2$$

$$\zeta_a = \psi_a + \theta\psi_1$$

This enables one to write a wavefunction for the donor mixed with the intermediate chromophore wavefunctions:

$$\zeta_a = \psi_a + \theta[\psi_1 + \sigma(\psi_2 + \rho\psi_3)] \tag{10.8}$$

That is,

$$\zeta_a = \psi_a + \lambda\psi_1 + \mu\psi_2 + \nu\psi_3 \tag{10.9}$$

where $\lambda > \mu > \nu$ are small coefficients (and may be given approximately by a perturbation treatment, see Chapter 3).

The initially-excited state is then given by a normalized, antisymmetrized excited ζ_a state (from Eq. 10.9). When interacted with the acceptor state ψ_b the overall electronic coupling matrix element is obtained, its algebraic form reflecting the pathway concepts connoted by the Lippmann-Schwinger approach (particularly when the mixing coefficients are given in terms of the perturbation representation).

Considerations such as these really underpin the superexchange mechanism, and emphasize its import. In fact, delocalization should be significant when the intermediate (relay) moieties are spectroscopically "donor-like". The immediate corollary being that a coherent (tunnelling) mechanism should feature in the energy migration process; acting to augment the interactions between non-nearest neighbour chromophores.

Rate expressions pertinent to energy migration and superexchange-mediated EET may be written explicitly as a sum over all possible pathways connecting donor to acceptor, as was done in Section VIII. However, for the general description of an arbitrary supramolecular system it may not be immediately obvious how to divide the spacer group so as to define the intermediate states. This has been addressed by Liang and Newton,[154] who discuss the cases of highly-localized intermediate states (which has the advantage that the model may be transferable among chemically related systems) and the other limit, where the intermediate state is taken to be the entire set of spacer eigenfunctions (which may improve convergence properties of the perturbation expansion).

The nature of such an expression allows us to approximate the description by neglecting terms in the expansion which one would expect to be small from physical considerations (e.g. terms involving long-distance excitation hops). In such a manner considerable physical insight into the importance of various pathways contributing to the energy migration process may be obtained. The concept of pathways has been invoked previously for the description of superexchange electron transfer processes.[103,111-113,155] In energy migration it is usually assumed that only nearest-neighbour excitation hops have significance. However, for such a weakly-attenuated interaction as dipole-allowed excitation transfer this will likely be a poor approximation. It is probable that considerable improvement would be gained by including the possibility of next-to-nearest-neighbour interactions.[156] Individually each contribution may have a small amplitude, but the summation for these higher-order contributions is over many terms.[†]

[†] Thus the total contribution may be significant, as has been emphasized (particularly by Feynman) for similar problems in quantum field theory.

XI. Conclusions

The goals of this Chapter were to provide a physical basis from which to gain a detailed insight into the microscopic mechanisms underlying electronic energy transfer (EET) phenomena, to provide a general understanding and theoretical formulation of the problem and a sound physical foundation which may be cast in a form related to experiment. Building upon this, superexchange and energy migration phenomena were unified therefore establishing their similarities and differences. The major question addressed was: What can we learn about energy migration? In particular, is it purely a manifestation of a random sequence of uncorrelated steps, or are coherent (superexchange-like) pathways available?

In the first part of the chapter it was shown that due to broadening of vibronic bands, particularly by fast non-radiative decay processes, an EET non-radiative transition may proceed from an initially prepared *non-stationary* state (of the complete system Hamiltonian) to a final state quasi-continuum, and is described by a Fermi Golden Rule rate expression. In general, such a rate expression applies as long as the donor-acceptor electronic coupling is less than the vibronic band width. This corresponds to the Simpson-Peterson weak coupling case, where EET is slow compared with vibrational relaxation. In this limit one should take care not to view resonance EET as a two-level process - in fact there is a near-infinite number of terms in the system Hamiltonian which need to be considered. In the strong coupling limit the emission properties of the supramolecular system are characteristic of excimer or exciplex emission, as appropriate.

In Section VIII a rate expression was developed enabling a juxtaposition of superexchange and energy migration in supramolecular systems. Quantum mechanical considerations of the problem were stressed. It was shown that the observed efficiency of energy migration in many systems may be attributed to contributions from coherent (electronic tunnelling) transfer steps in a non-

Markoffian random walk. Examination of the microscopic EET rate constant developed here related the coherent energy migration steps to superexchange-mediated EET, elucidated its mechanism and developed some expressions related to experiment. Thus, significantly higher rates of energy migration (to a trap) than predicted by a simple Markoffian random walk model are predicted by these considerations of the rate constant, primarily because the interactions are considered to have an effective longer range of influence. The weight of the coherent energy transport mechanism will be greater at lower temperatures, where many of the effects which lead to collapse of coherences to a diagonal matrix element (such as nuclear and solvent motion) occur on a slower time-scale.

Appendix A

The usual approach adopted to describe the time-development of a system is to partition the Hamiltonian into a time-independent part, H_0, and a small time-dependent perturbation, $V(t)$,

$$H = H_0 + \lambda V(t) \tag{A.1}$$

where the solutions to the eigenvalue problem $H_0 \psi_k = E_k \psi_k$ are known. The functions ψ_k form a complete set, so the general solution, Ψ, of the time-dependent Schrödinger equation

$$i\hbar \frac{\partial \Psi}{\partial t} = H\Psi \tag{A.2}$$

can be expressed as[157-159]

$$\Psi = \sum_k c_k(t) \psi_k \exp(-iE_k t / \hbar) \tag{A.3}$$

The unknown coefficients $c_k(t)$ depend on time, and since the eigenvectors ψ_k are orthonormal and provided Ψ is normalized to unity, $|c_k(t)|^2$ gives the probability of finding the system in state k at time t.

The perturbative solution of the time-dependent Schrödinger equation can be found using Dirac's variation-of-constants method,[20,156,159,160] which is particularly useful for describing resonance or transition phenomena requiring first- or second-order interactions between a finite number of states. If the perturbation λV is weak we can expand c_k in powers of λ:

$$c_k = c_k^{(0)} + \lambda c_k^{(1)} + \lambda^2 c_k^{(2)} + \ldots \tag{A.4}$$

where $c_k^{(0)}$ are the initial values of c_k. Now, to first order it can be shown that for a time-independent perturbation ("switched on" at $t = 0$) the transition probability from state a to b ($a \neq b$) is given by

$$P_{ba}(t) = |c_b^{(1)}(t)|^2 = \frac{2}{\hbar^2}|V_{ba}|^2 \frac{1 - \cos \omega t}{\omega^2} \tag{A.5}$$

where $\omega = (E_b - E_a)/\hbar$ and $V_{ba} = \langle b\,|V|\,a\rangle$.

Examination of the function $\frac{1 - \cos \omega t}{\omega^2}$ shows that transitions will mainly occur to those states whose energy, E_b, is contained in a band of width $\delta E \approx 2\pi\hbar/t$ about E_a. The area of this central band will approximate the total area for sufficiently large times, which leads to a transition proportional to time. Two cases of relevance are:

(a) The unperturbed energy is strictly conserved ($\omega = 0$), then

$$P_{ba}(t) = \frac{|V_{ba}|^2\, t^2}{\hbar^2} \tag{A.6}$$

This describes the behaviour of a simple two-level system.[161,162]

(b) The case of more physical interest, where there is a group of final states b having a range of energy levels E_b lying within an interval ($E_b - \delta$, $E_b + \delta$) centred on the energy of the initial state E_a. $P_{ba}(t)$ will obviously be large if $E_b \approx E_a$. We define the number of states b per unit energy as $\rho_b(E_b)$, the density of states.

Recalling that t must be small, so that the c_k don't vary much from $c_k^{(0)}$, we now add the further constraint that t must be much greater than one molecular period:

$$t >> \hbar/E_b \text{ or } \hbar/E_a \tag{A.7}$$

Then the average value of the oscillatory probability function is

$$P_{ba}(t) = \frac{2}{\hbar^2} \int_{E_b - \delta}^{E_b + \delta} |V_{ba}|^2 \frac{1 - \cos \omega t}{\omega^2} \rho_b(E_b)\, \mathrm{d}E_b \tag{A.8}$$

$$= \frac{2}{\hbar^2} |V_{ba}|^2 \rho_b(E_b) \int_{E_b - \delta}^{E_b + \delta} \frac{1 - \cos \omega t}{\omega^2} \, \mathrm{d}E_b \tag{A.9}$$

given that δ is small enough so that V_{ba} and ρ_b are nearly constant within the integration range. Now assume that t is large enough that $\delta >> 2\pi\hbar/t$ and extend the limits of integration from $-\infty$ to $+\infty$; thus

$$P_{ba}(t) = \frac{2\pi}{\hbar} |V_{ba}|^2 \rho_b(E_b)\, t \tag{A.10}$$

We may therefore obtain a probability coefficient, or rate, for the transition:

$$w_{ba} = \frac{\mathrm{d}\, P_{ba}(t)}{\mathrm{d}\, t} = \frac{2\pi}{\hbar} |V_{ba}|^2 \rho_b(E_b) \tag{A.11}$$

which is the well-known Fermi Golden Rule.[159,163,164]

References

(1) Th. Förster, Ann. Physik **2**, 55 (1948).

(2) Th. Förster, Z. Naturforsch **4a**, 321 (1949).

(3) Th. Förster, in *Modern Quantum Chemistry*, edited by O. Sinanoğlu (Academic Press, 1965) Vol. III.

(4) K. P. Ghiggino and T. A. Smith, Prog. Reaction Kinetics **18**, 375 (1993).

(5) S. H. Lin, W. Z. Xiao and W. Dietz, Phys. Rev. E **47**, 3698 (1993).

(6) S. H. Lin, Mol. Phys. **21**, 853 (1971).

(7) S. H. Lin, Proc. Royal Soc. London **A 335**, 51 (1973).

(8) D. L. Dexter, J. Chem. Phys. **21**, 836 (1953).

(9) A. I. Burshtein, Sov. JETP Phys. **35**, 882 (1972).

(10) S. G. Fedorenko and A. I. Burshtein, Chem. Phys. **98**, 341 (1985).

(11) W. T. Simpson and D. L. Peterson, J. Chem. Phys. **26**, 588 (1957).

(12) A. Witkowski and W. Moffitt, J. Chem. Phys. **33**, 872 (1960).

(13) R. L. Fulton and M. Gouterman, J. Chem. Phys. **35**, 1059 (1961).

(14) C. A. Langhoff and G. W. Robinson, Mol. Phys. **26**, 249 (1973).

(15) C. A. Langhoff and G. W. Robinson, Chem. Phys. **6**, 34 (1974).

(16) J. O. Berg, Chem. Phys. Letts. **41**, 547 (1976).

(17) M. Kasha, H. R. Rawls and M. A. El-Bayoumi, Pure Appl. Chem. **11**, 371 (1966).

(18) E. G. McRae and M. Kasha, in *Physical Processes in Radiation Biology*, edited (Academic Press, New York, 1964).

(19) R. McWeeny, *Methods of Molecular Quantum Mechanics;* 2nd ed. (Academic Press, 1992).

(20) D. P. Craig and T. Thirunamachandran, *Molecular Quantum Electrodynamics* (Academic Press, New York, 1984).

(21) R. Omnès, Rev. Mod. Phys. **64**, 339 (1992).

(22) D. P. Craig and T. Thirunamachandran, Chem. Phys. **167**, 229 (1992).

(23) G. W. Robinson and R. P. Frosch, J. Chem. Phys. **37**, 1962 (1962).

(24) G. W. Robinson, in *Excited States*, edited by E. C. Lim (Academic Press, New York, 1974) Vol. 1.

(25) W. Rhodes, J. Chem. Phys. **50**, 2885 (1969).

(26) W. Rhodes, B. R. Henry and M. Kasha, Proc. Natl. Acad. Sci. (USA) **63**, 31 (1969).

(27) T. Ikeda, B. Lee, S. Kurihara, S. Tazuke, S. Ito and M. Yamamoto, J. Am. Chem. Soc. **110**, 8299 (1988).

(28) T. Ikeda, B. Lee, H. Ushiki and K. Horie, J. Chem. Phys. **95**, 6877 (1991).

(29) A. D. McLachlan, Mol. Phys. **8**, 409 (1964).

(30) L. Pauling and E. B. Wilson, *Introduction to Quantum Mechanics* (McGraw-Hill, New York, 1935).

(31) G. V. Varanda and G. S. Agarwal, Phys. Rev. A **48**, 4062 (1993).

(32) G. W. Robinson, J. Molec. Spectrosc. **6**, 58, 509 (1961).

(33) G. W. Robinson and R. P. Frosch, J. Chem. Phys. **38**, 1187 (1963).

(34) C. A. Langhoff and G. W. Robinson, Mol. Phys. **29**, 613 (1975).

(35) G. W. Robinson and C. A. Langhoff, Chem. Phys. **5**, 1 (1974).

(36) D. P. Chock, J. Jortner and S. A. Rice, J. Chem. Phys. **49**, 610 (1968).

(37) M. Bixon and J. Jortner, J. Chem. Phys. **48**, 715 (1968).

(38) M. Bixon and J. Jortner, J. Chem. Phys. **50**, 4061 (1969).

(39) J. Jortner, S. A. Rice and R. M. Hochstrasser, Adv. Photochem. **7**, 149 (1969).

(40) G. W. Robinson and J. O. Berg, Can. J. Phys. **53**, 2068 (1975).

(41) O. K. Rice, Phys. Rev. **33**, 748 (1929).

(42) O. K. Rice, Phys. Rev. **34**, 1451 (1929).

(43) U. Fano, Phys. Rev. **124**, 1866 (1961).

(44) R. A. Harris, J. Chem. Phys. **39**, 978 (1963).

(45) M. L. Goldberger and K. M. Watson, *Collision Theory* (Wiley, 1964).

(46) A. Messiah, *Quantum Mechanics* (Wiley, New York, 1962).

(47) S. H. Lin, J. Chem. Phys. **44**, 3759 (1966).

(48) C. Bloch, Nucl. Phys. **6**, 329 (1958).

(49) T. Kato, Prog. Theor. Phys. **4**, 514 (1949).

(50) T. Kato, Prog. Theor. Phys. **5**, 95, 207 (1950).

(51) V. Hurtubise, J. Chem. Phys. **99**, 265 (1993).

(52) S. Mukamel and J. Jortner, J. Chem. Phys. **60**, 4760 (1974).

(53) S. Mukamel, Ph.D. Thesis, Tel-Aviv University, 1975.

(54) G. W. Robinson, J. Chem. Phys. **47**, 1967 (1967).

(55) J. Guillet, *Polymer Photophysics and Photochemistry* (Cambridge University Press, 1985).

(56) M. Gell-Mann and M. L. Goldberger, Phys. Rev. **91**, 398 (1953).

(57) S. Tomonaga, Prog. Theor. Phys. (Kyoto) **1**, 27 (1946).

(58) J. Schwinger, Phys. Rev. **74**, 1439 (1948).

(59) B. A. Lippmann and J. Schwinger, Phys. Rev. **79**, 469 (1950).

(60) R. Engleman and J. Jortner, Mol. Phys. **18**, 145 (1970).

(61) K. F. Freed and J. Jortner, J. Chem. Phys. **52**, 6272 (1970).

(62) G. R. Hunt, E. F. McCoy and I. G. Ross, Aust. J. Chem. **15**, 591 (1962).

(63) J. P. Byrne, E. F. McCoy and I. G. Ross, Aust. J. Chem. **18**, 1589 (1965).

(64) B. R. Henry and M. Kasha, Ann. Rev. Phys. Chem. **19**, 161 (1968).

(65) G. D. Scholes and K. P. Ghiggino, J. Phys. Chem. **98**, 4580 (1994).

(66) G. Juzeliūnas and D. L. Andrews, Phys. Rev. B **49**, 8751 (1994).

(67) M. Bixon and J. Jortner, J. Chem. Phys. **50**, 3284 (1969).

(68) (a) G. D. Scholes and K. P. Ghiggino, **101**, 1251 (1994).

(b) G. D. Scholes and K. P. Ghiggino, J. Chem. Phys. (submitted).

(69) C. J. Joachain, *Quantum Collision Theory;* 3rd ed. (Elsevier, Amsterdam, 1983).

(70) K. Huang and A. Rhys, Proc. Roy. Soc. (Lond.) **A204**, 406 (1950).

(71) R. Kubo, Phys. Rev. **86**, 929 (1952).

(72) R. Kubo and Y. Toyozawa, Prog. Theor. Phys. **13**, 160 (1955).

(73) Y. E. Perlin, Sov. Phys. Usp. **6**, 542 (1964).

(74) S. H. Lin and R. Bersohn, J. Chem. Phys. **44**, 3768 (1966).

(75) D. M. Burland and G. W. Robinson, Proc. Natl. Acad. Sci. (USA) **66**, 257 (1970).

(76) G. Fischer, *Vibronic Coupling: The Interaction between the Electronic and Nuclear Motions* (Academic Press, London, 1984).

(77) G. D. Scholes and K. P. Ghiggino, J. Photochem. Photobiol. A: Chem. **80**, 355 (1994).

(78) (a) R. D. Harcourt, G. D. Scholes and K. P. Ghiggino, J. Chem. Phys. (in press, December 1, 1994).

(b) G. D. Scholes, R. D. Harcourt and K. P. Ghiggino, J. Chem. Phys. (submitted).

(79) J. Ulstrup and J. Jortner, J. Chem. Phys. **63**, 4358 (1975).

(80) M. Bixon and J. Jortner, Faraday Disc. Chem. Soc. **74**, 17 (1982).

(81) I. Yamazaki, M. Mimuro, T. Murao, T. Yamazaki and Y. Fujita, Photochem. Photobiol. **39**, 233 (1984).

(82) I. Yamazaki, N. Tamai, T. Yamazaki, A. Murakami, M. Mimuro and Y. Fujita, J. Phys. Chem. **92**, 5035 (1988).

(83) D. L. Huber, Phys. Rev. B **20**, 2307, 5333 (1979).

(84) K. A. Peterson and M. D. Fayer, J. Chem. Phys. **85**, 4702 (1986).

(85) M. N. Berberan-Santos and J. M. G. Martinho, J. Phys. Chem. **94**, 5847 (1990).

(86) J. Jortner, S.-I. Choi, J. L. Katz and S. A. Rice, Phys. Rev. Lett. **11**, 323 (1963).

(87) J. Jortner, S. A. Rice, J. L. Katz and S.-I. Choi, J. Chem. Phys. **42**, 309 (1965).

(88) M. Levine, J. Jortner and A. Szöke, J. Chem. Phys. **45**, 1391 (1966).

(89) D. N. Beraton and J. J. Hopfield, J. Chem. Phys. **81**, 5753 (1984).

(90) M. A. El-Sayed, M. T. Wauk and G. W. Robinson, Mol. Phys. **5**, 205 (1962).

(91) H. Sternlicht, G. C. Nieman and G. W. Robinson, J. Chem. Phys. **38**, 1326 (1963).

(92) S. D. Colson and G. W. Robinson, J. Chem. Phys. **48**, 2550 (1968).

(93) G. L. Closs, P. Piotrowiak, J. M. MacInnis and G. R. Fleming, J. Am. Chem. Soc. **110**, 2652 (1988).

(94) G. L. Closs, M. D. Johnson, J. R. Miller and P. Piotrowiak, J. Am. Chem. Soc. **111**, 3751 (1989).

(95) H. Oevering, J. W. Verhoeven, M. N. Paddon-Row and E. Cotsaris, Chem. Phys. Letts. **143**, 488 (1988).

(96) J. Kroon, A. M. Oliver, M. N. Paddon-Row and J. W. Verhoeven, J. Am. Chem. Soc. **112**, 4868 (1990).

(97) G. D. Scholes, K. P. Ghiggino, A. M. Oliver and M. N. Paddon-Row, J. Am. Chem. Soc. **115**, 4345 (1993).

(98) G. D. Scholes, K. P. Ghiggino, A. M. Oliver and M. N. Paddon-Row, J. Phys. Chem. **97**, 11871 (1993).

(99) G. L. Closs and J. R. Miller, Science **240**, 440 (1988).

(100) G. L. Closs, L. T. Calcaterra, N. J. Green, K. W. Penfield and J. R. Miller, J. Phys. Chem. **90**, 3673 (1986).

(101) H. Oevering, M. N. Paddon-Row, M. Heppener, A. M. Oliver, E. Cotsaris, J. W. Verhoeven and N. S. Hush, J. Am. Chem. Soc. **109**, 3258 (1987).

(102) M. D. Newton, Chem. Rev. **91**, 767 (1991).

(103) K. D. Jordan and M. N. Paddon-Row, Chem. Rev. **92**, 395 (1992).

(104) A. Kuki, Structure and Bonding **75**, 49 (1991).

(105) J. R. Reimers and N. S. Hush, Chem. Phys. **134**, 323 (1989).

(106) J. R. Reimers and N. S. Hush, Chem. Phys. **146**, 89 (1990).

(107) P.-O. Löwdin, J. Mol. Spectrosc. **10**, 12 (1963).

(108) J. W. Evenson and M. Karplus, J. Chem. Phys. **96**, 5272 (1992).

(109) H. Haken and G. Strobl, in *The Triplet State,* edited by A. B. Zahlan (Cambridge University Press, Cambridge, Mass., 1967) p. 311.

(110) L. Mower, Phys. Rev. **165**, 145 (1968).

(111) J. N. Onuchic, P. C. P. der Andrade and D. N. Beraton, J. Chem. Phys. **95**, 1131 (1991).

(112) J. N. Onuchic and D. N. Beratan, J. Chem. Phys. **92**, 722 (1990).

(113) J. N. Onuchic and D. N. Beraton, J. Am. Chem. Soc. **109**, 6771 (1987).

(114) G. C. Nieman and G. W. Robinson, J. Chem. Phys. **37**, 2150 (1962).

(115) D. B. MacQueen, J. R. Eyler and K. S. Schanze, J. Am. Chem. Soc. **114**, 1897 (1992).

(116) N. Koga, K. Sameshima and K. Morokuma, J. Phys. Chem. **97**, 13117 (1993).

(117) M. A. Ratner, J. Phys. Chem. **94**, 4877 (1990).

(118) D. P. Craig and T. Thirunamachandran, Chem. Phys. **135**, 37 (1989).

(119) G. D. Scholes, A. H. A. Clayton and K. P. Ghiggino, J. Chem. Phys. **97**, 7405 (1992); (noting that the interference terms discussed here are different to those referred to in this work, the general conclusions, however, will be similar).

(120) S. H. Lin, J. Chem. Phys. **90**, 7103 (1989).

(121) Y. Fujimura and S. H. Lin, J. Chem. Phys. **70**, 247 (1979).

(122) R. Islampour and S. H. Lin, J. Phys. Chem. **95**, 10261 (1991).

(123) A. M. Kuznetsov and J. Ulstrup, J. Chem. Phys. **75**, 2047 (1981).

(124) J. E. Wessel and J. A. Syage, J. Phys. Chem. **94**, 737 (1990).

(125) K. F. Freed, J. Chem. Phys. **84**, 2108 (1986).

(126) G. W. Robinson, J. Chem. Phys. **46**, 572 (1967).

(127) R. P. Feynman, Phys. Rev. **76**, 769 (1949).

(128) I. Lindgren and J. Morrison, *Atomic Many-Body Theory;* 2nd ed. (Springer-Verlag, 1986).

(129) J. R. Reimers, N. S. Hush, D. M. Sammeth and P. R. Callis, Chem. Phys. Letts. **169**, 622 (1990).

(130) D. L. Andrews and G. Juzeliūnas, J. Chem. Phys. **96**, 6606 (1992).

(131) R. R. McLone and E. A. Power, Mathematika **11**, 91 (1964).

(132) V. M. Kenkre and R. S. Knox, Phys. Rev. Lett. **33**, 803 (1974).

(133) (a) T. S. Rahman, R. S. Knox and V. M. Kenkre, Chem. Phys. **44**, 197 (1979).

(b) R. S. Knox and D. Gülen, Photochem. Photobiol. **57**, 40 (1993).

(134) A. D. Stein, K. A. Peterson and M. D. Fayer, J. Chem. Phys. **92**, 5622 (1990).

(135) R. A. Marcus and R. Almeida, J. Phys. Chem. **94**, 2973, 2978 (1990).

(136) M. Kasha, A. Sytnik, J. C. Baum and R. L. Christiansen, XVIth International Conference on Photochemistry, Vancouver, Canada (August 1 – 6 1993).

(137) R. P. Feynman, Rev. Mod. Phys. **20**, 267 (1948).

(138) Y. Hu and S. Mukamel, J. Chem. Phys. **91**, 6973 (1989).

(139) S. Mukamel, Adv. Chem. Phys. **70**, 165 (1988).

(140) M. D. Todd, A. Nitzan and M. A. Ratner, J. Phys. Chem. **97**, 29 (1993).

(141) (a) R. A. Marcus, Chem. Phys. Lett. **133**, 471 (1987).

(b) R. A. Marcus, Chem. Phys. Lett. **146**, 13 (1988).

(142) H. Guo, L. Liu and K.-Q. Lao, Chem. Phys. Lett. **218**, 212 (1994).

(143) J. R. Reimers and N. S. Hush, J. Photochem. Photobiol. A: Chem. **82**, 31 (1994).

(144) H. Haken and G. Strobl, Z. Physik **262**, 135 (1973).

(145) H. Haken and P. Reineker, Z. Physik, **249**, 253 (1972).

(146) P. Reineker and R. Kühne, Z. Physik B **22**, 193 (1975).

(147) R. W. Munn and W. Siebrand, J. Chem. Phys. **52**, 47 (1970).

(148) T. Holstein, Ann. Phys. (N. Y.) **8**, 325 (1959).

(149) M. Grover and R. S. Silbey, J. Chem. Phys. **54**, 4843 (1971).

(150) (a) A. Katz, *Principles of Statistical Mechanics* (W. H. Freeman, 1967).

(b) I. Prigogine, *Non-Equilibrium Statistical Mechanics* (Wiley, 1962).

(151) V. M. Kenkre and R. S. Knox, Phys. Rev. B. **9**, 5279 (1974).

(152) R. Zwanzig, Physica **30**, 1109 (1964).

(153) L. van Hove, Physica **23**, 441 (1957).

(154) C. Liang and M. D. Newton, J. Phys. Chem. **97**, 3199 (1993).

(155) C. A. Naleway, L. A. Curtiss and J. R. Miller, J. Phys. Chem. **95**, 8434 (1991).

(156) B. H. Bransden and C. J. Jochain, *Physics of Atoms and Molecules* (Longman, N.Y. & London, 1983).

(157) E. Merzbacher, *Quantum Mechanics;* 2nd ed. (Wiley, 1970).

(158) W. Heitler, *The Quantum Theory of Radiation* (Clarendon, Oxford, 1954).

(159) (a) P. A. M. Dirac, Proc. Royal Soc. (Lond.) **A112**, 661 (1926).

(b) P. A. M. Dirac, Proc. Royal Soc. (Lond.) **A114**, 243, 710 (1927).

(c) P. A. M. Dirac, *The Principles of Quantum Mechanics;* 4th ed. (Clarendon, Oxford, 1958).

(160) P. W. Langhoff, S. T. Epstein and M. Karplus, Rev. Mod. Phys. **44**, 602 (1972).

(161) J.-M. Lopez-Castillo, A. Filali-Mouhim and J.-P. Jay-Gerin, J. Chem. Phys. **97**, 1905 (1992).

(162) Y. R. Kim, P. Share, M. Pereira, M. Sarisky and R. M. Hochstrasser, J. Chem. Phys. **91**, 7557 (1989).

(163) (a) G. Wentzel, Z. Phys. **43**, 524 (1927).

(b) G. Wentzel, Phys. Z. **29**, 333 (1928).

(164) E. Fermi, *Notes on Quantum Mechanics* (University of Chicago Press, Illinois, 1961).

CHAPTER THREE

Interchromophore Electronic Interactions

I. Preamble

Interactions between chromophores in an excited electronic state have import in a range of instances and applications. This work unifies the theories underlying excimer stability, molecular exciton interactions and electronic aspects of EET. The nature of interactions between a pair of chromophores (one excited) are discussed in detail and a general model, valid over all separations, is developed. Its usefulness is gauged with various *ab initio* calculations, which are presented as a paradigm for the trends in interchromophore interactions with varying separation, particularly exciton interactions and EET.

In Sections IV to VI a theory of the electronic factors in singlet-singlet and triplet-triplet EET is formulated such that a consistent physical picture emerges. This is achieved by the establishment and analysis of a new expression for the exciton resonance interaction term.

The magnitude of interchromophore electronic coupling - which largely governs the rate of EET - is then examined in detail, at separations from 4 Å to 40 Å. A quantitative description of higher multipole terms in the Coulombic interaction is presented, describing their distance and orientation dependencies, and their significance. Short-range effects are also studied and it is shown that, in general, they are not dominated by the Dexter exchange interaction, but that other terms are more important. The paralogism that the Coulombic and exchange mechanisms operate exclusively to each other in singlet-singlet EET is also explicated. Whilst this Chapter deals specifically with the interactions between two identical chromophores, its application to EET between two *different* chromophores is equally valid.

II. Interchromophore Interactions

The present inquiry deals with two major aspects of the interaction between a pair of chromophores which are in an excited electronic state. The first is the lowering of the excited state potential energy surface relative to the ground state by attractive forces in the dimer, which contributes significantly to the stability of excimers and plays an important role in mediating EET at close separations. The second aspect is the direct electronic coupling, or interaction matrix element, which is responsible for the splitting of the excited state potential energy surface (exciton splitting). In the weak coupling limit, both interactions contribute to mediate the transfer of electronic excitation (indicated by an asterisk) from being localized on one chromophore (A) to another (B):

$$|A^*B\rangle \rightarrow |AB^*\rangle$$

This time-dependent process may often be observed provided that excitation can be initially localized on one of the chromophores. The matrix element $\langle A^*B|H|AB^*\rangle$, where H is the complete Hamiltonian operator for the supermolecular system, reflects the magnitude of the coupling between the two (often non-stationary) states.

In the absence of direct interchromophore orbital overlap, two chromophores (one in an excited electronic state) interact via the long-range *Coulombic* mechanism. This interaction is specified by a two-electron integral which describes a simultaneous deexcitation and excitation of the two chromophores.[1] It thus denotes an interaction between the dipole transition moments of the two molecules, and is in many ways analogous to the classical electrostatic interaction between two real dipoles. The Coulombic potential may be conveniently represented as a multipolar expansion and in the usual treatment of molecular exciton interactions or EET just the dipole-dipole term is considered.[2] In Section XI the effect and moment of higher order terms are considered and the validity of the dipole approximation is critically examined.

When the electron density distributions (i.e. orbitals) on the two chromophores overlap, additional effects arise which can dramatically increase the electronic coupling. For example, as a consequence of the Pauli principle the Dexter exchange interaction - an integral directly related to the Coulombic integral by the antisymmetrizing operator - is often considered.[3,4] In the treatment of EET, two mechanisms are generally thought to prevail: the Coulombic interaction and this exchange interaction. Usually one is considered to act exclusively to the other, but in this work it is demonstrated that this is often a poor approximation. A unifying model is developed in Sections IV to VI, which accounts for both Coulombic interactions and contributions engendered by interchromophore orbital overlap. It is shown that other significant interactions depending on interchromophore orbital overlap should be considered in addition to the exchange term.

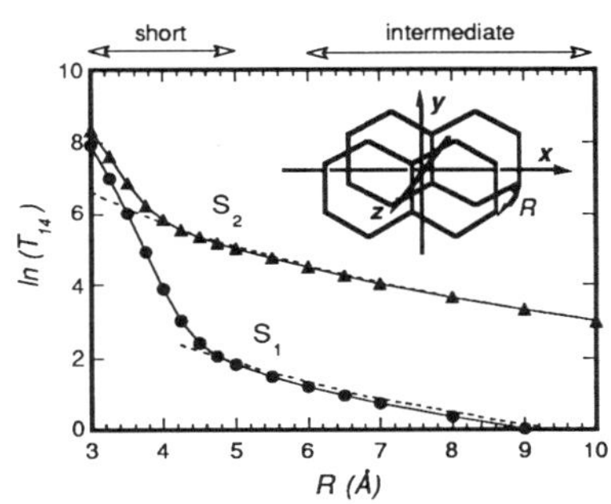

Figure 3.1. Variation of the resonance interaction, T_{14}, with interchromophore separation, R, for the two lowest singlet naphthalene electronic excited states. Solid lines depict data calculated by the CIS method for the D_{2h} symmetry dimer (STO-3g basis set), dashed lines indicate the interaction calculated via the dipole approximation. (From Ref. 17).

It is convenient to partition interchromophore interactions into three regimes; encompassing short-, intermediate-, and long-range interactions respectively. The boundary between short- and intermediate-range cases can be seen quite clearly in a log-scale plot of the calculated direct interaction matrix element, T_{14}, versus separation, R (Figure

3.1) as a marked change in gradient. Orbital overlap effects contribute only to the short-range limit, which therefore provides the most general expression for interchromophore interactions. It accounts for excimer stability as well as molecular exciton interactions and EET at close separations (approximately 3 to 6 Å for the naphthalene D_{2h} dimer configuration using this method and basis set). The intermediate regime (approximately 7 to 150 Å) is, in general, most commonly applicable to exciton interactions and EET. Overlap effects are insignificant and the Coulombic interaction mediates excitation resonance, simplifying to the leading (dipole-dipole) term at large separation.[2,5-8] The long-range interaction, which will be briefly considered in Section X, involves a modification to the dipole-dipole interaction by retardation effects.[9-11]

As discussed in Chapter 2, such electronic interactions may be efficiently mediated via relay chromophores, solvent, or connecting bonds, etc. Some electronic aspects of these superexchange interactions will be discussed in Section XIII.

This adumbration is elaborated upon in the following sections, whereupon calculations and a detailed interpretation of the results are provided (Sections XI and XII. A general discussion of the implications and application of this work is given in Section XIV.

III. Dimer Stationary States

It is desirable to express interchromophore interactions in terms of properties of the isolated constituent chromophores. If two chromophores, A and B, are weakly interacting, then their individuality is retained to a good approximation, and the total wavefunction may be expressed as a product of the A and B wavefunctions, Φ^A and

Φ^{B} respectively. The Pauli principle is satisfied by symmetry-adapting the total wavefunction (using the antisymmetrizer, A) then normalizing (by the constant N). The supermolecule wavefunction is then given by[1]

$$\Phi^{AB} = N\,\mathrm{A}\,(\Phi^{A}\Phi^{B}) \tag{3.1}$$

which, if the two chromophores were so well separated that the interaction terms were negligible, would yield an energy $E = E^{A} + E^{B}$ for a Hamiltonian $\mathrm{H} = \mathrm{H}^{A} + \mathrm{H}^{B}$.

Such a partitioning, however, prohibits the possibility of interchromophore electron transfer or exchange, since this would violate the orthogonality requirements imposed upon the generalized product. It is possible to account for non-orthogonality and short-range effects by using density functions rather than orbitals, however, we have chosen to pursue such an investigation using a variational treatment. The problem may then be stated in terms of a visualizable localized molecular orbital representation.

In the following exposition we will consider excitation resonance between a pair of model two-level systems. The hybrid wavefunction for the closed-shell ground state may be represented by the determinant (3.2), where unbarred spin orbitals (a and b) are associated with an electron of α spin, barred orbitals ($\bar{a}$ and $\bar{b}$) with an electron of β spin. M is a normalizing constant.

$$\psi_0 = M\left|ab\bar{a}\bar{b}\right|. \tag{3.2}$$

For the present treatment it is convenient to consider spin orbitals which are localized on the individual chromophores (which may be, for example, Hartree-Fock molecular orbitals for each isolated chromophore). In that case orbitals localized on the one chromophore are orthonormal (i.e., $\langle a|a'\rangle = \delta_{aa'}$), whereas interchromophore orbital overlap is accounted for (i.e., $\langle a|b\rangle = S_{ab}$).

In the limit of infinite separation the dimer has the degenerate excited states, representing localized excitations, $|A^*B\rangle$ and $|AB^*\rangle$. At smaller separations resonance forces arise which remove the degeneracy in the hybrid wavefunction. The correct zero-order excited state wavefunctions are then given by linear combinations of the two states: $\frac{1}{\sqrt{2}}\{|A^*B\rangle \pm |AB^*\rangle\}$. In such a way resonance produces a splitting of the potential energy surface, which is sometimes referred to as exciton splitting.

The correct zero-order exciton resonance (ER) states are therefore (where symmetry labels refer to the D_{2h} "sandwich" dimer with the x-axis defining the intermolecular axis and the y-axis parallel to the electronic dipole transition moment of one molecule):

$$\begin{aligned}\Phi^{ER}_{(B_{1g})} &= \frac{1}{\sqrt{2}}\{|A^*B\rangle + |AB^*\rangle\} \\ &= \frac{(\psi_1 + \psi_4)}{\sqrt{2 + 2\langle\psi_1 | \psi_4\rangle}} \\ &\approx \frac{1}{\sqrt{2}}\left\{\psi_1 + \psi_4\right\}\end{aligned} \tag{3.3a}$$

since $\langle\psi_1 | \psi_4\rangle \ll 1$, and similarly

$$\begin{aligned}\Phi^{ER}_{(B_{1u})} &= \frac{1}{\sqrt{2}}\{|A^*B\rangle - |AB^*\rangle\} \\ &\approx \frac{1}{\sqrt{2}}\left\{\psi_1 - \psi_4\right\} .\end{aligned} \tag{3.3b}$$

It has been postulated that the configuration interaction (CI) between exciton resonance (ER) and charge resonance (CR) states is important in the correct description of close range interactions, for example in excimers.[12-14] Thus, in addition to the excitation resonance, it may be necessary to consider the influence of ionic configurations, where the correct zero-order charge resonance (CR) states are

$$\Phi^{CR}_{(B_{3g})} \approx \frac{1}{\sqrt{2}}\{|A^-B^+\rangle + |A^+B^-\rangle\}$$
$$\approx \frac{1}{\sqrt{2}}\{\psi_2 + \psi_3\} \qquad (3.4a)$$

$$\Phi^{CR}_{(B_{2u})} \approx \frac{1}{\sqrt{2}}\{|A^-B^+\rangle - |A^+B^-\rangle\}$$
$$\approx \frac{1}{\sqrt{2}}\{\psi_2 - \psi_3\} \qquad (3.4b)$$

Mixing of the ER and CR states produces the final dimer stationary states shown in Figure (3.2).

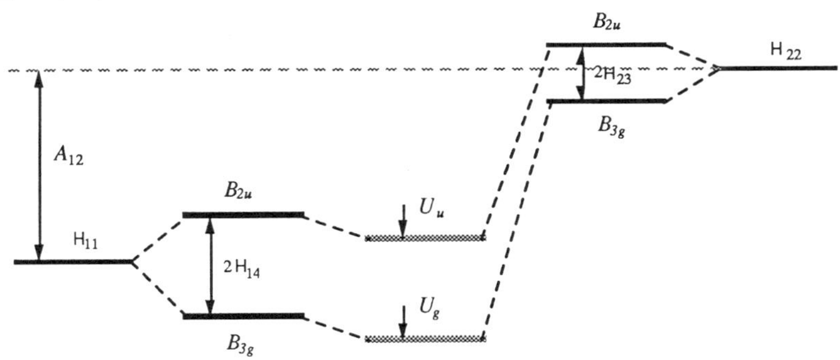

Figure 3.2. Schematic energy level diagram illustrating a dissection of the interactions in an excited state dimer (symmetry labels are for a D_{2h} "sandwich" dimer). Potential energy surfaces of the locally-excited (H_{11}) and ionic, or charge transfer, (H_{22}) configurations are split by the resonance interactions H_{14} and H_{23} respectively, to give the ER and CR wavefunctions. Interaction between these configurations (denoted by U) yields the final dimer states.

IV. The Model

To elucidate the nature of the electronic interactions promoting EET we consider the model involving two equivalent monomer chromophores, A and B, and the four orbital configurations depicted in Figure 3.3. Two of these configurations, $\psi_1(A^*B)$ and $\psi_4(AB^*)$, represent locally-excited configurations of the dimer. The other two, $\psi_2(A^+B^-)$ and $\psi_3(A^-B^+)$, are ionic configurations that can interact with ψ_1 and ψ_4 when the separation between the monomers is finite. The initial ("reactant") and final ("product") wavefunctions (Ψ_L and Ψ_R) then involve admixtures of either of (normalized) ψ_1 or ψ_4 together with both of ψ_2 and ψ_3:

$$\Psi_L \approx N(\psi_1 + \lambda\psi_2 + \mu\psi_3) \tag{4.1a}$$

$$\Psi_R \approx N(\psi_4 + \mu\psi_2 + \lambda\psi_3) \tag{4.1b}$$

with the normalizing factor

$$N = (1 + 2\lambda S_{12} + 2\mu S_{13} + \lambda^2 + \mu^2 + 2\lambda\mu S_{23})^{-1/2} \tag{4.2}$$

and overlap integrals defined by $S_{ij} = \langle\psi_i|\psi_j\rangle$. Such mixing between locally-excited and ionic configurations was found to be significant at close separations in the study of transannular interactions in ethene dimers by Koutecký and Paldus,[15] and in the correct description of excimer interactions.[12,13]

Other configurations, for example those that arise from $(b)^2 \rightarrow (b')^2$ and $(a)^2 \rightarrow (a')^2$ excitations for ψ_1 and ψ_4 respectively, have been excluded from the present treatment.

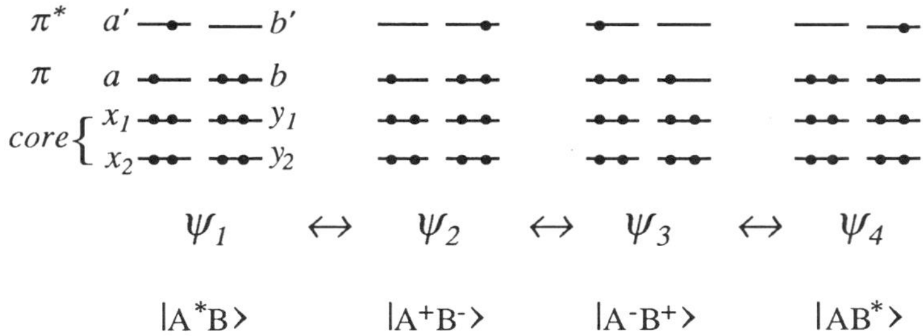

Figure 3.3. Representation of four possible electronic configurations of an excited state dimer. Core orbitals, x_l and y_k are kept doubly-occupied whilst properties of the configuration are determined by the active space electrons which occupy the π-electron orbitals a and a' (on molecule A) and b and b' (on molecule B). ψ_1 and ψ_4 represent locally excited configurations of molecule A (A*B) and B (AB*). ψ_2 and ψ_3 define the ionic configurations A^+B^- and A^-B^+ respectively (which are to be mixed with the locally excited configurations).

In order to ascertain the nature of the interactions that effect EET, we shall examine the Hamiltonian matrix element $\langle \Psi_L | H | \Psi_R \rangle$; the interaction that connects the two states Ψ_L and Ψ_R. The theory that follows is associated with the valence bond theory described previously[16] for O_2.

For the secular equations of Eq. (4.3),

$$\sum_J (H_{IJ} - ES_{IJ})C_J = 0, \quad I, J = L \text{ or } R \tag{4.3}$$

with $H_{IJ} = \langle \Psi_I | H | \Psi_J \rangle$ and $S_{IJ} = \langle \Psi_I | \Psi_J \rangle$, the overlap and off-diagonal Hamiltonian matrix elements are given by Eqs. (4.4) and (4.5) respectively,

$$S_{LL} = N^2\left\{1 + \lambda^2 + \mu^2 + 2\lambda S_{12} + 2\mu S_{13} + 2\lambda\mu S_{23}\right\} \tag{4.4a}$$

$$S_{LR} = N^2\left\{S_{14} + 2\lambda\mu + 2\mu S_{12} + 2\lambda S_{13} + (\lambda^2 + \mu^2)S_{23}\right\} \tag{4.4b}$$

$$H_{LR} = N^2\left\{H_{14} + 2\mu H_{12} + 2\lambda H_{13} + (\lambda^2 + \mu^2)H_{23} + \lambda\mu(H_{22} + H_{33})\right\} \quad (4.5)$$

in which use has been made of the equivalences $H_{12} = H_{34},\ H_{13} = H_{24},\ S_{12} = S_{34}$ and $S_{13} = S_{24}$.

To demonstrate how to simplify Eq. (4.5), the energies of ionic configurations will be explicitly written relative to that of the locally excited configuration ψ_1: $H_{22} = H_{11} + A_{12}$ and $H_{33} = H_{11} + A_{13}$, where A_{1n} is the energy gap between configurations 1 and n (i.e. $E_n - E_1$). The diagonal and off-diagonal Hamiltonian matrix elements $H_{LL} \equiv H_{RR}$ and H_{LR} from the secular equations of Eq. (4.3) are thus given by Eqs. (4.6) and (4.7).

$$\begin{aligned} H_{LL} &= H_{11} + N^2\left\{2\lambda(H_{12} - S_{12}H_{11}) + 2\mu(H_{13} - S_{13}H_{11}) + A_{12} + \mu^2 A_{13}\right. \\ &\qquad \left. + 2\lambda\mu(H_{23} - S_{23}H_{11})\right\} \qquad (4.6a) \\ &\equiv H_{11} + T_{LL} \qquad (4.6b) \end{aligned}$$

$$\begin{aligned} H_{LR} &= S_{LR}H_{11} + N^2\left\{(H_{14} - S_{14}H_{11}) + 2\mu(H_{12} - S_{12}H_{11}) + 2\lambda(H_{13} - S_{13}H_{11})\right. \\ &\qquad \left. + \lambda\mu(A_{12} + A_{13}) + (\lambda^2 + \mu^2)(H_{23} - S_{23}H_{11})\right\} \qquad (4.7a) \\ &\equiv S_{LR}H_{11} + T_{LR} \qquad (4.7b) \end{aligned}$$

where we have introduced the notation $T_{LL} = H_{LL} - S_{LL}H_{11}$ and $T_{LR} = H_{LR} - S_{LR}H_{11}$.

The secular equations (Eq. 4.3) may then be redefined according to Eq. (4.8),

$$\sum_J (T_{IJ} - \varepsilon S_{IJ})C_J = 0 \qquad (4.8)$$

with $\varepsilon = E - H_{11}$, and it is seen that the transfer matrix element T_{LR} of Eq. (4.7) is then the most convenient form of Hamiltonian matrix element which connects Ψ_L and Ψ_R. If S_{LR} is omitted, as occurs in the subsequent treatment, it may be interpreted as the stabilization energy of Ψ_L relative to ψ_1 when Ψ_L interacts with Ψ_R (i.e., yielding the final states represented in Figure 3.2).

Thus, T_{LR} ($= H_{LR} - S_{LR}H_{11}$), rather than H_{LR}, is the correct interaction matrix element for EET. As will become evident, this form of the interaction matrix element is essential for the correct description of the orbital overlap-dependent contributions to this electronic coupling.

It is evident from Eq. (4.7) that

$$T_{LR} = N^2\{(H_{14} - S_{14}H_{11}) + 2\mu(H_{12} - S_{12}H_{11}) + 2\lambda(H_{13} - S_{13}H_{11}) + \lambda\mu(A_{12} + A_{13}) + (\lambda^2 + \mu^2)(H_{23} - S_{23}H_{11})\} \tag{4.9a}$$

or equivalently, in the abbreviated notation,

$$\boxed{T_{LR} = N^2\{T_{14} + 2\mu T_{12} + 2\lambda T_{13} + \lambda\mu(A_{12} + A_{13}) + (\lambda^2 + \mu^2)T_{23}\}} \tag{4.9b}$$

Now, by invoking perturbation theory (since the third order contribution is negligible), the mixing coefficients are given by Eq. (4.10)

$$\lambda \approx \frac{-H_{12} + S_{12}H_{11}}{A_{13}} \tag{4.10a}$$

$$\mu \approx \frac{-H_{13} + S_{13}H_{11}}{A_{12}} \tag{4.10b}$$

Then retaining terms which contribute up to second order in orbital overlap (and therefore putting $N^2 = 1$), the T_{LR} of Eq. (4.9) may be written, to a good approximation, as Eq. (4.11)* .

$$T_{LR} \approx T_{14} - T_{12}T_{13}\left(\frac{1}{A_{12}} + \frac{1}{A_{13}}\right) \tag{4.11}$$

with
$$T_{ij} = H_{ij} - S_{ij}H_{11} \tag{4.12}$$

Similarly,

$$T_{LL} \approx -\frac{T_{12}^2}{A_{12}} - \frac{T_{13}^2}{A_{13}} \tag{4.13}$$

The transfer matrix element T_{LR} of Eq. (4.11) is a simplified form of T_{LR} in Eq. (4.9) and gives clearly the primary interaction terms that promote donor-acceptor EET. These are:

(a) A *direct exciton resonance interaction*, T_{14}, which is associated with a two-electron transfer between ψ_1 and ψ_4.

(b) The *through-configuration exciton resonance interaction* of Eq. (4.14),

$$-(H_{12} - S_{12}H_{11})(H_{13} - S_{13}H_{11})\left(\frac{1}{A_{12}} + \frac{1}{A_{13}}\right) \equiv -\frac{T_{12}T_{24}}{A_{12}} - \frac{T_{13}T_{34}}{A_{13}} \tag{4.14}$$

* The result may also be obtained by putting $H_{LR} = N \langle \Psi_L |H| \psi_4 \rangle$, which, proceeding as above, gives Eq. (4.11).

which are associated with two equivalent one-electron transfer superexchange processes mediated via the bridging, ionic configurations, namely $\psi_1 \rightarrow \psi_2 \rightarrow \psi_4$ and $\psi_1 \rightarrow \psi_3 \rightarrow \psi_4$.

We begin by examining the electronic coupling matrix elements within a 4-orbital, 4-electron framework (i.e. Figure 3.3, excluding the core electrons). In Sections VIII and IX we will inquire as to the applicability of these results to many-electron calculations.

V. The Direct Exciton Transfer Interaction Matrix Element

The nature of the interactions which *directly* promote donor-acceptor EET is elucidated here by considering the direct interaction matrix element which connects the two configurations ψ_1 and ψ_4, that is, T_{14} (in Eqs. 4.9 and 4.11).

The interaction matrix element for a two-electron transfer between a pair of model, two-level, systems is first considered. The model is a useful paradigm for a many-electron system in that it treats only the "active" outer valence electrons, keeping the comparatively unperturbed core electrons frozen. The cases of singlet-singlet and triplet-triplet EET are compared.

Orbitals a (a') and b (b') represent the π-electron HOMO (LUMO) MOs on molecules A and B respectively (Figure 3.3), then orbitals a and a' (b and b') are orthogonal. The relevant Slater determinantal wavefunctions for the singlet-singlet ($S = M_S = 0$) and triplet-triplet ($S = 1, M_S = 0$) cases are:

$$^{1,3}\psi_1 = N_1 \left(\left| a\bar{a}'b\bar{b} \right| \pm \left| a'\bar{a}b\bar{b} \right| \right) \tag{5.1}$$

$$^{1,3}\psi_4 = N_4 \left(\left| a\bar{a}b\bar{b}' \right| \pm \left| a\bar{a}b'\bar{b} \right| \right) \tag{5.2}$$

where the absence or presence of a bar over the fragment MO indicates that the associated spin wavefunction is α or β respectively. The normalizing coefficients are (up to second order in orbital overlap):

$$N_1 = N_4 = (2 - 2S_{ab}^2 - 2S_{a'b'}^2)^{-1/2} \tag{5.3}$$

Hence, by expanding the Slater determinants of Eqs. (5.1) and (5.2), it is seen that the electronic coupling matrix elements have the form (retaining terms to second order in orbital overlap, and therefore setting normalizing constants to $1/\sqrt{2}$):

$$\begin{aligned} {}^{1,3}H_{14} &= \langle a\bar{a}'b\bar{b} \mid \mathrm{H} \mid a\bar{a}b\bar{b}' - b\bar{a}a\bar{b}' - a\bar{b}'b\bar{a} + b\bar{b}'a\bar{a} \rangle \\ &\quad \pm \langle a\bar{a}'b\bar{b} \mid \mathrm{H} \mid a\bar{a}b'\bar{b} - b'\bar{a}a\bar{b} - a\bar{b}b'\bar{a} + b'\bar{b}a\bar{a} \rangle \\ &= J - Z \pm J - P' \end{aligned} \tag{5.4}$$

in which $\mathrm{H} = \sum_{\mu} \mathrm{h}(\mu) + \sum_{\mu<\nu} r_{\mu\nu}^{-1}$, with $\mathrm{h}(\mu) = -\frac{1}{2}\nabla^2(\mu) - \sum_n \frac{Z_n}{r_{n\mu}}$ in atomic units. Core (1-electron) integrals are specified by $h_{ij} \equiv \langle i(\mu) | \mathrm{h}(\mu) | j(\mu) \rangle$ and two-electron integrals are written in the notation $(ik|jl) \equiv \langle i(\mu)\, j(\nu)\, | r_{\mu\nu}^{-1} |\, k(\mu)\, l(\nu) \rangle$, which highlights their orbital overlap dependencies, and

$$J = (a'a|bb') \tag{5.5a}$$

$$Z = (a'b'|ba) \tag{5.5b}$$

$$\begin{aligned} P' &= h_{aa}S_{ab}S_{a'b'} + h_{bb}S_{ab}S_{a'b'} + h_{a'b'}S_{ab} + h_{ab}S_{a'b'} + (aa|bb)S_{ab}S_{a'b'} \\ &\quad + (aa|a'b')S_{ab} + (bb|a'b')S_{ab} + (ab|aa)S_{a'b'} + (ab|bb)S_{a'b'} \end{aligned} \tag{5.5c}$$

Similarly we have

$$^{1,3}S_{14} \approx -S_{ab}S_{a'b'} \tag{5.6}$$

$$^{1,3}H_{11} \approx h_{aa} + 2h_{bb} + h_{a'a'} + 2(aa|bb) + (aa|a'a') + 2(a'a'|bb) + (bb|bb) \pm J_0 \tag{5.7}$$

where H_{11} has been given such that $S_{14}H_{11}$ contributes to second order in overlap and $J_0 = (a'a|aa')$, accounting for the primary energy difference between singlet and triplet monomer excited states.

Now combining Eqs. (5.4), (5.6) and (5.7) we obtain

$$\boxed{^{1,3}T_{14} = {}^{1,3}H_{14} - {}^{1,3}S_{14}{}^{1,3}H_{11} \approx J - Z \pm J - {}^{1,3}P} \tag{5.8}$$

where, by employing the Mulliken approximation for two electron integrals which involve orbital overlap charge densities, i.e., $ab \approx \frac{1}{2}S_{ab}(aa + bb)$, etc.,

$$P \equiv P' - S_{14}H_{11} \approx S_{a'b'}\beta_{ab} + S_{ab}\beta_{a'b'} - S_{ab}S_{a'b'}\{(a'a'|bb) \pm J_0\} \tag{5.9}$$

in which the "bond integrals" are defined according to Eq. (5.10).

$$\beta_{ab} = h_{ab} - S_{ab}h_{aa}\,, \quad \beta_{a'b'} = h_{a'b'} - S_{a'b'}h_{a'a'} \tag{5.10}$$

The $2J$ term, the well-known Coulombic interaction (u^{Coul}), is an interaction between charge transition densities of the two chromophores,[1,17] as can be seen by expanding the two-electron repulsion operator in a Taylor series (multipole expansion). The leading non-zero term for this integral is the dipole-dipole interaction (proportional to $1/R^3$), which has a significant magnitude over large separations if the $a' \rightarrow a$ and $b \rightarrow b'$ transitions are electric dipole-allowed. It will be examined in detail in Sections X and XI.

The Z term is an exchange "correction" to J, manifested by the Pauli antisymmetry principle, which has been previously supposed to make the dominant contribution to the matrix element promoting EET at close range (i.e., the Dexter exchange term). The P term, which involves one-electron penetration terms, is opposite in sign to Z and is of greater magnitude, as can be seen by application of the Mulliken approximation to Z, and combining Eqs. (5.8) and (5.9):

$$^{1,3}T_{14} \approx J \pm J - S_{a'b'}\beta_{ab} - S_{ab}\beta_{a'b'} + \tfrac{1}{2}S_{ab}S_{a'b'}\{(aa|b'b') - (aa|a'a')\} \pm S_{ab}S_{a'b'}J_0 \tag{5.11}$$

VI. Through-Configuration Interaction Matrix Elements

The role of the bridging configurations ψ_2 and ψ_3 in mediating a through-configuration interaction (cf. Eq. 4.14) will now be elucidated. The singlet and triplet Slater determinantal wavefunctions representing the ionic configurations are:

$$^{1,3}\psi_2 = \left(|ab\bar{b}'\bar{b}| \pm |b'b\bar{a}\bar{b}|\right)/\sqrt{2} \tag{6.1}$$

$$^{1,3}\psi_3 = \left(|ab\bar{a}\bar{a}'| \pm |aa'\bar{a}\bar{b}|\right)/\sqrt{2} \tag{6.2}$$

where normalizing constants are again set to $1/\sqrt{2}$. Retaining terms up to first order in overlap, we have

$$^{1,3}H_{12} \equiv {}^{1,3}H_{34} \approx -\langle ab\bar{a}'\bar{b}|\mathrm{H}|ab\bar{b}'\bar{b}\rangle$$

$$\approx -\left[h_{a'b'} + 2(a'b'|bb) + (aa|a'b') + S_{a'b'}\{h_{aa} + 2h_{bb} + (bb|bb) + 2(aa|bb)\}\right] \tag{6.3}$$

$$^{1,3}H_{13} \equiv {}^{1,3}H_{24} \approx -\langle ab\bar{a}'\bar{b}|\mathrm{H}|-ab\bar{a}'\bar{a}\rangle$$

$$\approx h_{ab} + (ab|aa) + (ab|bb) + (ab|a'a')$$
$$+ S_{ab}\{\, h_{aa} + h_{a'a'} + h_{bb} + (a'a'|aa) + (a'a'|bb) + (aa|bb)\} \qquad (6.4)$$

$$^{1,3}H_{22} = {}^{1,3}H_{33} \approx h_{aa} + 2h_{bb} + h_{b'b'} + 2(aa|bb) + 2(bb|b'b') + (aa|b'b') + (bb|bb) + (b'b|bb') \qquad (6.5)$$

$$^{1,3}S_{12} \approx -S_{a'b'}\,,\ {}^{1,3}S_{13} \approx S_{ab} \qquad (6.6)$$

Thus, by employing the Mulliken approximation, Eqs. (6.3 - 6.6) give Eqs. (6.7 - 6.8):

$$^{1,3}T_{12} = {}^{1,3}H_{12} - {}^{1,3}S_{12}{}^{1,3}H_{11}$$
$$\approx -\left[\beta_{a'b'} + \tfrac{1}{2}S_{a'b'}\{(aa|a'a') - (a'a'|bb) + 2J_0 \mp 2J_0\}\right] \qquad (6.7)$$

and

$$^{1,3}T_{13} = {}^{1,3}H_{13} - {}^{1,3}S_{13}{}^{1,3}H_{11}$$
$$\approx \beta_{ab} + \tfrac{1}{2}S_{ab}\{(aa|a'a') - (a'a'|bb) + 2J_0 \mp 2J_0\} \qquad (6.8)$$

Such matrix elements contribute second order interactions (Eq. 4.14) which are proportional to overlap squared (as are the first order contributions, cf. Eq. 5.8). Also note that $\{(aa|a'a') - (a'a'|bb) - J_0 \mp J_0\} \approx {}^{1,3}A_{12}$. We shall now define the through-configuration interaction by combining Eqs. (4.14, 6.7 and 6.8):

$$^{1,3}U \approx -T_{12}T_{13}\left(\frac{1}{A_{12}} + \frac{1}{A_{13}}\right) = \frac{2[\beta_{a'b'} + \frac{1}{2}S_{a'b'}(A_{12} + W)][\beta_{ab} + \frac{1}{2}S_{ab}(A_{12} + W)]}{A_{12}} \qquad (6.9)$$

where $W = 2J_0 \mp 2J_0$. The magnitude of U is moderated by its energy gap dependence, but even so, it would be expected to dominate over $-Z$-P since it involves contributions such as $h_{ab}h_{a'b'}$ rather than $h_{ab}S_{a'b'}$. From analogous considerations it is inferred that ${}^{1,3}U$ and $-Z-P$ have almost identical distance-dependencies.

The total coupling, ${}^{1,3}T_{LR}$, in Eq. (4.11) is obtained by combining Eqs. (5.8) and (6.9) to give Eq. (6.10).

$$ {}^{1,3}T_{LR} \approx J - Z \pm J - P + {}^{1,3}U \tag{6.10} $$

Eq. (6.10) may be further simplified, giving Eq. (6.11), wherein it is evident that the uncancelled orbital overlap-dependent terms are associated entirely with the through-configuration interaction.

$$ {}^{1}T_{LR} \approx J \pm J + 2\beta_{ab}\beta_{a'b'}/{}^{1}A \tag{6.11a} $$

$$ {}^{3}T_{LR} \approx 2\beta_{ab}\beta_{a'b'}/{}^{3}A + (2J_0/A)\left\{ \tfrac{1}{2}S_{ab}S_{a'b'}({}^{3}A + 2J_0) + (S_{ab}\beta_{a'b'} + S_{a'b'}\beta_{ab}) \right\} \tag{6.11b} $$

VII. Model Localized Molecular Orbital (LMO) Calculations

By inspection of Eq. (5.8) one would anticipate that the dominant orbital overlap-dependent integrals in the expression for T_{14} would be those involving one-electron integrals (the penetration integrals). To test this supposition a simple dimer system was considered, consisting of two ethene "fragments" positioned in a sandwich configuration at a separation of 4 Å. An RHF calculation, using Gaussian-92,[18] on one of the ethene fragments found that the relevant HOMO and LUMO MOs were given by $a = 0.6356(\, p_1 + p_2 \,)$ and $a' = 0.8099(\, p_1 - p_2 \,)$ respectively (where p_1 denotes a $2p_y$ atomic orbital located on carbon 1, cf. Figure 3.4a). Core, overlap and

Table 3.1. Estimated values of integrals composing J, Z and P (cf. Eq. 5.8) for STO-6g basis functions centred on the ethene dimer at 4 Å separation. Core integrals were determined for carbon centres with +1 charge, as were overlap integrals, S_{ab} = -0.009789 and $S_{a'b'}$ = -0.004875 (see Appendix B). Thus β_{ab} = +0.3072 and $\beta_{a'b'}$ = +0.1989.

[a]P^{Mull} denotes the value of P determined using the Mulliken approximation to simplify two-electron integrals (Eq. 5.9).

[b]P^{Full} refers to that obtained by complete evaluation of all two-electron integrals, i.e., where $S_{ab}S_{a'b'}(\text{"full"}) \equiv 2S_{ab}(aa|a'b') + 2S_{a'b'}(aa|ab) - S_{a'b'}S_{ab}\{(bb|bb)+(aa|a'a')+(aa|bb)+2(a'a'|bb)\}$.

integral	calculated value (eV)	interaction component (eV)
$(a'a\|bb')$	0.12439	$2J \approx 0.24878$
$(a'b'\|ba)$	0.00047724	$-Z \approx -0.0004772$
$S_{a'b'}\,\beta_{ab}$	-0.0014975	
$S_{ab}\beta_{a'b'}$	-0.0019465	
$\mp S_{ab}S_{a'b'}J_0$	$\mp$0.0002047	
$- S_{ab}S_{a'b'}(a'a'\|bb)$	-0.00017295	$_{-1,3}P^{Mull} \approx 0.003822$[a] 0.003412
$- S_{ab}S_{a'b'}(\text{"full"})$	-0.0004398	$_{-1,3}P^{Full} \approx 0.004089$[b] 0.003679

two-electron integrals were taken from the 4-electron localized molecular orbital (LMO) calculation described next, so they relate to charges of +1 on each carbon centre (see Appendix B). Utilizing these results the interaction matrix elements pertinent to Eq. (5.8) were determined (Table 3.1). The importance of terms involving the core integrals is highlighted.

It should be emphasized that this calculation is not intended to be a reliable *quantitative* study of the ethene dimer; values of the integrals were obtained simply to give an indication of the relative significance of each contribution to the electronic coupling.

An important result of this simple dissection of the T_{14} interaction is the obvious significance of the terms involving one electron integrals (P) as compared to the two electron Dexter exchange integral (Z). The magnitude of the one electron integrals is determined primarily by electron-nuclear attractions (hence their negative sign) as well as a kinetic energy contribution (positive in sign), which means that P not only has a greater magnitude, but has the opposite sign compared to the Dexter exchange interaction, Z.

Values of ($-Z-{}^{1,3}P$) may be compared in order to check the usefulness of the Mulliken approximation in this instance. Full evaluation of $-Z$ - (P' - $S_{14}H_{11}$) (i.e., without the Mulliken approximation) gives 0.003612 eV and 0.003202 eV for singlet and triplet states respectively. Using the Mulliken approximation for P only (i.e. Eq. 5.9) gives 0.003345 eV and 0.002935 eV (for singlet and triplet interactions), whilst application of the Mulliken approximation to *all* two-electron integrals (Eq. 5.11) gives 0.003419 eV and 0.003009 eV. All these approximate results are in reasonable agreement with the "exact" treatment summarized in Table 3.2, which gave $-Z-P$ = 0.0037355 eV and 0.0033168 eV.

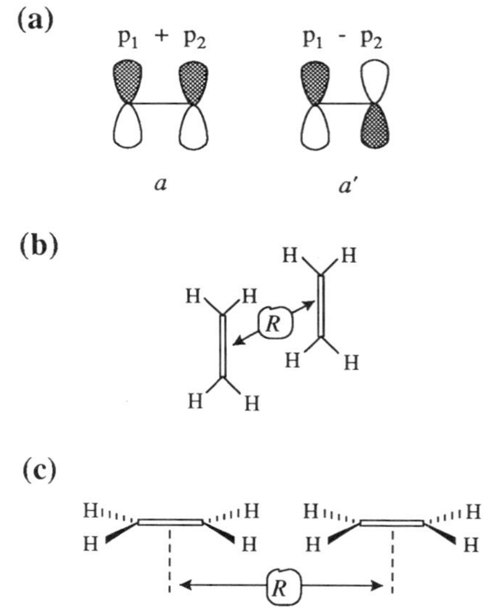

Figure 3.4. (a) Depiction of the π- and π*-orbital (*a* and *a'* respectively) description used in the LMO calculations; (b) the ethene sandwich geometry; and (c) the head-to-tail geometry.

Model 4-electron LMO calculations (STO-6g basis set) were undertaken using the method detailed in Section VII and Appendix A. Four carbon centres, each with +1 charge and a p atomic orbital modelled the active space electrons (c.f. Figure 3.3 without the core electrons) of a 4 Å separation ethene sandwich dimer (Figure 3.4b). Thus, the validity of σ–π separability is assumed (cf. Ref. 19,20). The results are summarized in Table 3.2. It is seen that the interactions calculated in this way agree with the results of the simple treatment summarized in Table 3.1.

In addition, it is evident that U makes the greatest contribution to the overall orbital overlap-dependent electronic coupling, as expected from considerations of the previous section. Thus, while the distance dependencies of the Z, P and U terms are similar (they are determined by the same overlap-dependencies), their magnitudes differ appreciably. The differences between the coefficients λ and μ for the singlet-singlet and triplet-triplet cases is explained using Eqs. (4.10) and comparing ${}^{(1)}A_{12} \approx 1$ eV with ${}^{(3)}A_{12} \approx 10$ eV.

Table 3.2. Matrix elements (eV) and normalized coefficients from the 4-electron LMO calculations, together with a corresponding dissection of the total electronic coupling (T_{LR}) into Coulombic (J) and various short-range (overlap-dependent) interactions, see Eqs. (4.10, 4.11 and 6.10).

	singlet	triplet
H_{12}	0.089669	0.099775
H_{13}	-0.050875	-0.11716
H_{14}	0.25099	0.0024398
H_{11}	26.9577	18.3771
H_{22} (= H_{33})	28.0221	28.0214
S_{12}	-0.0048755	"
S_{13}	+0.0097887	"
S_{14}	-4.7724 × 10^{-5}	"
S_{23}	-4.7724 × 10^{-5}	"
λ	-0.18603	-0.019614
μ	0.26444	0.030753
$2J$	0.24854	-
$-Z - P$	0.0037355	0.0033168
μT_{12}	0.058468	0.0058238
λT_{13}	·0.058554	0.0058263
T_{LR}	0.36672	0.014967

VIII. Inclusion of the Core Electrons

The derivation of Eq. (6.11) did not take account of the σ-core electrons. To examine how the inclusion of these electrons affects the magnitudes of the H_{ij} and T_{ij}, we shall include the doubly-occupied x and y (σ) orbitals of Figure 3.3 in the Slater determinants of ψ_1 to ψ_4. Upon introduction of the approximations of Eqs. (8.1a) to (8.1d),

$$h_{ab}^{\sigma\pi} + 2(ab|xx) + 2(ab|yy) - (ax|xb) - (ay|yb) \approx h_{ab}^{\pi} \tag{8.1a}$$

$$h_{a'b'}^{\sigma\pi} + 2(a'b'|xx) + 2(a'b'|yy) - (a'x|xb') - (a'y|yb') \approx h_{a'b'}^{\pi} \tag{8.1b}$$

$$h_{aa}^{\sigma\pi} + 2(aa|xx) + 2(aa|yy) - (ax|xa) - (ay|ya) \approx h_{aa}^{\pi} \equiv h_{bb}^{\pi} \tag{8.1c}$$

$$h_{a'a'}^{\sigma\pi} + 2(a'a'|xx) + 2(a'a'|yy) - (a'x|xa') - (a'y|ya') \approx h_{a'a'}^{\pi} \equiv h_{b'b'}^{\pi}, \tag{8.1d}$$

it may be deduced that

$$H_{ij}^{\sigma\pi} \approx H_{ij}^{\pi} + S_{ij}E^{\sigma} \tag{8.2}$$

in which those matrix elements labelled with a π superscript pertain to the 4-electron calculation, and those labelled $\sigma\pi$ refer to an all-electron calculation. Therefore

$$T_{ij}^{\sigma\pi} \approx T_{ij}^{\pi}, \tag{8.3}$$

Thus it is seen that, although the magnitudes of the H_{ij} matrix elements depend on the number of σ-core electrons that are included explicitly in the Slater determinants, the magnitudes of the T_{ij} matrix elements are almost invariant with respect to this number. This conclusion will be further demonstrated via the results of the *ab initio* studies reported in the next section for the ethene dimer with all electrons included in the calculations.

IX. All-Electron LMO Calculations

Matrix elements connecting the structures 1 to 4 were obtained for various separations of the D_{2h} (sandwich) configuration ethene dimer using all-electron *ab initio* LMO calculations, as detailed in Appendix A. The active space of localized π and π^* orbitals was represented by symmetric and antisymmetric combinations respectively of C(2p) orbitals (as defined in the previous section). As may be gathered from the simple description of each ethene excited state configuration these calculations are not meant to reproduce exactly the interactions in a "real" ethene dimer. They should, however, quantify the EET matrix elements relevant to the model shown in Figure 3.3 and the associated theory of Section IV.

Overlap integrals and matrix elements between structures 1 to 4 are collected in Table 3.3 for various separations, together with the coefficients c_1, c_2 and c_3 which indicate the relative admixture of functions ψ_1, ψ_2 and ψ_3 respectively in Ψ_L (Eq. 4.1). The VB structural weights for the configuration which includes all four structures are given in Table 3.4. A decomposition of the calculated interaction matrix element T_{LR} into $2J$ and $-Z$-P, plus second order matrix elements which involve the mixing of locally-excited and ionic configurations (via H_{12} and H_{13} - a through-configuration interaction), U, are also given (cf. Eqs. 4.11, 5.8 and 6.10).

Rather than using Eq, (4.11) to determine T_{LR}, the, more accurate, Eq. (9.1) has been used, where the coefficients were determined variationally (although the two methods were found to be in close agreement),

$$T_{LR} \approx T_{14} + \mu T_{12} + \lambda T_{13} \tag{9.1}$$

where μ and λ from Equations (4.10) are written in terms of the VB coefficients as:

$$\lambda = \frac{c_2}{c_1}, \qquad \mu = \frac{c_3}{c_1} \tag{9.2}$$

The obvious point to note from these calculations is the close agreement with the predictions of the simple four electron model. At close separations the influence of the through-configuration interaction, U, becomes significant, but in the singlet-singlet case the Coulombic term, $2J$, dominates. The short-range interactions are attenuated rapidly with separation (approximately as the square of interchromophore overlap), whereas the Coulombic interaction is seen to diminish more slowly with increasing separation (as expected). As the electronic coupling becomes negligibly small the excitation is best described as being localized on one of the chromophores, as evidenced by the triplet dimer at a separation of 6 Å.

The distance-dependence of the diagonal matrix elements H_{11} and H_{22} is also quite drastic. This is thought to be primarily due to variation in the magnitude of one electron integrals, for example h_{aa}, as attractions with the approaching nuclei in the B molecule obtain increasing significance.

Table 3.3. Normalized overlap and Hamiltonian matrix elements (eV) between configurations 1 to 4 (Figure 3.3) for various separations of the ethene dimer, together with the normalized coefficients c_1, c_2 and c_3 indicating the relative admixture of structures 1, 2 and 3 respectively in Ψ_L (Eqs. 4.1 & 9.2).

R (Å)	4.0		5.0		6.0	
	singlet state	triplet state	singlet state	triplet state	singlet state	triplet state
S_{11}	1.0000	"	1.0000	"	1.0000	"
S_{12}	-0.0047586	"	-0.00042904	"	-0.0000283	"
S_{13}	0.0095544	"	0.00097843	"	0.0000709	"
S_{14}	-0.0000454	"	-4.1978×10^{-7}	"	-2.0057×10^{-9}	"
H_{12}	33.011	33.020	2.9076	2.9083	0.18873	0.18875
H_{13}	-66.200	-66.265	-6.6221	-6.6295	-0.47200	-0.47257
H_{14}	0.56439	0.31615	0.12464	0.0028524	0.06869674	0.00001342
H_{23}	0.31614	0.31615	0.0028531	0.0028524	0.0000135	0.0000134
H_{11}	-6913.90	-6922.53	-6750.51	-6759.13	-6637.48	-6646.10
H_{22}	-6912.85	-6912.85	-6748.72	-6748.72	-6635.21	-6635.21
c_1	0.98492	0.99972	0.99992	1.0000	1.0000	1.0000
c_2	-0.099905	-0.0081122	-0.0063847	-0.0007998	-0.00039434	-0.00006219
c_3	0.12878	0.012882	0.0096407	0.0015557	0.0006894	0.00014013

Table 3.4. The VB structural weights for the configuration including all four structures, together with a decomposition of the total calculated excitation transfer matrix element, T_{LR}, into $2J$, $-Z$-P and the through-configuration interaction, U (see Eqs. 4.11, 5.8, 6.10 & 9.1). $\lambda = c_2/c_1$ and $\mu = c_3/c_1$ (where the coefficients are listed in Table 3.3).

R (Å)	4.0		5.0		6.0	
	singlet	triplet	singlet	triplet	singlet	triplet
W_1	0.48252	0.49963	0.49988	0.499996	0.499999	1.00000
W_2	0.017478	0.0037004	0.0001216	0.000004	0.000001	0.00000
W_3	0.017478	0.0037004	0.0001216	0.000004	0.000001	0.00000
W_4	0.48252	0.49963	0.49988	0.499996	0.499999	0.00000
$2J$	0.24825	-	0.12179	-	0.068683	-
$-Z - P$	0.002249	0.0018671	1.936×10^{-5}	1.504×10^{-5}	1.773×10^{-7}	9.005×10^{-8}
μT_{12}	0.014450	0.0010109	1.095×10^{-4}	1.343×10^{-5}	6.131×10^{-7}	9.324×10^{-8}
λT_{13}	0.014386	0.0010093	1.098×10^{-4}	1.293×10^{-5}	5.531×10^{-7}	8.467×10^{-8}
T_{LR}	0.27934	3.887×10^{-2}	0.12203	4.140×10^{-5}	0.068684	2.680×10^{-7}

In Figure 3.5 each component contributing to the total singlet-singlet electronic coupling calculated by the all-electron *ab initio* LMO method is plotted as a function of separation of the two ethene molecules. The relatively weak attenuation of the Coulombic interaction, J, with separation (going approximately as R^{-3}) is evident. In contrast, the overlap-dependent short-range interactions, $-Z - P$ and U, strongly diminish with increasing separation. Their attenuation is close to exponential, with exponent approximately 5.0 in all cases; only the pre-exponential factor is varying between interactions. The plots highlight the significance of the U

interaction compared with the $-Z\text{-}P$ interaction (at least for the ethene dimer). The consequences on calculated rates for electronic energy transfer will be dramatic, as is evidenced even in the (what we now know to be very conservative) estimates of a previous contribution.[21] Figure 3.6 compares the distance dependencies of the orbital overlap-dependent components of both singlet-singlet and triplet-triplet interactions.

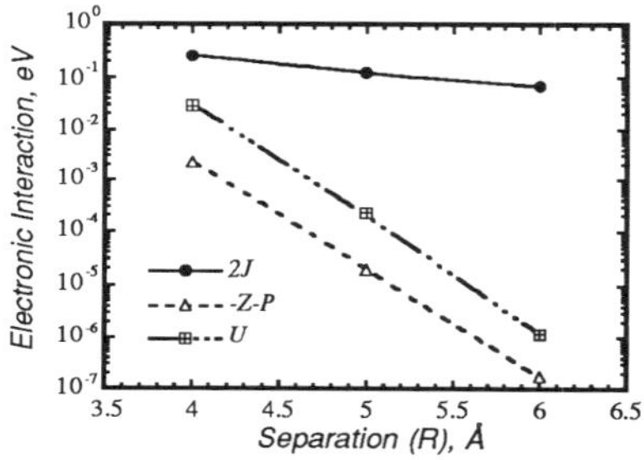

Figure 3.5. Comparison of distance dependencies of electronic coupling components for the ethene singlet state, calculated by the all electron LMO method.

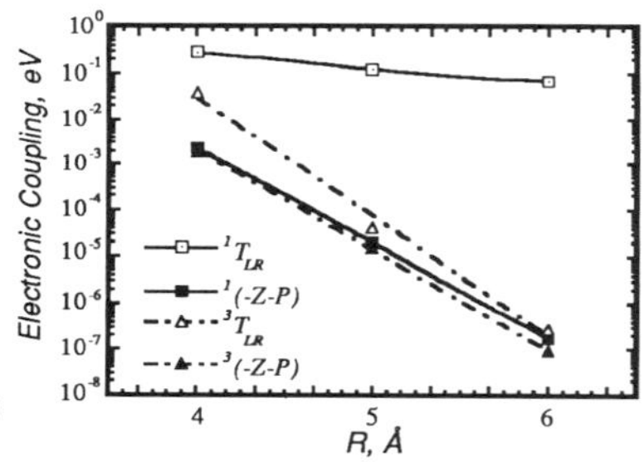

Figure 3.6. Comparison of the total electronic coupling matrix element, T_{LR}, and the direct short range interactions, $-Z\text{-}P$, for the ethene singlet and triplet states, calculated by the all electron LMO method.

After looking at the form of the Coulombic and short-range interactions one may conjecture as to what would happen if the dimer is arranged in such a way that the Coulombic interaction is negative in sign. Would the short-range interactions then interfere destructively with the Coulombic interaction? To investigate this supposition an all-electron LMO calculation was undertaken on a head-to-tail configuration ethene dimer (with a 5 Å separation between molecular centres, see

Figure 3.4c). Allowance was not made for polarization of the π orbitals due to the now non-equivalent p atomic orbitals.

Table 3.5. Results (analogous to Tables 3.3 and 3.4) for a head-to-tail dimer configuration, with ethene centres separated by 5 Å. The calculation was for the singlet state only.

matrix elements (eV)	overlap integrals	coefficients	electronic coupling (eV)
H_{12} = 8.8496	S_{12} = -0.0012947	λ = -0.018494	$2J$ = -0.22209
H_{13} = 5.7493	S_{13} = -0.00084213	μ = -0.0079455	$-Z\text{-}P$ = 0.00004583
H_{14} = -0.22945	S_{14} = 1.0903×10^{-6}		λT_{13} = -0.00025221
H_{23} = 0.0074927	S_{23} = -1.0903×10^{-6}		μT_{12} = -0.00025051
H_{11} = -6810.899			
H_{22} = -6809.195			T_{LR} = -0.22253

The results, summarized in Table 3.5, are interesting. They show that, indeed, the $-Z$ - P interaction does oppose the (now negative in sign) Coulombic interaction. The U term nonetheless still has the same sign as the Coulombic interaction and still makes the dominant contribution to the short range (orbital overlap-dependent) interactions. We therefore conclude that it is unlikely there would be a configuration of the dimer in which the overall electronic coupling vanishes due to an interference between its component interactions.

Before proceeding, it would be worthwhile to recap on the preceding results and foreshadow some of the work to be presented in the remaining sections of this chapter. This is done diagrammatically in Chart 3.1, below.

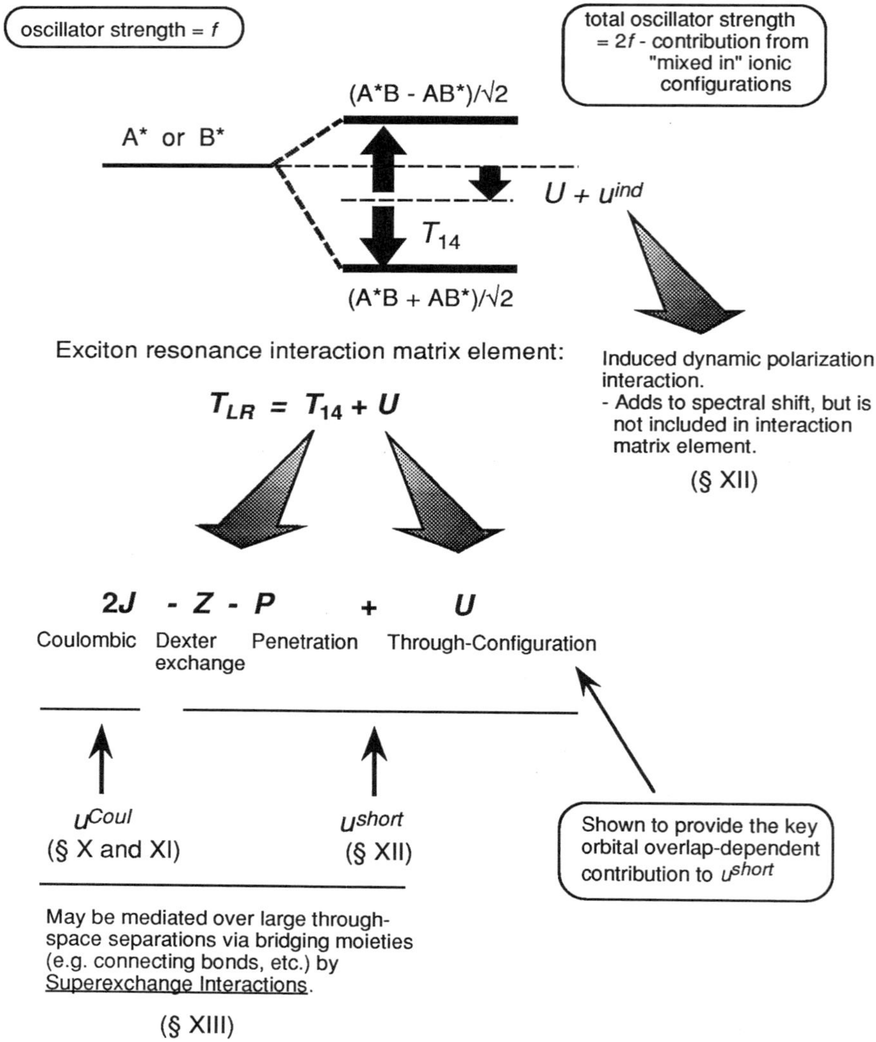

Chart 3.1. Diagrammatic summary of the Chapter so far and its connection with following sections.

X. The Quantum Electrodynamical Approach

So as to describe a somewhat more flexible and complete theory, another expression for the Hamiltonian is now considered. The short-range (interchromophore orbital overlap-dependent) effects are described as before, but the intermediate-range, Coulombic, interaction will now represented in a quantum electrodynamical framework. We shall write

$$\mathrm{H} = \mathrm{H}_{\mathrm{mol}} + \mathrm{H}_{\mathrm{rad}} + \mathrm{H}_{\mathrm{int}} \tag{10.1}$$

where the unperturbed molecular Hamiltonian is given by

$$\mathrm{H}_{\mathrm{mol}} = \sum_{N,n,i\neq i'} \left\{ \frac{-\hbar^2}{2m} \nabla_i^2(N) + \mathrm{V}_{ni}(N) + \mathrm{V}_{ii'}(N) \right\} \tag{10.2}$$

for electrons i and nuclei n of molecule N, and where V is the Coulomb operator.

$$\mathrm{H}_{\mathrm{rad}} = \frac{1}{2} \int \left(\frac{\boldsymbol{d}^{\perp 2}}{\varepsilon_0} + \varepsilon_0 c^2 \boldsymbol{b}^2 \right) \mathrm{d}^3\boldsymbol{r} \tag{10.3}$$

is the Hamiltonian for the radiation field, and the interaction Hamiltonian is written as Eq. (10.4). The first two terms in Equation (10.4) define the core (one-electron) Hamiltonian operator and the third term the modified two-electron repulsion operator (such that it excludes regard of the Coulomb integral, J). The fourth term is only important in self-energy calculations and may be ignored in the present context. The remaining terms in the equation couple molecule N to the electromagnetic field. In this way the intermolecular Coulombic interactions can be mediated entirely by the exchange of transverse photons.

$$\mathrm{H_{int}} = \sum_{N,n,i} \left\{ \frac{-\hbar^2}{2m} \nabla_i^2(N') + \mathrm{V}_{nj}(N,N') + \mathrm{V}_{ij}(N,N') \right\} + \frac{1}{2} \int \left| \boldsymbol{p}_A^{\perp}(\boldsymbol{r}) \right|^2 \mathrm{d}^3\boldsymbol{r}$$

$$+ \sum_N \left\{ -\frac{1}{4\pi\varepsilon_0} \mu_\alpha(N).\boldsymbol{d}_\alpha^{\perp}(\boldsymbol{R}_N) - \frac{1}{4\pi\varepsilon_0} \Theta_{\alpha\beta}(N)\, \nabla_\alpha\, \boldsymbol{d}_\beta^{\perp}(\boldsymbol{R}_N) \right.$$

$$\left. - \frac{1}{4\pi\varepsilon_0} \Omega_{\alpha\beta\gamma}(N)\, \nabla_\alpha \nabla_\beta\, \boldsymbol{d}_\gamma^{\perp}(\boldsymbol{R}_N) - \frac{1}{4\pi} \mathrm{m}_\alpha(N).\boldsymbol{b}_\alpha(\boldsymbol{R}_N) + \ldots \right\} \qquad (10.4)$$

Nuclei n and electrons i are associated with molecule N, while electrons j are associated with the different molecule, N'. $\boldsymbol{p}^{\perp}(\boldsymbol{r})$ is the transverse component of the electric polarization field and $\mu_\alpha(N)$, $\Theta_{\alpha\beta}(N)$, and $\Omega_{\alpha\beta\gamma}(N)$ are the electric dipole, quadrupole, and octopole transition moment tensor operators respectively (for molecule N), while $\mathrm{m}_\alpha(N)$ is the magnetic dipole transition moment operator. In Cartesian tensor notation they are defined as

$$\mu_\alpha(N) = \sum_i e_i(\boldsymbol{r}_i - \boldsymbol{R}_N)_\alpha \qquad (10.5\text{a})$$

$$\Theta_{\alpha\beta}(N) = \frac{1}{2!}\sum_i e_i(\boldsymbol{r}_i - \boldsymbol{R}_N)_\alpha(\boldsymbol{r}_i - \boldsymbol{R}_N)_\beta \qquad (10.5\text{b})$$

$$\Omega_{\alpha\beta\gamma}(N) = \frac{1}{3!}\sum_i e_i(\boldsymbol{r}_i - \boldsymbol{R}_N)_\alpha(\boldsymbol{r}_i - \boldsymbol{R}_N)_\beta(\boldsymbol{r}_i - \boldsymbol{R}_N)_\gamma \qquad (10.5\text{c})$$

$$\mathrm{m}_\alpha(N) = \sum_i m_i(\boldsymbol{r}_i - \boldsymbol{R}_N)_\alpha \qquad (10.5\text{d})$$

$\boldsymbol{d}^{\perp}$ is the transverse displacement vector relating to the fundamental electric field, $\boldsymbol{b}$ is the fundamental magnetic field (defined as in Eqs. 10.6 and 10.7 respectively), while ε_0 is the vacuum permittivity and c is the speed of light *in vacuo*. Outside the source the transverse displacement vector, $\boldsymbol{d}^{\perp}(\boldsymbol{R}_N)$, equals the total electric field at $\boldsymbol{R}_N$ (the position vector of molecule N), since $\boldsymbol{d}^{\parallel}(\boldsymbol{r})$ is zero for a neutral system.

$$\boldsymbol{d}^{\perp}(\boldsymbol{r}) = \mathrm{i}\sum_{k,\lambda} \left(\frac{2\pi\hbar ck\varepsilon_0}{V}\right)^{1/2} \left\{ e^{(\lambda)}(\boldsymbol{k})a^{(\lambda)}(\boldsymbol{k})e^{\,\mathrm{i}\boldsymbol{k}.\boldsymbol{r} - \mathrm{i}\omega t} - \bar{e}^{(\lambda)}(\boldsymbol{k})a^{\dagger(\lambda)}(\boldsymbol{k})e^{\,-\mathrm{i}\boldsymbol{k}.\boldsymbol{r} + \mathrm{i}\omega t} \right\} \qquad (10.6)$$

$$\boldsymbol{b}(\boldsymbol{r}) = \mathrm{i}\sum_{\boldsymbol{k},\lambda}\left(\frac{2\pi\hbar k}{c\varepsilon_0 V}\right)^{1/2}\left\{ b^{(\lambda)}(\boldsymbol{k})a^{(\lambda)}(\boldsymbol{k})e^{\,\mathrm{i}\boldsymbol{k}.\boldsymbol{r}-\mathrm{i}\omega t} - \bar{b}^{(\lambda)}(\boldsymbol{k})a^{\dagger(\lambda)}(\boldsymbol{k})e^{-\mathrm{i}\boldsymbol{k}.\boldsymbol{r}+\mathrm{i}\omega t}\right\} \tag{10.7}$$

with $$b^{(\lambda)}(\boldsymbol{k}) = \hat{k}\times e^{(\lambda)}(\boldsymbol{k}) \tag{10.8}$$

This molecular quantum electrodynamical model is a very general framework within which to treat the interactions between a molecule and the (quantized) electromagnetic field, as well as between molecules themselves.[5-8,10,11,22-26] Such a partitioning is not essential for the evaluation of the electronic coupling, but is capable of accounting for retarded interactions[10,11] if necessary and, in the opinion of the author, provides a more lucid physical picture of the Coulombic interaction.

Now, using Rayleigh-Schrödinger perturbation theory we can write down an expression for the interaction matrix element, where the Coulombic interaction, mediated via the radiation field, is described by the second order term:

$$^1T_{LR} = -Z - {}^1P - {}^1U - \sum_I \frac{\langle f|\mathrm{H_{int}}|I\rangle\langle I|\mathrm{H_{int}}|i\rangle}{E_I - E_i} \tag{10.9a}$$

$$= u^{short} + u^{Coul} \tag{10.9b}$$

The first three terms in Eq.(10.9a) define u^{short}: those interactions whose magnitude depends upon the degree of interchromophore orbital overlap. It encompasses the $-Z$-P and U terms given in Eq. 6.10 and contributes at small separations only. It is discussed further in Section XII. The second term describes the Coulombic interaction, $2J$ (or u^{Coul}), which is independent of interchromophore orbital overlap and acts over all separations, via the radiation field.

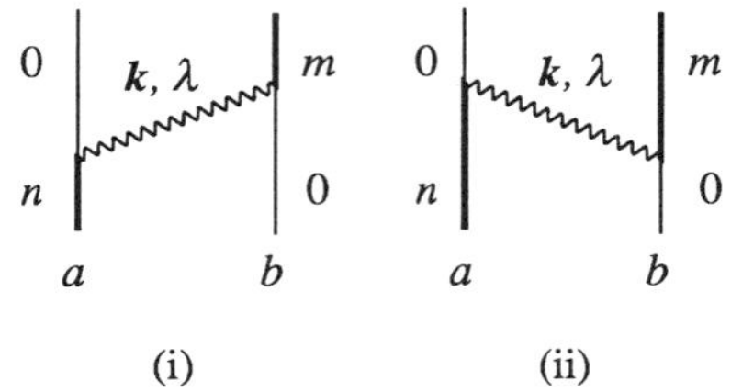

Figure 3.7. Time-ordered diagrams depicting the two contributions to the electronic coupling matrix element which describes u^{Coul}. Time proceeds upwards, from the initial state to the final state, via the intermediate state.

In the multipolar formalism all intermolecular interactions are mediated by the exchange of transverse (virtual) photons. Energy is conserved between initial (i) and final (f) states, but not between initial and intermediate states. This Coulombic term reduces to the Coulomb's law result at the limit of short to intermediate-range separations. For the present purposes just two possible intermediate states, I, in (10.9) are considered: one where both molecules are in their ground states with one virtual photon present, and another where both molecules are in excited states, again, with one virtual photon present (Figure 3.7). Thus, for this part of Eq.(10.9), the important terms from the interaction Hamiltonian, Eq.(10.4), are those effecting an interaction between molecule and field.

From Eq.(10.9), we may write:

$$
\begin{aligned}
u^{Coul} = & \sum_{k,\lambda} \frac{\hbar ck}{2\varepsilon_0 V} \left[\mu_\alpha^{0n}(a)\mu_\beta^{m0}(b) + \mu_\alpha^{0n}(a)\Theta_{\beta\gamma}^{m0}(b)\nabla_\gamma + \Theta_{\alpha\gamma}^{0n}(a)\mu_\beta^{m0}(b)\nabla_\gamma \right. \\
& \left. + \Theta_{\alpha\gamma}^{0n}\Theta_{\beta\delta}^{m0}\ \nabla_\gamma\nabla_\delta + \mu_\alpha^{0n}(a)\Omega_{\beta\gamma\delta}^{m0}(b)\nabla_\gamma\nabla_\delta + \Omega_{\alpha\gamma\delta}^{0n}(a)\mu_\beta^{m0}(b)\nabla_\gamma\nabla_\delta + \ldots \right] \\
& \times \left[e_\alpha^{(\lambda)}(k)\bar{e}_\beta^{(\lambda)}(k)\frac{e^{ik.R}}{E_{n0}-\hbar ck} - e_\beta^{(\lambda)}(k)\bar{e}_\alpha^{(\lambda)}(k)\frac{e^{-ik.R}}{E_{n0}+\hbar ck} \right] \\
& + \sum_{k,\lambda} \frac{\hbar k}{2\varepsilon_0 V} \left\{ \mu_\alpha^{0n}(a)m_\beta^{m0}(b) \left[\bar{e}_\alpha^{(\lambda)}(k)b_\beta^{(\lambda)}(k)\frac{e^{ik.R}}{E_{n0}-\hbar ck} - e_\alpha^{(\lambda)}(k)\bar{b}_\beta^{(\lambda)}(k)\frac{e^{-ik.R}}{E_{n0}+\hbar ck} \right] \right. \\
& \left. + m_\alpha^{0n}(a)\mu_\beta^{m0}(b) \left[\bar{b}_\alpha^{(\lambda)}(k)e_\beta^{(\lambda)}(k)\frac{e^{ik.R}}{E_{n0}-\hbar ck} - b_\alpha^{(\lambda)}(k)\bar{e}_\beta^{(\lambda)}(k)\frac{e^{-ik.R}}{E_{n0}+\hbar ck} \right] \right\} \\
& + \sum_{k,\lambda} \frac{\hbar k}{2\varepsilon_0 cV} m_\alpha^{0n}(a)m_\beta^{m0}(b) \left[b_\alpha^{(\lambda)}(k)\bar{b}_\beta^{(\lambda)}(k)\frac{e^{ik.R}}{E_{n0}-\hbar ck} - b_\beta^{(\lambda)}(k)\bar{b}_\alpha^{(\lambda)}(k)\frac{e^{-ik.R}}{E_{n0}+\hbar ck} \right]
\end{aligned}
\tag{10.10}
$$

We now sum over polarizations using:

$$\sum_{\lambda} e_{\alpha}^{(\lambda)}(\boldsymbol{k})\bar{e}_{\beta}^{(\lambda)}(\boldsymbol{k}) = (\delta_{\alpha\beta} - \hat{k}_{\alpha}\hat{k}_{\beta}) \tag{10.11a}$$

$$\sum_{\lambda} e_{\alpha}^{(\lambda)}(\boldsymbol{k})\bar{b}_{\beta}^{(\lambda)}(\boldsymbol{k}) = \varepsilon_{\alpha\beta\gamma}\hat{k}_{\gamma}) \tag{10.11b}$$

$$\sum_{\lambda} b_{\alpha}^{(\lambda)}(\boldsymbol{k})\bar{b}_{\beta}^{(\lambda)}(\boldsymbol{k}) = (\delta_{\alpha\beta} - \hat{k}_{\alpha}\hat{k}_{\beta}) \tag{10.11c}$$

and convert the $\boldsymbol{k}$-sum to an integral with the relation:

$$\frac{1}{V}\sum_{\boldsymbol{k}} \Rightarrow \int \frac{d^3\boldsymbol{k}}{(2\pi)^3} \tag{10.12}$$

Then, putting $\kappa = (E_n - E_0)/\hbar c = (E_m - E_0)/\hbar c$, we have

$$\begin{aligned} u^{Coul} = \Big\{ & \big[\mu_{\alpha}^{0n}(a)\mu_{\beta}^{m0}(b) + \mu_{\alpha}^{0n}(a)\Theta_{\beta\gamma}^{m0}(b)\nabla_{\gamma} + \Theta_{\alpha\gamma}^{0n}(a)\mu_{\beta}^{m0}(b)\nabla_{\gamma} \\ & + \Theta_{\alpha\gamma}^{0n}\Theta_{\beta\delta}^{n0}\ \nabla_{\gamma}\nabla_{\delta} + \mu_{\alpha}^{0n}(a)\Omega_{\beta\gamma\delta}^{m0}(b)\nabla_{\gamma}\nabla_{\delta} + \Omega_{\alpha\gamma\delta}^{0n}(a)\mu_{\beta}^{m0}(b)\nabla_{\gamma}\nabla_{\delta} + \dots\big] \\ & + \frac{1}{c^2}m_{\alpha}^{0n}(a)m_{\beta}^{m0}(b)\Big\} \times \int d^3\boldsymbol{k}\,\frac{k}{2\varepsilon_0(2\pi)^3}(\delta_{\alpha\beta} - \hat{k}_{\alpha}\hat{k}_{\beta})\left\{\frac{e^{i\boldsymbol{k}.\boldsymbol{R}}}{\kappa - k} - \frac{e^{-i\boldsymbol{k}.\boldsymbol{R}}}{\kappa + k}\right\} \\ & + \big[\mu_{\alpha}^{0n}(a)m_{\beta}^{m0}(b) + m_{\beta}^{0n}(a)m_{\alpha}^{m0}(b)\big]\int d^3\boldsymbol{k}\,\frac{k}{2\varepsilon_0 c(2\pi)^3}\varepsilon_{\alpha\beta\gamma}\hat{k}_{\gamma}\left\{\frac{e^{i\boldsymbol{k}.\boldsymbol{R}}}{\kappa - k} - \frac{e^{-i\boldsymbol{k}.\boldsymbol{R}}}{\kappa + k}\right\} \end{aligned} \tag{10.13}$$

The integrations are performed as shown in Appendix C, to give the retarded interaction matrix element:

$$\begin{aligned} u^{Coul(ret)} = \Big\{ & \big[\mu_{\alpha}^{0n}(a)\mu_{\beta}^{m0}(b) + \mu_{\alpha}^{0n}(a)\Theta_{\beta\gamma}^{m0}(b)\nabla_{\gamma} + \Theta_{\alpha\gamma}^{0n}(a)\mu_{\beta}^{m0}(b)\nabla_{\gamma} \\ & + \Theta_{\alpha\gamma}^{0n}\Theta_{\beta\delta}^{n0}\ \nabla_{\gamma}\nabla_{\delta} + \mu_{\alpha}^{0n}(a)\Omega_{\beta\gamma\delta}^{m0}(b)\nabla_{\gamma}\nabla_{\delta} + \Omega_{\alpha\gamma\delta}^{0n}(a)\mu_{\beta}^{m0}(b)\nabla_{\gamma}\nabla_{\delta} + \dots\big] \\ & + \frac{1}{c^2}m_{\alpha}^{0n}(a)m_{\beta}^{m0}(b)\Big\}\frac{1}{4\pi\varepsilon_0 R^3}\big[(\delta_{\alpha\beta} - 3\hat{R}_{\alpha}\hat{R}_{\beta})(\cos\kappa R + \kappa R\sin\kappa R) \\ & - (\delta_{\alpha\beta} - \hat{R}_{\alpha}\hat{R}_{\beta})\,\kappa^2R^2\cos\kappa R\big] + \mathrm{Im}\big[\mu_{\alpha}^{0n}(a)m_{\beta}^{m0}(b) + m_{\beta}^{0n}(a)m_{\alpha}^{m0}(b)\big] \\ & \times \varepsilon_{\alpha\beta\gamma}\frac{1}{4\pi\varepsilon_0 c}\frac{\hat{R}_{\gamma}}{R^3}(\kappa R\cos\kappa R + \kappa^2R^2\sin\kappa R)\,. \end{aligned} \tag{10.14}$$

In the near zone (i.e., at close separations, where $\kappa R \ll 1$) Eq.(10.14) reduces to a form reminiscent of the interaction between two static dipoles:

$$
\begin{aligned}
u^{Coul(nz)} \approx & - \left[\mu_\alpha^{0n}(a)\mu_\beta^{m0}(b) + \frac{1}{c^2} m_\alpha^{0n}(a) m_\beta^{m0}(b) \right] T_{\alpha\beta} \\
& + \mathrm{Im} \left[\mu_\alpha^{0n}(a) m_\beta^{m0}(b) + m_\beta^{0n}(b)\mu_\alpha^{m0}(a) \right] \frac{\varepsilon_{\alpha\beta\gamma}}{c} \kappa T_\gamma \\
& + \left[\mu_\alpha^{0n}(a)\Theta_{\beta\gamma}^{m0}(b) + \Theta_{\beta\gamma}^{0n}(a)\mu_\alpha^{m0}(b) \right] T_{\alpha\beta\gamma} \\
& - \left[\Theta_{\alpha\gamma}^{0n}(a)\Theta_{\beta\delta}^{m0}(b) + \mu_\alpha^{0n}(a)\Omega_{\beta\gamma\delta}^{m0}(b) + \Omega_{\beta\gamma\delta}^{0n}(a)\mu_\alpha^{m0}(b) \right] T_{\alpha\beta\gamma\delta} + \ldots
\end{aligned} \tag{10.15}
$$

where $\varepsilon_{\alpha\beta\gamma}$ is the Levi-Civita unit tensor, and the interaction tensors pertaining to the Cartesian multipole expansion are given by $T_{\alpha\beta\gamma\ldots\nu} = (4\pi\varepsilon_0)^{-1} \nabla_\alpha\nabla_\beta\nabla_\gamma\ldots\nabla_\nu R^{-1}$: Tensors which are proportional to $R^{-(n+1)}$. They are symmetric with respect to the interchange of any pair of suffixes and reduce to zero on contraction. Specifically, we have:

$$T_\alpha = (4\pi\varepsilon_0)^{-1} R_\alpha R^{-3} \tag{10.16a}$$

$$T_{\alpha\beta} = (4\pi\varepsilon_0)^{-1} (3R_\alpha R_\beta - R^2\delta_{\alpha\beta}) R^{-5} \tag{10.16b}$$

$$T_{\alpha\beta\gamma} = (4\pi\varepsilon_0)^{-1} (-3) \left[5R_\alpha R_\beta R_\gamma - R^2 (R_\alpha\delta_{\beta\gamma} + R_\beta\delta_{\gamma\alpha} + R_\gamma\delta_{\alpha\beta}) \right] R^{-7} \tag{10.16c}$$

$$
\begin{aligned}
T_{\alpha\beta\gamma\delta} = & (4\pi\varepsilon_0)^{-1}\ 3\ [35R_\alpha R_\beta R_\gamma R_\delta - 5R^2 (R_\alpha R_\beta\delta_{\gamma\delta} + R_\alpha R_\gamma\delta_{\beta\delta} + R_\alpha R_\delta\delta_{\beta\gamma} \\
& + R_\beta R_\gamma\delta_{\alpha\delta} + R_\beta R_\delta\delta_{\alpha\gamma} + R_\gamma R_\delta\delta_{\alpha\beta}) + R^4 (\delta_{\alpha\beta}\delta_{\gamma\delta} + \delta_{\alpha\gamma}\delta_{\beta\delta} + \delta_{\alpha\delta}\delta_{\beta\gamma}) \big] R^{-9}
\end{aligned} \tag{10.16d}
$$

At very large donor-acceptor separations (i.e., in the wave zone, with $\kappa R \gg 1$) one may envisage two principal forms of the interaction matrix element. Firstly for an electric dipole-allowed transition (as previously discussed[5,8,10,11]) :

$$u^{Coul(wz;1)} \approx \frac{1}{4\pi\varepsilon_0} \mu_\alpha^{0n}(a)\mu_\beta^{m0}(b) \left(\hat{R}_\alpha\hat{R}_\beta - \delta_{\alpha\beta} \right) \frac{\kappa^2 R^2}{R^3} e^{i\kappa R} . \tag{10.17}$$

This is inferred by rewriting the leading term of Eq. (10.14) in the form

$$u^{Coul(ret,d\text{-}d)} \approx \mu_\alpha^{0n}(a)\mu_\beta^{m0}(b)\ \sigma_{\alpha\beta}(\kappa,R) \tag{10.18}$$

where

$$\sigma_{\alpha\beta}(\kappa,R) = \frac{\kappa^3}{4\pi\varepsilon_0}\left\{(\delta_{\alpha\beta} - 3\hat{R}_\alpha\hat{R}_\beta)\left(\frac{\cos\kappa R}{\kappa^3 R^3} + \frac{\sin\kappa R}{\kappa^2 R^2}\right) - (\delta_{\alpha\beta} - \hat{R}_\alpha\hat{R}_\beta)\frac{\cos\kappa R}{\kappa R}\right\} \tag{10.19}$$

and noting that

$$\sigma_{\alpha\beta} = \text{Re}\ \theta_{\alpha\beta} \tag{10.20a}$$

and $\tau_{\alpha\beta} = \text{Im}\ \theta_{\alpha\beta}$

(10.20b)

where the interaction tensor is

$$\theta_{\alpha\beta} = \left\{(\delta_{\alpha\beta} - 3\hat{R}_\alpha\hat{R}_\beta)\left(\frac{1}{\kappa^3 R^3} + \frac{i}{\kappa^2 R^2}\right) - (\delta_{\alpha\beta} - \hat{R}_\alpha\hat{R}_\beta)\frac{1}{\kappa R}\right\}\kappa^3 e^{i|k|R} \tag{10.21}$$

So that the correct nonoscillatory and smoothly varying R-dependence is retrieved from the Fermi Golden Rule rate expression.

Secondly, for an electric dipole-forbidden transition (in the absence of vibronic coupling) we have:

$$u^{Coul(2)} \approx \left[\frac{1}{c^2}m_\alpha^{0n}(a)m_\beta^{m0}(b) + \Theta_{\alpha\gamma}^{0n}\Theta_{\beta\delta}^{n0}\ \nabla_\gamma\nabla_\delta\right]\theta_{\alpha\beta}(k,\boldsymbol{R}) \tag{10.22}$$

therefore the higher multipole terms also reduce to a κ^2/R distance dependence in the wave zone:

$$u^{Coul(wz;2)} \approx \frac{1}{4\pi\varepsilon_0 c^2}m_\alpha^{0n}(a)m_\beta^{m0}(b)\left(\hat{R}_\alpha\hat{R}_\beta - \delta_{\alpha\beta}\right)\frac{\kappa^2R^2}{R^3}e^{i\kappa R} + \frac{1}{4\pi\varepsilon_0}\Theta_{\alpha\gamma}^{0n}\Theta_{\beta\delta}^{n0}\left(\delta_{\alpha\beta}\hat{R}_\alpha\hat{R}_\beta - \hat{R}_\alpha\hat{R}_\beta\hat{R}_\gamma\hat{R}_\delta\right)\frac{\kappa^4R^4}{R^5}e^{i\kappa R} \tag{10.23}$$

XI. The Coulombic Interaction

In this section, the properties and convergence of the multipole expansion of the Coulombic interaction ($2J$ in Eq.(5.8)) is examined. The theory is a continuation of Section X, in the near zone limit, and thus relates to previous work on interacting static charge distributions.[27,28] Specifically, distance and orientation dependencies of the near zone Coulombic interaction, Eq.(10.15) will be presented and investigated computationally for model systems.

A multipole expansion about a point P converges formally (and then at best asymptotically) only at distances farther from P than the most extreme part of the charge distribution.[29] Quantum mechanical charge clouds have no strict boundary surface, so extra care must be taken when using a multipole expansion for molecular problems to ensure that the expansion is converging and enough terms are included. In particular, for the situation where the leading term (dipole-dipole) is extremely small, it is essential to consider higher terms in the expansion. For example, the Coulombic interaction in a pair of D_{2h} symmetry chromophores with dipole-allowed electronic transitions, dipole-dipole, dipole-octopole and octopole-octopole interactions will be the leading non-zero terms in the multipole expansion which should need to be considered. Electric dipole-quadrupole, quadrupole-quadrupole and magnetic dipole-dipole terms are zero since they represent completely symmetric transitions in such chromophores.[30,31]

Rather than use the Cartesian tensor notation of Eqs. (10.5), it is more convenient to employ a notation deriving from an expansion of the Coulombic interaction potential in terms of spherical harmonics. In this way compact expressions for the interaction, referring to local (molecule-fixed) frames, may be obtained. However, when evaluating orientation dependencies for higher transition moments, such expressions are still somewhat cumbersome. To illustrate the ideas involved, an evaluation of Eq. (10.15) for the simple case of a dipole-allowed transition in a molecule of $D_{\infty h}$ symmetry will be presented. More general expressions will be discussed after that.

Axially Symmetric Polarized Transitions. If $\hat{r}_\alpha$ is a component of the unit vector along the axis of symmetry, then the expressions obtained in the previous section simplify considerably since we have (cf. Buckingham[32] for a discussion of analogous, static charge distributions):

$$\mu_\alpha = \mu \hat{r}_\alpha \tag{11.1a}$$

$$\Theta_{\alpha\beta} = \frac{1}{2}\Theta (3 \hat{r}_\alpha \hat{r}_\beta - \delta_{\alpha\beta}) \tag{11.1b}$$

$$\Omega_{\alpha\beta\gamma} = \frac{1}{2}\Omega (5 \hat{r}_\alpha \hat{r}_\beta \hat{r}_\gamma - \hat{r}_\alpha\delta_{\beta\gamma} - \hat{r}_\beta\delta_{\gamma\alpha} - \hat{r}_\gamma\delta_{\alpha\beta}) \tag{11.1c}$$

That is, for an axially symmetric polarized electronic transition, each multipole transition moment is determined by a single *scalar* quantity. In the classical analogue these scalars refer to the multipole moments Q_{10}, Q_{20} and Q_{30}, defined (in spherical tensor notation) by

$$Q_{10} = \sum_i e_i z_i \tag{11.2a}$$

$$Q_{20} = \sum_i e_i \frac{1}{2} (3z_i^2 - r_i^2) \tag{11.2b}$$

$$Q_{30} = \sum_i e_i \frac{1}{2} (5z_i^3 - 3z_i r_i^2) \tag{11.2c}$$

They are given in Cartesian tensor notation as μ_z, Θ_{zz} and Ω_{zzz} respectively. Thus, in this instance there is a direct correspondence between Cartesian and Spherical tensor notations.

Using Eq. (10.15) with (11.1) we find that the leading terms in the multipole expansion which are important for the electronic coupling between dipole-allowed electronic transitions (the dipole-dipole, dipole-octopole and octopole-octopole interactions) in chromophores with high axial symmetry (e.g. $D_{\infty h}$, D_{6h}, C_{3h}

molecular point groups) as well as the magnetic dipole-dipole interaction ($u^{\text{m-m}}$, important for totally symmetric electronic transitions) are:

$$u^{\text{d-d}} = \frac{1}{4\pi\varepsilon_0 R^3}\,\mu^{(a)}\mu^{(b)}\,(2\cos\theta_a\cos\theta_b + \sin\theta_a\sin\theta_b\cos\varphi) \tag{11.3a}$$

$$u^{\text{m-m}} = \frac{1}{4\pi c^2\varepsilon_0 R^3}\,m^{(a)}m^{(b)}\,(2\cos\theta_a\cos\theta_b + \sin\theta_a\sin\theta_b\cos\varphi) \tag{11.3b}$$

$$u^{\text{d-o}} = \frac{1}{4\pi\varepsilon_0 R^5}\Big[\mu^{(a)}\Omega^{(b)}\Big(10\cos\theta_a\cos^3\theta_b - 6\cos\theta_a\cos\theta_b + \tfrac{15}{2}\cos^2\theta_b\sin\theta_a\sin\theta_b\cos\varphi - \tfrac{3}{2}\sin\theta_a\sin\theta_b\cos\varphi\Big) + \mu^{(b)}\Omega^{(a)}\Big(10\cos^3\theta_a\cos\theta_b - 6\cos\theta_a\cos\theta_b + \tfrac{15}{2}\cos^2\theta_a\sin\theta_a\sin\theta_b\cos\varphi - \tfrac{3}{2}\sin\theta_a\sin\theta_b\cos\varphi\Big)\Big] \tag{11.3c}$$

$$u^{\text{o-o}} = \frac{1}{4\pi\varepsilon_0 R^7}\,\Omega^{(a)}\Omega^{(b)}\Big[5\cos\theta_a\cos\theta_b(5\cos^2\theta_a - 3)(5\cos^2\theta_b - 3) + \tfrac{45}{8}(5\cos^2\theta_a - 1)(5\cos^2\theta_b - 1)\sin\theta_a\sin\theta_b\cos\varphi - \tfrac{45}{2}\cos\theta_a\cos\theta_b\sin^2\theta_a\sin^2\theta_b\cos2\varphi + \tfrac{5}{8}\sin^3\theta_a\sin^3\theta_b\cos3\varphi\Big] \tag{11.3d}$$

where the angles defining the relative orientation of transition moments in molecules A and B are shown in Figure 3.8. The higher multipole terms may thus either increase or decrease u^{Coul} relative to the limiting dipole-dipole term, depending upon the relative orientations of the chromophores.

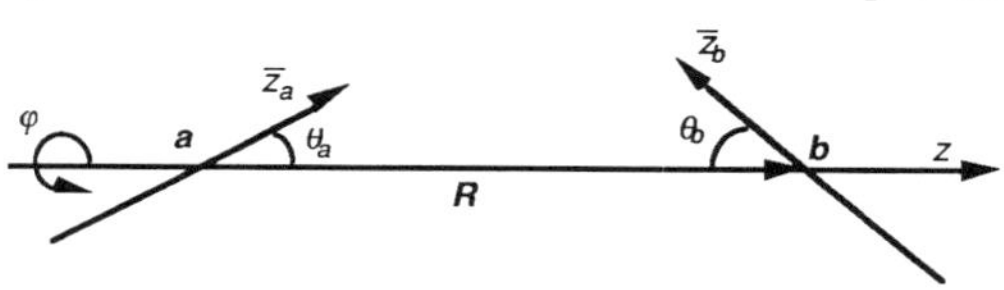

Figure 3.8. Angles defining the orientation between the dipole transition moments of molecules a and b, separated by a distance R. The z axis is in the laboratory reference frame, while the $\bar{z}_i$ axes are in the molecular reference frames.

The orientation-dependencies of the higher multipole terms follow similar forms to the dipole-dipole term, but are more drastic, as is seen for example in Figure 3.9. Their importance varies, primarily depending on the magnitude of the dipole transition moment: they have a significant relative contribution to a weak electronic transition, but generally make a less significant contribution for a very intense transition. For randomly oriented appropriate orientation factors of $\frac{1}{3}$, $\frac{1}{5}$ and $\frac{1}{7}$ for dipole-dipole, dipole-octopole and octopole-octopole interactions respectively[5,33] tend to decrease the relative importance of the higher multipole terms. This was also implied by a computational study of energy transfer in chlorophylls by Chang.[34]

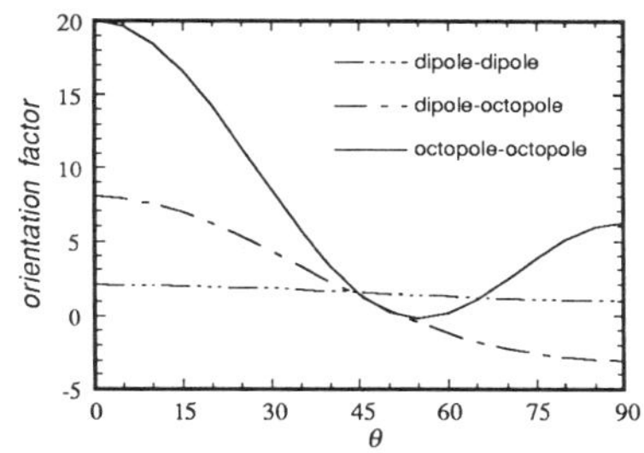

Figure 3.9. Illustration of the marked orientation dependence of the higher multipole terms compared to the dipole-dipole term. The angle $\theta = \theta_a = \theta_b$ (Figure 3.8).

Alternatively one may employ the analogy of a general "two-centre" expansion of the potential energy of interaction of two charge distributions by expanding the interaction potential in terms of products of surface harmonics.[35-37] This method is valid at all separations, but will be presented here for non-overlapping charges. The multipole moments embedded in the charge distributions are related to each other by Wigner coefficients, as shown by Hirschfelder, et al.,[37] so that the potential energy of interaction between two real dipoles a and b of length l_a and l_b is given by:

$$u_{ab} = e_a e_b \sum_{n_a, n_b, m} (-1)^{\{n_b + m\}} \frac{[1-(-1)^{n_a}][1-(-1)^{n_b}]\,(n_a + n_b)!\,(\frac{1}{2} l_a)^{n_a} (\frac{1}{2} l_b)^{n_b}}{(n_a + |m|)!\,(n_b + |m|)!} \times \frac{P_{n_a}^{m}(\cos\theta_a)\, P_{n_b}^{m}(\cos\theta_b)\, \exp(im\phi_b)}{R_{ab}^{n_a+n_b+1}} \quad (11.4)$$

with the associated Legendre polynomials being defined as

$$P_l^m(x) = (1 - x^2)^{\frac{|m|}{2}} \frac{d^{|m|}}{dx^{|m|}} P_l(x) \;. \quad (11.5)$$

Table 3.6. Multipole Moments to rank 3: spherical tensors versus Cartesian tensors (from Stone[27]).

Spherical Tensor	Cartesian Multipole Equivalent
Q_{10} (Q_z)	μ_z
Q_{11c} (Q_x)	μ_x
Q_{11s} (Q_y)	μ_y
Q_{20}	Θ_{zz}
Q_{21c}	$\sqrt{4/3}\ \Theta_{xz}$
Q_{21s}	$\sqrt{4/3}\ \Theta_{yz}$
Q_{22c}	$\sqrt{1/3}(\Theta_{xx} - \Theta_{yy})$
Q_{22s}	$\sqrt{4/3}\ \Theta_{xy}$
Q_{30}	Ω_{zzz}
Q_{31c}	$\sqrt{3/2}\ \Omega_{xzz}$
Q_{31s}	$\sqrt{3/2}\ \Omega_{yzz}$
Q_{32c}	$\sqrt{3/5}(\Omega_{xxz} - \Omega_{yyz})$
Q_{32s}	$2\sqrt{3/5}\ \Omega_{xyz}$
Q_{33c}	$\sqrt{1/10}(\Omega_{xxx} - 3\Omega_{xyy})$
Q_{33s}	$\sqrt{1/10}(3\Omega_{xxy} - \Omega_{yyy})$

In this representation n_a (and n_b) define the order of the basic zonal harmonic $P_l(x)$ (tabulated, for example in Ref. 38), and hence the corresponding multipole moment (derived from the associated Legendre polynomials). Thus, for example, putting n_a (n_b) = 1 and n_b (n_a) = 5 would define the next term in the expansion: the dipole-(32-pole) interaction, which has an R^{-7} distance dependence.

Coulombic Interactions in the General Case. The near zone limit of the Coulombic interaction (Eq. 10.15) can be written in terms of the spherical tensor formulation as:*

$$u^{Coul} \approx Q_t^{0n}(a)T_{tu}^{ab}Q_u^{0m}(b) + M_t^{0n}(a)T_{tu}^{ab}M_u^{0m}(b) \qquad (11.6)$$

where Q and M represent the electric and magnetic multipole transition moments respectively. The T_{tu}^{ab} are functions of the relative orientations of chromophores a and b. The notation of Stone[27] has been used in the subscripts to label the multipole transition moments. For completeness, those relevant to this work are collected in Table 3.6.

---

* See Ref. 27 for an excellent discussion of spherical tensors and electrostatic interactions.

Table 3.7. Transition moments corresponding to some common molecular point groups.

Mol. Pt. Group	Transition Symmetry	Spherical Tensor Transition Moments: Dipole	Quadrupole	Octopole
C_s	A"	Q_z, M_x, M_y	Q_{21c}, Q_{21s}	Q_{30}, Q_{32c}, Q_{32s}
C_2	A	Q_z, M_Z	Q_{20}, Q_{22c}, Q_{22s}	Q_{30}, Q_{32c}, Q_{32s}
C_{2v}	A_1	Q_z	Q_{20}, Q_{22c}	Q_{30}, Q_{32c}
	A_2	M_z	Q_{22s}	Q_{32s}
	B_1	Q_x, M_y	Q_{21c}	Q_{31c}, Q_{33c}
C_{3v}	A_1	Q_z	Q_{20}	Q_{30}, Q_{33c}
	A_2	M_z		Q_{33s}
	E	$(Q_x,Q_y), (M_x,M_y)$	(Q_{22c},Q_{22s}),	(Q_{31c},Q_{31s}),
			(Q_{21c},Q_{21s})	(Q_{32s},Q_{32c})
C_{4v}	A_1	Q_z	Q_{20}	Q_{30}
	A_2	M_z		
	B_2		Q_{22s}	Q_{32s}
	E	$(Q_x,Q_y), (M_x,M_y)$	(Q_{22c},Q_{22s}),	(Q_{31c},Q_{31s}),
			(Q_{21c},Q_{21s})	(Q_{33c},Q_{33s})
C_{2h}	A_g	M_z	Q_{20}, Q_{22c}, Q_{22s}	
	A_u	Q_z		Q_{30}, Q_{32c}, Q_{32s}
D_{2h}	A_g		Q_{20}	
	B_{1g}	M_z	Q_{22s}	
	A_u			Q_{32s}
	B_{1u}	Q_z		Q_{30}, Q_{32c}
D_{3h}	A'_1		Q_{20}	Q_{33c}
	A'_2	M_z		Q_{33s}
	E'	(Q_x,Q_y)	(Q_{22c},Q_{22s})	(Q_{31c},Q_{31s})
	A''_2	Q_z		Q_{30}
	E"	(M_x,M_y)	(Q_{21c},Q_{21s})	(Q_{32s},Q_{32c})
D_{4h}	A_{1g}		Q_{20}	
	A_{2g}	M_z		
	B_{1g}		Q_{22c}	
	E_g	(M_x,M_y)	(Q_{21c},Q_{21s})	
	A_{2u}	Q_z		Q_{30}
	B_{1u}			Q_{32s}
	E_u	(Q_x,Q_y)		(Q_{31c},Q_{31s}),
				(Q_{33c},Q_{33s})

Which of these transition moments should be considered will depend upon the symmetries of donor and acceptor chromophores. Using standard character tables, the relevant information for some common molecular point groups are collected in Table 3.7. A general simplification has been employed, such that if a transition is dipole-allowed, the local axis system should be rotated so that the transition is aligned along the *z*-axis.

It is seen immediately that more transition moments enter the equation for lower symmetry chromophores. For example, even for a molecule of C_{2v} symmetry, in addition to the symmetry-allowed electric dipole transition moment, one may need to consider two quadrupole and two octopole transition moments. Whereas for higher symmetry chromophores, the simple model of Eqs. (11.3) suffices.

For such chromophores, then, an expression for the Coulombic (2*J*) interaction for the case of electric dipole-allowed (symmetrically-polarized) electronic transitions may be inferred from Eqs. (11.3):

$$u^{Coul} \approx u^{\text{d-d}} + u^{\text{d-o}} + u^{\text{o-o}} \tag{11.7}$$

Such an expression has been advocated previously by Craig and Walsmley to describe exciton splittings in crystal spectra of very weak transitions in crystalline naphthalene.[39,40] The computational results to be described next indicate that the Coulombic interaction should be taken at least to the dipole-octopole term even for moderately-allowed dipole transitions.

Conversely, an expression for totally-symmetric (electric dipole-forbidden) transitions cannot be cast in such a simple form.

Computational Analysis. In Figure 3.10 the calculated (CIS) T_{14} interaction matrix element for ethene singlet and triplet B_{1u} and B_{1g} electronic states is plotted as a function of separation. The plot shows that the major contributor to this direct exciton resonance interaction matrix element at all separations, for the electronic dipole-allowed ethene $^1B_{1u}$ state, is the Coulombic interaction (given the substantial difference between singlet-singlet and triplet-triplet interactions). Conversely, for the magnetic-dipole and electric quadrupole-allowed ethene $^1B_{1g}$ state, the short-range $-Z$-P interactions dominate at closer separations. At larger separations, however, the magnetic dipole-dipole interaction provides the bulk of the electronic coupling (as will be shown presently).

Using Eq. (5.9) the triplet-triplet interactions may be related to the short range component of the corresponding singlet-singlet interaction via the relation (11.8).

$$^1T_{14} \approx {}^3T_{14} + S_{14}\Delta E_{(\text{singlet-triplet})} + 2J_0 \tag{11.8}$$

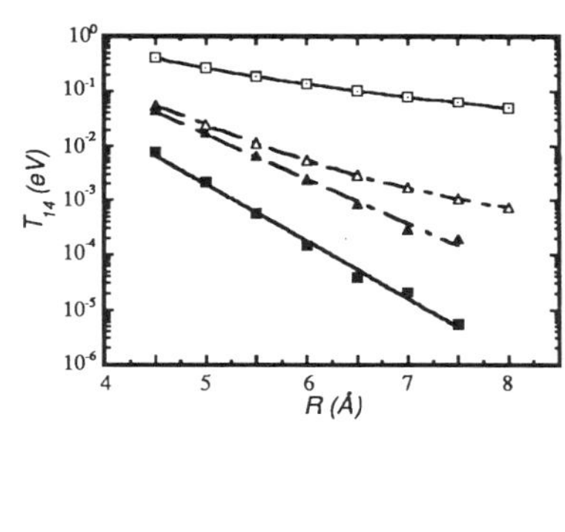

Figure 3.10. A plot of the calculated (CIS) T_{14} interaction matrix element for ethene singlet (open symbols) and triplet (filled symbols) B_{1u} (squares) and B_{1g} (triangles) electronic states as a function of separation.

Thus, using the calculated singlet-triplet energy gap, $\Delta E_{(\text{singlet-triplet})} \approx$ 4.1685 eV, and an approximate expression for the distance dependence of the orbital overlap (determined from the attenuation of $^3T_{14}$), $S_{14} \propto \exp(-2.408R)$, the total Coulombic interaction, $2J$, for the ethene B_{1u} state may be obtained as a function of interchromophore separation. The static dipole transition moment for the ethene B_{1u} state was calculated to be $\mu =$

1.4060×10^{-29} Cm (1.6583 a.u.), so that the electric dipole-dipole interaction is well-characterized. The calculated u^{Coul} (i.e., $2J$) data were fit by a function of the form:

$$u^{Coul} = \frac{11.0887}{R^3} + \frac{5.6093 \times 10^{78}(-6\mu Q_{30} + \sqrt{15}Q_{32c})}{R^5} + \frac{5.6093 \times 10^{98}X}{R^7} \text{ eV} \tag{11.9}$$

where the interchromophore separation is in units of Å and orientation factors are given for the first three terms as 1, -3 and $\frac{\sqrt{15}}{2}$ respectively. Transition moments and orientation factors for the final term are collectively labelled X, which contains contributions from both octopole-octopole and dipole-(32-pole) terms.

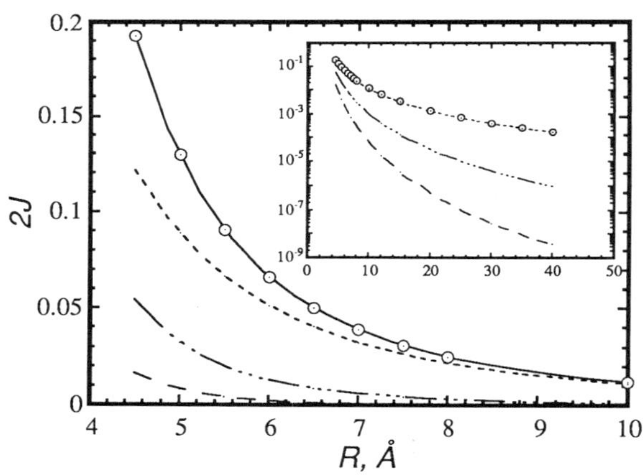

Figure 3.11. Calculated (CIS) Coulombic interaction for the ethene $^1B_{1u}$ electronic state at various separations (circles and solid line fit). Contributions from dipole-dipole (small dashes), dipole-octopole (dash-dots) and R^{-7} (long dashes) interactions are indicated. The inset (plotted on a log scale) shows the data at larger separations.

The results, presented in Figure 3.11, indicate that the contribution from octopole transition moments (i.e. $\sqrt{15}Q_{32c}$ $-6Q_{30}$) is 1.27×10^{-48} Cm^3 and the X term is approximately 1.1×10^{-96} C^2m^6. It is therefore apparent that the dipole-octopole interaction is important at close separations.

A similar analysis is not so straightforward for the ethene B_{1g} transition (Figure 3.12) due to the very small electronic couplings involved. It is evident, however, that the interaction exhibits a long range R^{-3} behaviour (with $m_z = 1.606 \times 10^{-22}$ Am^2 (0.0632 a.u.)) as well as a short range quadrupole-quadrupole component, varying as R^{-5}.

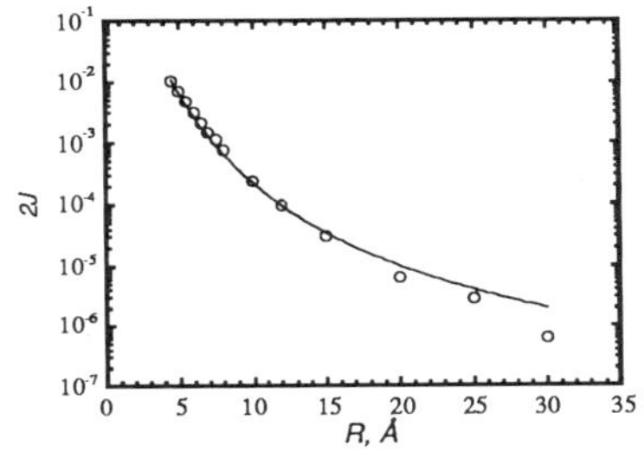

Figure 3.12. Calculated (CIS) Coulombic interaction for the ethene $^1B_{1g}$ electronic state at various separations (open circles). The solid line is a fit of the data, accounting for magnetic dipole-dipole and electric quadrupole-quadrupole interactions.

XII. The Short-range Interactions

In this section the distance dependence of the short-range, orbital overlap-dependent, interaction terms (*-Z-P* and *U,* cf. Eq. 6.10) will be examined. Both simplistic, but nevertheless useful, as well as some more proper, or flexible, models will be investigated.

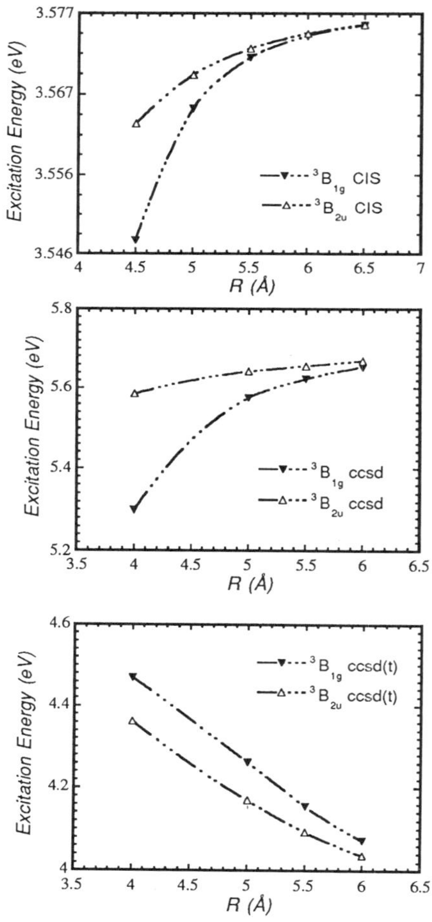

Figure 3.13. Excitation energies of the ethene $^3B_{1u}$ state at various separations of the D_{2h} dimer. The upper panel depicts the CIS results, the middle panel the CCSD results, and the CCSD(T) results are displayed in the lower panel.

Excitation energies of the split ethene $^3B_{1u}$ excited (triplet) state† at various separations of the D_{2h} "sandwich" dimer are shown in Figure 3.13. At the CIS level of theory the model described in Section IV should be represented. Double excitations are included via the CCSD method; adding correlation, particularly to the closed-shell ground state. The depression of both potential energy

† Labelled B_{1g} and B_{2u} because of a change in axis convention for the dimer.

surfaces with decreasing separation is attributed to the U interaction, as has been shown previously.[13,15,41,42]

Inclusion of triple substitutions (CCSD(T)) considerably alters the trends observed using the CIS and CCSD methodologies. This is possibly because quadruple substitutions have not been included in the description of the ground state wavefunction: as the separation between the molecules becomes larger, and the molecular orbitals obtain a more localized character, quadruple excitations may effectively provide double substitutions on each monomer. Since the effects of electron correlation are not considered in the model of Section IV, most of the analysis presented in this Section will deal with the CIS data.

Similar potential energy surfaces for the ethene singlet $^1B_{1u}$ (electric dipole-allowed) transitions are presented in Figure 3.14. The notably larger splittings between the $^1B_{1g}$ and $^1B_{2u}$ dimer states compared with those of the triplet state are due to the appreciable Coulombic interaction. In the absence of interchromophore orbital overlap the oscillator strength of the allowed supermolecular state is twice that of the corresponding monomer transition for this dimer geometry. Such a state is said to be superradiant.[41,43] As first noted by Koutecký and Paldus,[15] the dependence of oscillator strength of electronically-allowed transitions on interchromophore separation in the region of non-negligible orbital overlap reflects the mixing of locally-excited and ionic configurations. Thus, in this case, the intensity of this allowed band of the "monomer" decreases as the degree of mixing with the ionic configurations (cf. Eq. 4.1) increases. Oscillator strengths of the transitions into the corresponding ionic states increases concommitantly. The mixing appears to begin increasing significantly with decreasing separation at a critical distance of about 5.5 Å.

It is evident from Sections V and VI that the most significant orbital overlap-dependent terms (U and $-Z\text{-}P$; hitherto collectively referred to as u^{short}) are governed by the product of two intermolecular orbital overlap integrals. Thus one would expect that the distance-dependence of such interactions would be of the form:

$$u^{short} \propto S_{ab}S_{a'b'} \tag{12.1}$$

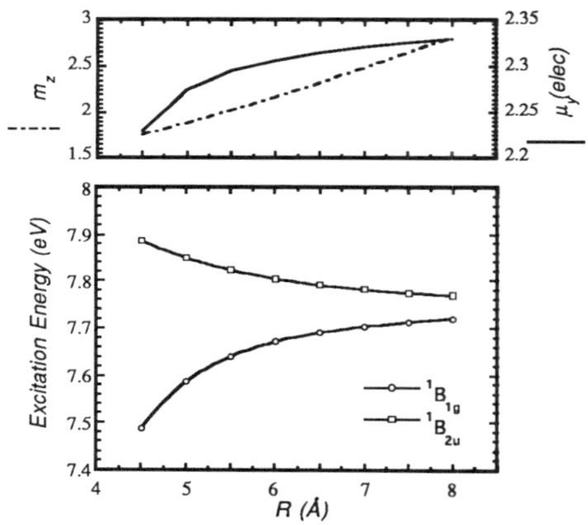

Figure 3.14. Lower panel: Excitation energy (CIS) versus separation for the ethene $^1B_{1u}$ state in a D_{2h} sandwich dimer configuration. Upper panel: Corresponding variation of the electric dipole transition moment of the $^1B_{1u}$ dimer transition (solid line), and the magnetic dipole transition moment of the $^1B_{1g}$ dimer transition (dashed line).

The dependence of u^{short} on interchromophore separation, for the triplet-triplet interaction, is depicted in Figures 3.15 and 3.16. It has been fit by two functions, each approximating the attenuation of interchromophore orbital overlap with increasing separation: u^{exp}, a simple exponential function (Eq. 12.2), and u^M, a function utilizing the Mulliken form of the 2pσ-2pσ overlap integral[44] (Eq. 12.3).

$$u^{exp} = A \exp\left(-2\alpha R\right) \tag{12.2}$$

$$u^M = A\, S(p)S(p') \tag{12.3}$$

with

$$S(p) = e^{-p}\left(-1 - p - \tfrac{1}{5}p^2 + \tfrac{2}{15}p^3 + \tfrac{1}{15}p^4\right) \tag{12.4}$$

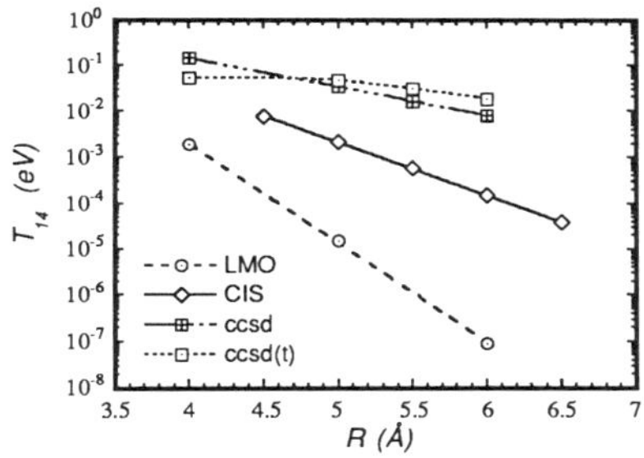

Figure 3.15. Distance dependence of the direct exciton resonance interaction matrix element for the ethene $^3B_{1u}$ state, calculated by the all-electron LMO method (Section IX) as well as the CIS, CCSD and CCSD(T) methods.

where $p = \xi R$, with R (taken to be the centre-to-centre distance) in a.u. and the A are scalars which normalize u^{exp} or u^M to the calculated u^{short}. The results are collected in Table 3.8.

The distance dependence of $^3T_{14}$, determined by various methods, is shown in Figure 3.15. Notable is the over one order of magnitude increase in going from the CIS to the CCSD method. In Figure 3.16 $^3T_{14}$ and ^3U (CIS method) for the $^3B_{2u}$ and $^3B_{1g}$ ethene states are compared. They have been fit by Eq. (12.3), as detailed in Table 3.6. Note the weaker attenuation of the U interaction with separation compared to T_{14}. Such a comparison of T_{14} to U is not reported for the CCSD calculations because of ambiguity involved in determining U due to unbalanced inclusion of correlation compared to the monomer.

Table 3.8. Fitted distance dependencies (using Eqs. 12.2. and 12.3) of the CIS data for the triplet state ethene dimer at close separations.

Equation	State	T_{14}		U	
		exponent (Å^{-1})	A (eV)	exponent (Å^{-1})	A (eV)
12.2	B_{1u}	1.33	1200	0.706	11.8
	B_{1g}	1.00	374	0.580	13.8
12.3	B_{1u}	1.18, 1.03	3.26	0.781, 0.749	2.26
	B_{1g}	1.01, 0.727	10.8	0.752, 0.480	1.87

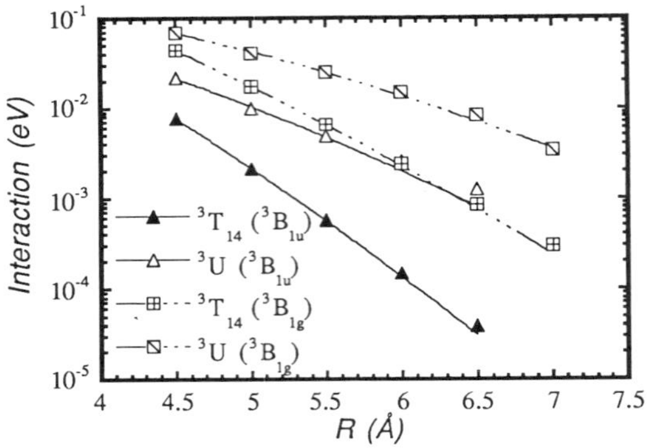

Figure 3.16. Comparison of the direct exciton resonance interaction, $^3T_{14}$, and the through-configuration exciton resonance interaction, ^3U, for the ethene $^3B_{1u}$ (triangles) and $^3B_{1g}$ (squares) states.

An advantage of calculating u^{short} via u^M is that the orientation dependence of the Mulliken overlap integral may be determined (see e.g. Ref. 46). Thus it would be useful to be able to determine A, the basic form of which has been detailed by Mason and Hirschfelder.[45] Thus, a quantitative evaluation of u^{short} may be possible by using the Mulliken or Sklar approximations to reduce the problem to a number of classical electrostatic interactions. Such a procedure has been previously described[45] for short-range intermolecular forces.

An interesting additional correction to U, possibly relevant to exciton interactions and the stabilization of excimers, is the induction, or dynamic polarization, interaction. Following Power and Thirunamachandran,[47] this interaction may be perceived as the coupling between transition moments in each molecule, induced by vacuum fluctuations of the electromagnetic field. The induced (transition) moment for a molecule at $\boldsymbol{r}$, with isotropic polarizability $\alpha(k)$ is:

$$\mu_\alpha^{\text{ind}}(\boldsymbol{k}) = \alpha(k)\, d_\alpha^{\perp}(\boldsymbol{r}) \tag{12.5}$$

which can be used to give a fully retarded dispersion interaction (Casimir-Polder potential) between two molecules (cf. Section X). When one molecule (e.g. A) is in an excited state, an additional contribution arises such that the product of the two induced moments have a resonance at $k = k_a$ (specifying the deexcitation energy of A). Thus, for an interacting molecular pair, where one molecule is in an electronically excited state, the leading interaction (independent of orbital overlap) is

$$u^{\text{ind}} = \frac{-k_a^4}{(4\pi\varepsilon_0)^2R^2}[\mu_\alpha(a)]^2\alpha_{\beta\gamma}(b;k_a)\left(\delta_{\alpha\beta} - 3\hat{R}_\alpha\hat{R}_\beta\right)\left(\delta_{\alpha\gamma} - 3\hat{R}_\alpha\hat{R}_\gamma\right) \times \left(1 + \frac{1}{k_a^2R^2} + \frac{1}{k_a^4R^4}\right) \tag{12.6}$$

where $\alpha_{\beta\gamma}(b;k_a)$ is the dynamic polarizability of molecule B.

This interaction may be viewed as a coupling between the transition moments, induced by vacuum fluctuations. The leading contribution arises from an excited state to ground state transition of the donor chromophore, and a corresponding dynamic polarization of the acceptor. As such, it contributes to the stabilization of excimers, but does not effect EET. We term this the *induced dynamic polarization interaction,* u^{ind}. It has an R^{-6} distance-dependence in the near zone limit, and is independent of orbital overlap.

XIII. Superexchange-Mediated Interactions

Through-Bond Propagated Short-Range Interactions. Electronic coupling need not be realized solely by a direct interaction between donor and acceptor chromophores. If there are additional components present in the system, $\{C_n\}$, which couple to A and B via the interaction Hamiltonian, then excitation transfer from A to B may be mediated indirectly - it becomes a multistep process, being relayed by $\{C_n\}$. Such a mechanism is known to promote long-range electron transfer,[48-54] and may similarly enhance rates of EET between A and B relative to that due to direct coupling only. Typical relay units include σ bond "spacer" groups, intervening chromophoric substituents, a lipid bilayer, or crystalline "host" molecules.

Relayed excitation transfer may be thought of as a multistep mechanism where each step proceeds by the *same* mechanism as for direct excitation transfer (i.e., Coulombic and short range interactions):

$$|A^*C_1C_2...C_mB\rangle \rightarrow |AC_1^*C_2...C_mB\rangle \rightarrow ... \rightarrow |A^*C_1C_2...C_i^*...C_mB\rangle$$
$$\rightarrow ... \rightarrow |AC_1C_2...C_mB^*\rangle$$

In this picture, the complete relayed interaction is described by a sum over all possible pathways leading from the initial to the final state. An important aspect of a relayed mechanism is that energy does not need to be conserved in each step, since states representing localization of excitation on a relay unit are *virtual* states. The interaction is formally made possible by the Heisenberg (time-energy) uncertainty principle, but it means that energies of the intermediate states can be entirely different from that of the initial and final states so long as there is an overall conservation of energy between initial and final states of the transition.

In such a way the short range, orbital overlap-dependent, interactions may be promoted over large through-space separations; essentially by a delocalization (or mixing) of the donor and acceptor wavefunctions into the bridge orbitals. Donor and acceptor wavefunctions may then overlap appreciably, even at separations far exceeding their van der Waals radii. The net through-bond EET process may thus be represented as[55]

$$|A^*C_1C_2 \ldots C_mB\rangle \rightarrow |AC_1C_2 \ldots C_mB^*\rangle$$

where $C_1, C_2, \ldots, C_m$ correspond to m bridging units. This eigenvalue problem for the $(m+2) \times (m+2)$ Hamiltonian may be reduced to an eigenvalue problem for the 2×2 Hamiltonian (13.1) by using the Löwdin partitioning technique[56,57] (Eq. 13.1) whereby the mixed donor-bridge and acceptor-bridge wavefunctions are interacted.

$$\mathbf{H'} = \begin{pmatrix} \varepsilon'_a & V_{ab} \\ V_{ab} & \varepsilon'_b \end{pmatrix} \qquad (13.1)$$

ε' is thus the *effective* donor and acceptor energy, and V_{ab} is the *effective* electronic coupling between donor and acceptor. As described by Evenson and Karplus[58] this partitioning scheme leads to the effective coupling

$$V_{ab} = -\sum_{ij} V_{ai}\, \mathbf{G}_{ij}(\varepsilon)\, V_{jb} \qquad (13.2)$$

where the Green's function for the bridge is $\mathbf{G}(\varepsilon) = (\mathbf{H}_{\text{bridge}} - \varepsilon)^{-1}$, with $\mathbf{H}_{\text{bridge}}$ being the Hamiltonian for the bridge and ε is an eigenvalue of the full Hamiltonian for both donor and acceptor states. V_{ai} and V_{jb} define the electronic coupling between donor and bridge, and bridge and acceptor, respectively.

If one employs a tight-binding Hamiltonian and there is a large energy gap between bridge states and donor/acceptor states as well as weak coupling between bridge units, the McConnell formula[59,60] may be obtained for identical donor and acceptor chromophores. Then, for m bridge units, the through-bond mediated electronic coupling, $u^{t\text{-}b}$, is given by (Eq. 13.3).

$$u^{t\text{-}b} \approx \left(\frac{T^2}{-A}\right)\left(\frac{t}{-A}\right)^{m-1} \tag{13.3}$$

If diagonalization of the $m \times m$ bridge block of the Hamiltonian yields eigenfunctions ψ_i^c, $(i = 1,2,...,m)$ and eigenvalues ε_i^c then we have:

$T = T_{ai} = \langle \psi^a | \mathrm{H} - H_{aa} | \psi_i^b \rangle = T_{jb}$ = chromophore-bridge coupling,

$t = T_{i(i+1)} = \langle \psi_i^c | \mathrm{H} - H_{aa} | \psi_{i+1}^c \rangle$ = bridge-bridge coupling,

$A = \varepsilon_i^c - \varepsilon_a$ = energy gap between chromophore and bridge.

For long bridges, such a model predicts an exponential attenuation of the magnitude of electronic coupling with increasing bridge length (Eq. 13.4),

$$u^{t\text{-}b} \approx V_0 \exp\left[-\beta(m-1)\right] \tag{13.4}$$

where V_0 is the through-bond electronic coupling for $m = 1$ and β is the effective decay constant (per bond). A detailed analysis of the application and validity of the McConnell model has been given by Jordan and Paddon-Row.[54]

The Relayed Coulombic Interaction. Eqs. (10.15) and (11.3) strictly describe the direct Coulombic interaction between two isolated chromophores A and B. At the other limit, where there is a high density of a third body C (such as solvent molecules) between A and B, the interaction is viewed as occurring in a dielectric medium. In this case one simply changes ε_0 to ε, the medium dielectric constant (cf. Förster's Equation,[2] Eq. 2.6.13). However, when there are only few C molecules between A and B then the direct Coulombic interaction can be accompanied by three-body processes, leading to a modification in electronic coupling and a change in distance dependence.[61]

It is also possible that the connecting bridge between the two chromophores may act as the third body and mediate the EET process by the exchange of virtual excitation. Craig and Thirunamachandran have derived an expression for this relayed dipole-dipole interaction,[61] Eq. (13.5), where R_{AC} and R_{BC} define the separations between the respective chromophores and the third body C, while $\alpha_{kl}(c)$ is the dynamic polarizability of C.

$$u^{rel} \approx \frac{-1}{(4\pi\varepsilon_0)^2 R_{AC}^3 R_{CB}^3}\, \mu_\alpha(a)\mu_\beta(b)\alpha_{\gamma\delta}(c)\left(\delta_{\alpha\delta} - 3\hat{R}_\alpha\hat{R}_\delta\right)\left(\delta_{\beta\gamma} - 3\hat{R}_\beta\hat{R}_\gamma\right) \qquad (13.5)$$

Resonant EET in condensed matter has also been discussed by Juzeliūnas and Andrews.[11,62]

XIV. Discussion

The exponential attenuation with distance of the orbital overlap-dependent component of the interaction matrix element was first recognized by Dexter.[3] Dexter was concerned with the problem of sensitized luminescence in the solid state. In particular, the situation where an impurity atom in a crystalline host may be sensitized by energy transfer (from another impurity atom or from the host) even though its electronic transition is electric dipole-forbidden. The "many-body" nature of solid state problems was emphasized, but Dexter's interest was in the mechanism of the energy transfer process in such situations.

Dexter's approach was to write out properly antisymmetrized initial and final state wavefunctions of the donor-acceptor pair. He then considered only two-electron integrals and showed that the Coulomb interactions (incorporated in the dipole approximation in Förster's theory[2]) could be obtained without including the antisymmetry requirements of the initial and final state dimer wavefunctions. However, with antisymmetry considerations another interaction, the exchange integral, was recovered. The magnitude of this interaction is critically dependent upon the degree of overlap between donor and acceptor wavefunctions (i.e., they must be closely separated) and it was stated that exchange effects become important for magnetic dipole-allowed (and higher order) transitions.

The most important result of Dexter's work is that the exchange interaction (or more generally, as is seen in this work, interactions depending upon orbital overlap between donor and acceptor chromophores) are attenuated approximately exponentially with increasing separation. Experimentally, the Dexter exchange interaction is, however, difficult to quantify *a priori*.

What is seldom recognized when Dexter's work is used to interpret interactions between chromophores with electric dipole *allowed* electronic transitions (remember, his paper discussed the sensitization of *forbidden* transitions) is that the exchange and Coulombic interactions act in conjunction, *not* exclusively.[17] In some cases this may appear to lead to a paradox - where the interaction vanishes at a certain close separation between donor and acceptor. However, as demonstrated by the more detailed theory presented in this work (where the origins of "Dexter-type" EET are shown to be quite different than originally outlined), such a situation would be exceptional.

A more rigorous derivation of the interactions responsible for mediating singlet-singlet and triplet-triplet interactions was later presented by Merrifield,[4] but still orbital overlap-dependent interactions were attributed solely to a two-electron exchange integral. As pointed out by Nieman and Robinson[63] such treatments omit other kinds of intermolecular overlap effects, such as those corresponding to interactions between electrons and screened nuclei (penetration integrals). These terms were recovered in an investigation of the interaction between a ground state and an excited helium atom by Buckingham and Dalgarno.[64] Mulliken also examined the interactions between differently excited, like atoms,[65] and found that they consisted of a first-order dispersion energy, falling off approximately as $1/R^3$, and "valence-force" integrals, attenuated exponentially with separation (R).

Recently, a study has been presented which determines an excitation transfer matrix element applicable for any multiconfigurational wavefunctions,[66] but again the sole orbital overlap-dependent term is the two-electron exchange contribution. An elucidation of short- and intermediate-range electronic coupling in singlet-singlet EET was recently given by the present author,[17] as was the more detailed theory given in Sections IV to VI[67].

The current work is a thorough examination and juxtaposition of interactions responsible for promoting singlet-singlet and triplet-triplet EET. The approach was to allow ionic configurations of the donor-acceptor pair to mix with the initial and final locally excited configurations (Eq. 4.1 and Figure 3.3). The Hamiltonian interaction matrix element is then given by Eq. (4.7), however the electronic coupling matrix element connecting donor and acceptor states is obtained from the secular equations of Eq. (4.8), and is given by $T_{LR} = \langle \Psi_L | \mathrm{H} - E_1 | \Psi_R \rangle$. This theory is easily extended to the instance where donor and acceptor are different. To do this Eqs. (4.1) should be rewritten as Eqs. (14.1).

$$\Psi_L \approx N_L(\psi_1 + \lambda_L \psi_2 + \mu_L \psi_3) \tag{14.1a}$$

$$\Psi_R \approx N_R(\psi_4 + \mu_R \psi_2 + \lambda_R \psi_3) \tag{14.1b}$$

Proceeding as with the identical chromophore case, we obtain the Eq. (14.2), which is equivalent to Eq. (4.11), but for the case of non-identical donor and acceptor chromophores. Note that, in accord with a "standard" perturbation theory approach, it involves only energy differences relative to the monomer.

$$T_{LR} \approx T_{14} - (T_{12}T_{24})/A_{12} - (T_{13}T_{34})/A_{13} \tag{14.2}$$

The electronic coupling was later rewritten as Eq. (6.10), which clearly involves contributions from three types of term. The direct interaction, T_{14}, contributed two of these terms. The J-term, the well-known Coulombic interaction, represents an interaction between transition moments of the two chromophores and acts over *all* separations (cf. Sections X and XI). Note that for triplet-triplet EET the J's cancel each other, which means that a first order term involving a coherent deexcitation/excitation process is actually *absent*, not just small because it is spin-forbidden. The Z term, by analogy to J, is called the Dexter exchange interaction and, together with the penetration contribution, P, it is termed a short-range interaction[17] because of its dependence upon the degree of interchromophore orbital overlap. The approximate calculations on a model ethene dimer (Section VII) elucidated the expected dominance of these penetration interactions which involve products of one-electron integrals and overlap integrals, over the Dexter exchange interaction. Thus it was found that the direct interactions engendered by interchromophore orbital overlap are primarily determined by P.

The through-configuration interaction (Eq. 4.14) contributed the third term to the overall electronic coupling. This term, U, is a second-order contribution to the electronic coupling which involves successive virtual one-electron transfers mediated via the ionic bridging configurations (which places the relationship between electron and energy transfer[50,51] on a more rigorous foundation). It actually makes the dominant contribution to the interchromophore orbital overlap-dependent electronic coupling (provided that there is no symmetry restrictions on the configuration interaction[13]).

The U term leads to a lowering of the centre of gravity of the supermolecule's excited state potential energy surface. This, in turn, is split relative to the single molecule excited state by twice the T_{14} term ($2J$ - Z - P), as discussed in Section III.

The cognate computational investigation presented in Sections IX, XI and XII, provides valuable insight into the electronic coupling between two chromophores. At intermediate interchromophore separations (generally specified by distances of ca. 6 to 150 Å between the closest atomic centres) the electronic coupling is given primarily by u^{Coul}. For the interaction between electronically allowed transitions in chromophores u^{Coul} is given accurately by the leading terms in the multipole expansion (Section XI). Note that the dipole-quadrupole, quadrupole-quadrupole and magnetic dipole-electric dipole terms are zero for electric dipole-allowed transitions in molecules with high axial symmetry, whereas the higher order dipole-octopole and octopole-octopole terms may be important.

In Section XII it was shown that u^{short} is proportional to the product of two intermolecular orbital overlap integrals, and is well approximated by a simple exponential distance dependence (Eq. 12.2). A more rigorous (but still approximate) treatment, from which orientation dependencies may be determined was also considered, Eq. (12.3).

An extensive and systematic study of through-space "exchange-mediated" EET (i.e., based on the Dexter model[3]) has been undertaken by Speiser, et al.[68-70] By investigating EET in a series of macrocyclic α-diketones incorporating an aryl group (specifically *p*- and *m*-cyclophanes and the ortho analogues, benzocycloalkenediones) they obtained experimentally, for the first time, the parameters incorporated in Dexter's theory.[70] This was made possible by firstly ensuring poor spectral overlap between donor and acceptor chromophores and that the electronic transition of the donor was forbidden (so that the dipole-dipole interaction is negligible) and, secondly, employing semi-rigid geometries of the bichromophores so as to enable distance dependencies to be elucidated. A parallel study on triplet-triplet EET in these molecules was unable to provide quantitative correlations between EET efficiency and interchromophore separation, although the EET efficiency was shown to be sensitive to molecular geometry.

In a more recent study Levy, et al.[68] developed a model to account for the observed structural and temperature dependencies of the rate for Dexter-type intramolecular EET. The temperature effect was attributed to conformational changes induced by temperature variations. This study, together with a cognate, primarily computational, investigation[69] demonstrated that it is important to account for orientation effects on the magnitude of the short-range interactions, although the simple Dexter (exponential) distance-dependence relation holds for symmetric systems.

Similarly, a close-to exponential attenuation of the short-range component of the electronic coupling with increasing separation is observed for the calculations in Section XII in this *symmetric dimer configuration.* However, as described and demonstrated experimentally by Levy et al.,[68,69] orbital overlap-dependent interactions are critically dependent upon orientation of the interacting chromophores.

Whilst the ethene dimer is used in this paper to illustrate and quantify the theory presented herein, the essential results and conclusions of this work are transferable to larger molecular systems. The energy gaps between locally-excited and ionic (charge transfer) configurations should be of the same order, however the representation of excited states will be more complex. Vibronically induced spectral intensity may also need to be considered.[71]

The treatment presented in this Chapter examines the electronic aspects of interacting, discrete, Born-Oppenheimer levels. This is, of course, a somewhat artificial representation of a "real" system, but serves to illustrate the nature of the electronic component of the interaction matrix element which promotes excitation transfer or excitation delocalization. A complete description of a "real" system may be obtained by considering, in unison, the theory presented here and theories of non-radiative processes (cf. Chapter 2).

The illustrative MO-CI calculations on the ethene dimer in Sections XI and XII involved more than one HOMO and LUMO, via the CIS description. However, this does not imply that a four-level model need be invoked to account for the excitation transfer since the problem reduces to a linear combination of two-level systems, with relative weights given by the CIS coefficients.

In reality, the electronic overlaps implicit in each term of the interaction matrix elements are associated with vibrational overlap (Franck-Condon) factors, summed over all possible vibrational modes. Electronic coupling terms are appreciable for the "promoting modes";[72] for which the Franck-Condon factors are related to the electronic integrals. For example, the important vibrational overlap factors associated with the Coulombic interaction are the products[26] $\langle\chi_{a'}|\chi_a\rangle\langle\chi_{b'}|\chi_b\rangle$, while those for the orbital overlap-dependent interactions are[73] $\langle\chi_a|\chi_b\rangle\langle\chi_{a'}|\chi_{b'}\rangle$. Sums of such integrals are usually factored out of the interaction matrix element, assuming approximately constant and weak electronic coupling over nuclear coordinates (see Chapter 2 and references therein). In some cases this may alter the significance of U in comparison with $2J$.

Additional mechanisms that arise in complex systems include superexchange mediated EET, in which the electronic coupling (particularly via the short-range interactions) is promoted over large through-space separations via virtual interactions with intervening chemical entities (such as connecting σ bonds - through-bond interactions).[48,54,74-76] The present work is also relevant to the understanding of such phenomena, since each virtual electronic coupling matrix element in the superexchange interaction has an identical form to those donor-acceptor interactions addressed in this contribution. Vibronically-induced interactions may also be important in both direct[71] and superexchange-mediated[77] electronic interactions.

XV. Conclusions

The electronic interactions which promote singlet-singlet and triplet-triplet electronic excitation (energy) transfer (EET) were investigated in detail. Although this work specifically considered the electronic interaction between two *identical* chromophores, the conclusions are equally applicable to the case of two interacting *non-identical* chromophores (A and B). The approach was to allow the ionic configurations $\psi_2(A^+B^-)$ and $\psi_3(A^-B^+)$ to mix with the initial (A*B) and final (AB*) locally excited configurations via the small mixing coefficients λ and μ (Eq. 4.1 and Figure 3.3):

$$\Psi_L \approx N(\psi_1 + \lambda\psi_2 + \mu\psi_3)$$
$$\Psi_R \approx N(\psi_4 + \mu\psi_2 + \lambda\psi_3)$$

The interaction matrix element connecting these two states, T_{LR} , was obtained via the secular equations (4.3, 4.8) and given by Eq. (4.7):

$$T_{LR} = N^2\{T_{14} + 2\mu T_{12} + 2\lambda T_{13} + \lambda\mu(A_{12} + A_{13}) + (\lambda^2 + \mu^2)T_{23}\}$$

where $T_{ij} = H_{ij} - S_{ij}H_{11}$, then with Eqs. (4.10) it was found (Eq. 4.11) that (to second order in overlap):

$$T_{LR} \approx T_{14} - T_{12}T_{24}/A_{12} - T_{13}T_{34}/A_{12}$$

Singlet-singlet (1) and triplet-triplet (3) forms of T_{LR} were later rewritten as

$${}^{1,3}T_{LR} \approx J - Z \pm J - {}^{1,3}P + {}^{1,3}U$$

(Eq. 6.10), which clearly involves contributions from two types of term:

(a) The direct interaction $^{1,3}T_{14}$ ($= J - Z \pm J - P$), consisting of a long-range Coulombic interaction, $2J$, (for singlet-singlet EET only) and the short-range (orbital overlap-dependent) penetration, P, and Dexter exchange interactions, Z, (refer to Section V and Eqs. 5.5, 5.8 & 5.9);

(b) The through-configuration interaction $^{1,3}U$ ($= \mu T_{12} + \lambda T_{13}$), associated with two equivalent one electron transfer processes involving the ionic, bridging, configurations (refer Section IV, Eq. 4.14 and Section VI, Eqs. 6.8, 6.9 and 6.11).

The associated theory and model LMO calculations on ethene dimers (cf. Figure 3.4) showed that:

(i) Penetration terms, P, are more significant than the Dexter exchange contribution, Z (cf. Table 3.1, Section VII);

(ii) The through-configuration interaction generally dominates the short-range interactions (cf. Tables 3.2 and 3.4, Figures 3.5 and 3.6);

(iii) U and J have the same sign, even when $-Z$ does not (cf. Table 3.5);

(iv) The core electrons (refer to Section VIII) have little bearing on the calculated matrix elements.

Whilst the ethene dimer was used in this work to illustrate and quantify the theory presented herein, the essential results and conclusions of the work are transferable to any molecular system. The results imply that the electronic coupling between two chromophores can be given, over all separations, as a sum of the Coulombic interaction ($2J$, or u^{Coul}) and interactions dependent upon the degree of interchromophore orbital overlap ($-Z-P$ and U, or u^{short}).

u^{Coul} was described by a multipole expansion in the framework of quantum electrodynamics (Section X). It was shown that, in addition to the leading dipole-dipole interaction, higher multipole contributions may often need to be considered at separations of less than ca. 20 to 30 Å (Section XI).

u^{short} was shown to contain two important terms: those arising from direct donor-acceptor interactions (-*Z*-*P*) and those mediated via the ionic configurations (*U*). Generally, the *U* term acts to reinforce u^{Coul}, while -*Z*-*P* may sometimes oppose it. It was found that the distance dependence of u^{short} can be described by a simple exponential fall-off or, alternatively, in terms of the product of overlap integrals between Slater carbon 2pσ orbitals: Eqs. (12.2 - 12.4). An additional interaction, u^{ind}, the Coulombic induction interaction, arises in the dimer due to an induced dynamic polarization of the acceptor caused by the dimer transition moment (Eq. 12.6). It does not, however, effect EET.

It was also shown that it is not generally correct to consider exclusively a Förster[2] *or* a Dexter[3] mechanism for singlet-singlet interchromophore electronic energy transfer at close separations. If interchromophore orbital overlap effects, given by u^{short}, are significant then they must add to the Coulombic interaction, u^{Coul}. Furthermore, it is concluded that use of the Dexter exchange integral to describe short-range interactions is specious.

Appendix A. Calculations

The *ab initio* localized molecular orbital (LMO) calculations on the ethene dimer were undertaken using Roso's valence bond (VB) program[78,79] and an STO-6g basis set with exponents of 5.6727 for C(1s) AO's, 1.64 for C(2s), 1.63 for C(2p), and 1.20 for H(1s). Polarized C-H sp hybrid σ bonds were included in the closed-shell 14-orbital core,[80] together with C(1s) and C-C 2s-2p hybrid σ bond orbitals.

The experimental geometry of planar (D_{2h}) ethene given by Harmony et al.[81] was employed in all calculations: R(C) = 1.339, R(H) = 1.085, $\angle$(HCC) = 117.8°. Symmetry labels for the single ethene molecule refer to Mulliken's axis convention, where the x-axis is taken as that normal to the plane of the molecule and the z-axis as that through the two carbon centres.

Calculations on various naphthalene dimer configurations were undertaken using a minimal basis set (STO-3G) and the *ab initio* CI-singles (CIS; configuration interaction with single excitations from a Hartree-Fock reference determinant) method in Gaussian 92.[18] The naphthalene chromophore geometry was taken from crystal structure data.[82,83] In the D_{2h} symmetry dimer configuration the x-axis is taken along the long molecular axis, the y-axis along the short molecular axis, and the z-axis connects the centres of the two chromophores.

The CIS method was also used for calculations on the ethene dimers. A near triple zeta quality basis set was used: the 6-311+g(2d) basis set, where a diffuse p-type function and two d-type polarization functions have been added. To quantify the effects of electron correlation in the ground, triplet and singlet excited states an augmented coupled cluster method was employed. The converged coupled cluster wavefunction, which included all single and double substitutions (CCSD) was used in the evaluation of the contribution from triple substitutions (CCSD(T)). The resulting energy is correct to fourth order in a Møller-Plessett perturbation expansion, but contains significant fifth and higher order terms. An important advantage of the coupled cluster ansatz (and perturbative methods) over configuration interaction is its size-consistency.

The poor correlation of calculated excitation energies with experiment using the CIS method is not problematic since we are interested in a qualitative gauge of variation in the *splittings and shifts* of the naphthalene excited states, rather than absolute vertical excitation energies. It is emphasized, however, that the calculations are employed to illustrate the theory rather than to quantify actual electronic couplings. The higher-order methods give good agreement with experimentally-measured vertical excitation energies, but have some problems associated with "unbalanced" inclusion of correlation effects.

Basis set superposition error (BSSE) will have an effect on the calculations, but should be small compared to the large spectral perturbations being determined at closer interchromophore separations. The interactions reported in this work are of the order of > 0.01 Hartree at close separations, compared with BSSE of the order of < 10^{-4} Hartree.[84]

Appendix B. Calculated LMO Integrals

Integrals relevant to the ethene sandwich dimer described in Section V were calculated via four-electron LMO calculations with STO-6g basis functions centred on four carbon centres, each with +1 core charge.

Two-electron integrals (not printed by the program we used) were obtained indirectly by interacting a series of two-electron configurations, for example:

$$(aa|a'\,a') = \langle a\bar{a}'|\mathrm{H}|a\bar{a}'\rangle - h_{aa} - h_{a'a'}$$

(where the barred spin orbital contains an electron of β spin. Relevant integrals are tabulated below (Table B1). Further details are provided in Ref 67b.

Table B1. Two-electron integrals for the ethene dimer at 4 Å and 5 Å separations using STO-6g basis functions.

integral	4 Å dimer (eV)	5 Å dimer (eV)
$(aa|aa)$	13.2918	13.3201
$(a'a'|a'a')$	13.6530	13.6842
$(aa|a'a')$	13.2686	13.2995
$(aa'|aa')$	4.29007	4.31038
$(aa|bb)$	3.6654	2.91400
$(a'a'|bb)$	3.62421	2.89349
$(a'a'|b'b')$	4.05829	2.72308
$(aa|ab)$	0.07116	0.005856
$(a'a'|ab)$	0.06929	0.005719
$(aa|a'b')$	0.03344	0.002442
$(ab|a'b')$	0.00047724	3.8147×10^{-6}
$(ab|ab)$	0.0010558	9.7453×10^{-6}
$(a'b'|a'b')$	0.0002340	1.6348×10^{-6}
$(a'b|a'b)$	0.0001236	
$(ab'|a'b)$	0.0001196	
$(a'a|ab')$	0.006482	0.0003855

Appendix C. *k*-Integrations

The method for the **k**-integrations, so as to obtain Eq. (10.14) from Eq. (10.13), is demonstrated here. First Eq. (10.13) is re-written as Eq. (C1):

$$u^{Coul} = \frac{1}{2\varepsilon_0 V}\Bigg\{ \mu_\alpha^{0n}(a)\mu_\beta^{m0}(b) + \mu_\alpha^{0n}(a)\Theta_{\beta\gamma}^{m0}(b)\nabla_\gamma + \Theta_{\alpha\gamma}^{0n}(a)\mu_\beta^{m0}(b)\nabla_\gamma$$
$$+ \Theta_{\alpha\gamma}^{0n}\Theta_{\beta\delta}^{n0}\ \nabla_\gamma\nabla_\delta + \mu_\alpha^{0n}(a)\Omega_{\beta\gamma\delta}^{m0}(b)\nabla_\gamma\nabla_\delta + \Omega_{\alpha\gamma\delta}^{0n}(a)\mu_\beta^{m0}(b)\nabla_\gamma\nabla_\delta + \ldots$$
$$+ \frac{1}{c^2}m_\alpha^{0n}(a)m_\beta^{m0}(b)\Bigg\} \times \int \frac{k^2 \mathrm{d}k\mathrm{d}\Omega}{(2\pi)^3}\frac{k^2}{\kappa^2 - k^2}(\delta_{\alpha\beta} - \hat{k}_\alpha\hat{k}_\beta)\,(e^{i\mathbf{k}.\mathbf{R}} - e^{-i\mathbf{k}.\mathbf{R}})$$

$$+ \frac{1}{2\varepsilon_0 c}\left[\mu_\alpha^{0n}(a)m_\beta^{m0}(b) + m_\beta^{0n}(a)m_\alpha^{m0}(b)\right]\int \frac{k^2 \mathrm{d}k\mathrm{d}\Omega}{(2\pi)^3}\frac{k^2}{\kappa^2 - k^2}\varepsilon_{\alpha\beta\gamma}\hat{k}_\gamma\,(e^{i\mathbf{k}.\mathbf{R}} - e^{-i\mathbf{k}.\mathbf{R}})$$

(C1)

where only terms of the integrand even in k remain. Now, using the relations (C2) and (C3),

$$\frac{1}{4\pi}\int e^{\pm i\mathbf{k}.\mathbf{R}}\,\mathrm{d}\Omega = \frac{\sin kR}{kR} \tag{C2}$$

$$\frac{1}{k}\nabla_\alpha\int e^{\pm i\mathbf{k}.\mathbf{R}}\,\mathrm{d}\Omega = \pm\int \hat{k}_\alpha e^{\pm i\mathbf{k}.\mathbf{R}}\,\mathrm{d}\Omega\,, \tag{C3}$$

one obtains (cf. Craig and Thirunamachandran[5] pp. 148 and 169) Eqs. (C4) and (C5).

$$\frac{1}{4\pi}\int(\delta_{\alpha\beta} - \hat{k}_\alpha\hat{k}_\beta)\,e^{\pm i\mathbf{k}.\mathbf{R}}\,\mathrm{d}\Omega =$$
$$(\delta_{\alpha\beta} - \hat{R}_\alpha\hat{R}_\beta)\frac{\sin kR}{kR} + (\delta_{\alpha\beta} - 3\hat{R}_\alpha\hat{R}_\beta)\left(\frac{\cos kR}{k^2R^2} - \frac{\sin kR}{k^3R^3}\right) = \tau_{\alpha\beta}(kR) \tag{C4}$$

$$\frac{1}{4\pi}\int \hat{k}_\gamma e^{\pm i\mathbf{k}.\mathbf{R}}\,\mathrm{d}\Omega = \mp i\left(\frac{\cos kR}{kR} - \frac{\sin kR}{k^2R^2}\right)\hat{R}_\gamma\,. \tag{C5}$$

The final k-integration is carried out using the contour-integral method, for example (since the integrand is an even function of k):

$$\frac{1}{R}\int_0^\infty \mathrm{d}k\,\frac{k^3 \sin kR}{\kappa^2 - k^2} = \frac{1}{2iR}\int_0^\infty \mathrm{d}k\,\frac{k^3\, \mathrm{e}^{ik.R}}{\kappa^2 - k^2} \tag{C6}$$

$$= \frac{1}{2iR}\lim_{\eta\to 0}\int_C \mathrm{d}k\, k^3\, \mathrm{e}^{ik.R}\left[\frac{1}{-(k-\kappa+i\eta)(\kappa+k)} + \frac{1}{(k-\kappa)(k+\kappa-i\eta)}\right] \tag{C7}$$

where η is an infinitesimal positive, real quantity.

Regarding the integrand as a function of the complex variable k (cf. Sakurai[85] p. 209), we note that it has poles at two points:

$$k = \kappa - i\eta$$
$$k = -\kappa + i\eta$$

Thus we find that Eq. (C7) may be given as Eq. (C8).

$$\frac{1}{2iR}\left[\frac{-i\pi\kappa^3\, \mathrm{e}^{i\kappa R}}{-2\kappa} + \frac{i\pi(-\kappa)^3\, \mathrm{e}^{-i\kappa R}}{-2\kappa}\right] = -\frac{\pi}{2}\frac{\kappa^2}{R}\cos\kappa R\ . \tag{C8}$$

References

(1) R. McWeeny, *Methods of Molecular Quantum Mechanics;* 2nd ed. (Academic Press, 1992).

(2) Th. Förster, in *Modern Quantum Chemistry*, edited by O. Sinanoğlu (Academic Press, 1965) Vol. III.

(3) D. L. Dexter, J. Chem. Phys. **21**, 836 (1953).

(4) R. E. Merrifield, J. Chem. Phys. **23**, 402 (1955).

(5) D. P. Craig and T. Thirunamachandran, *Molecular Quantum Electrodynamics* (Academic Press, New York, 1984).

(6) D. P. Craig and T. Thirunamachandran, Acc. Chem. Res. **19**, 10 (1986).

(7) D. P. Craig and T. Thirunamachandran, Adv. Quantum Chem. **16**, 97 (1982).

(8) D. L. Andrews and B. S. Sherborne, J. Chem. Phys. **86**, 4011 (1987).

(9) R. R. McLone and E. A. Power, Mathematika **11**, 91 (1964).

(10) D. L. Andrews, Chem. Phys. **135**, 195 (1989).

(11) D. L. Andrews and G. Juzeliunas, J. Chem. Phys. **96**, 6606 (1992).

(12) J. N. Murrell and J. Tanaka, Mol. Phys. **7**, 363 (1964).

(13) S. P. McGlynn, A. T. Armstrong and T. Azumi, in *Modern Quantum Chemistry*, edited by O. Sinanoğlu (Academic Press, 1965) Vol. III.

(14) T. Azumi and S. P. McGlynn, J. Chem. Phys. **42**, 1675 (1965).

(15) J. Koutecký and J. Paldus, Theoret. Chim. Acta (Berl.) **1**, 268 (1963).

(16) R. D. Harcourt, J. Phys. Chem. **96**, 7616 (1992).

(17) G. D. Scholes and K. P. Ghiggino, J. Phys. Chem. **98**, 4580 (1994).

(18) M. J. Frisch, M. Head-Gordon, G. W. Trucks, J. B. Foresman, H. B. Schlegel, K. Ragavachari, M. Robb, J. S. Binkley, C. Gonzalez, D. J. DeFrees, D. J. Fox, R. A. Whiteside, R. Seeger, C. F. Melius, J. Baker, R. L. Martin, L. R. Kahn, J. J. P. Stewart, S. Topiol and J. A. Pople, Gaussian, Inc., (Pittsburgh PA, 1992).

(19) P. G. Lykos and R. G. Parr, J. Chem. Phys. **24**, 211 (1956).

(20) K. Jug and M. Köster, J. Am. Chem. Soc. **112**, 6772 (1990).

(21) G. D. Scholes, A. H. A. Clayton and K. P. Ghiggino, J. Chem. Phys. **97**, 7405 (1992).

(22) E. A. Power and T. Thirunamachandran, Phys. Rev. A **28**, 2649 (1983).

(23) E. A. Power and T. Thirunamachandran, Phys. Rev. A **28**, 2663 (1983).

(24) E. A. Power and T. Thirunamachandran, Phys. Rev. A **28**, 2671 (1983).

(25) D. P. Craig and T. Thirunamachandran, Chem. Phys. **167**, 229 (1992).

(26) D. L. Andrews and G. Juželiunas, J. Chem. Phys. **95**, 5513 (1991).

(27) A. J. Stone, in *Theoretical Models of Chemical Bonding*, edited by Z. B. Maksić (Springer-Verlag, 1991) Part 4.

(28) A. D. Buckingham, in *Intermolecular Forces - from Diatomics to Biopolymers*, edited by B. Pullman (Wiley, New York, 1978).

(29) A. J. Stone, Chem. Phys. Letts. **83**, 233 (1981).

(30) H. Sponer and E. Teller, Rev. Mod. Phys. **13**, 75 (1941).

(31) G. Herzberg, *Electronic Spectra and Electronic Structure of Polyatomic Molecules* (Van Nostrand, Princeton, N.J., 1950).

(32) A. D. Buckingham, Quarterly Reviews (Chem. Soc., London) **13**, 189 (1959).

(33) D. L. Andrews and T. Thirunamachandran, J. Chem. Phys. **67**, 5026 (1977).

(34) J. C. Chang, J. Chem. Phys. **67**, 3901 (1977).

(35) R. J. Buehler and J. O. Hirschfelder, Phys. Rev. **83**, 628 (1951).

(36) R. J. Buehler and J. O. Hirschfelder, Phys. Rev. **85**, 149 (1952).

(37) J. O. Hirschfelder, C. F. Curtis and R. B. Bird, *Molecular Theory of Gases and Liquids,* pp. 26-27, 843-846 (John Wiley and Sons, Inc., New York, 1954).

(38) W. E. Byerly, *An Elementary Treatise on Fourier Series and Sperical, Cylindrical, & Ellipsoidal Harmonics* (Ginn & Co., Boston, 1893).

(39) D. P. Craig, Mol. Phys. **4**, 113 (1961).

(40) D. P. Craig and S. H. Walmsley, *Excitons in Molecular Crystals* (Benjamin, NY, 1968).

(41) J. Koutecký and J. Paldus, Coll. Czech. Chem. Commun. **27**, 599 (1962).

(42) R. Polák and J. Paldus, Theoret. Chim. Acta (Berl.) **4**, 37 (1966).

(43) R. H. Dicke, Phys. Rev. **93**, 99 (1954).

(44) R. S. Mulliken, C. A. Rieke and H. Orloff, J. Chem. Phys. **17**, 1248 (1949).

(45) E. A. Mason and J. O. Hirschfelder, J. Chem. Phys. **26**, 173 (1957).

(46) G. D. Scholes, K. P. Ghiggino and G. J. Wilson, Chem. Phys. **155**, 127 (1991).

(47) E. A. Power and T. Thirunamachandran, Phys. Rev. A **48**, 4761 (1993).

(48) M. D. Newton, Chem. Rev. **91**, 767 (1991).

(49) G. L. Closs, L. T. Calcaterra, N. J. Green, K. W. Penfield and J. R. Miller, J. Phys. Chem. **90**, 3673 (1986).

(50) G. L. Closs, P. Piotrowiak, J. M. MacInnis and G. R. Fleming, J. Am. Chem. Soc. **110**, 2652 (1988).

(51) G. L. Closs, M. D. Johnson, J. R. Miller and P. Piotrowiak, J. Am. Chem. Soc. **111**, 3751 (1989).

(52) A. Kuki, Structure and Bonding **75**, 49 (1991).

(53) (a) P. Bertrand, Chem. Phys. Lett. **140**, 57 (1987).
(b) P. Bertrand, Structure and Bonding **75**, 1 (1991).

(54) K. D. Jordan and M. N. Paddon-Row, Chem. Rev. **92**, 395 (1992).

(55) S. H. Lin, W. Z. Xiao and W. Dietz, Phys. Rev. E **47**, 3698 (1993).

(56) P.-O. Löwdin, J. Mol. Spectrosc. **10**, 12 (1963).

(57) S. Larsson, J. Am. Chem. Soc. **103**, 4034 (1981).

(58) J. W. Evenson and M. Karplus, J. Chem. Phys. **96**, 5272 (1992).

(59) K. Hertzfeld, J. Chem. Phys. **10**, 508 (1942).

(60) H. M. McConnell, J. Chem. Phys. **35**, 508 (1961).

(61) D. P. Craig and T. Thirunamachandran, Chem. Phys. **135**, 37 (1989).

(62) G. Juželiunas and D. L. Andrews, Phys. Rev. B **49**, 8751 (1994).

(63) G. C. Nieman and G. W. Robinson, J. Chem. Phys. **39**, 1298 (1963).

(64) R. A. Buckingham and A. Dalgarno, Proc. Roy. Soc. (London) **A213**, 327 (1952).

(65) R. S. Mulliken, Phys. Rev. **120**, 1674 (1960).

(66) H. Nagae, T. Kakitani, T. Katoh and M. Mimuro, J. Chem. Phys. **98**, 8012 (1993).

(67) (a) R. D. Harcourt, G. D. Scholes and K. P. Ghiggino, J. Chem. Phys. (in press, December 1, 1994).

(b) G. D. Scholes, R. D. Harcourt and K. P. Ghiggino, J. Chem. Phys. (submitted).

(68) S.-T. Levy, M. D. Rubin and S. Speiser, J. Am. Chem. Soc. **114**, 10747 (1992).

(69) S.-T. Levy and S. Speiser, J. Chem. Phys. **96**, 3585 (1992).

(70) S. Hassoon, H. Lustig, M. B. Rubin and S. Speiser, J. Phys. Chem. **88**, 6367 (1984).

(71) G. D. Scholes and K. P. Ghiggino, J. Chem. Phys. **101**, 1251 (1994).

(72) M. Bixon and J. Jortner, J. Chem. Phys. **48**, 715 (1968).

(73) J. Ulstrup and J. Jortner, J. Chem. Phys. **63**, 4358 (1975).

(74) G. D. Scholes and K. P. Ghiggino, J. Photochem. Photobiol. A: Chem., **80**, 355 (1994).

(75) G. D. Scholes, K. P. Ghiggino, A. M. Oliver and M. N. Paddon-Row, J. Phys. Chem. **97**, 11871 (1993).

(76) G. D. Scholes, K. P. Ghiggino, A. M. Oliver and M. N. Paddon-Row, J. Am. Chem. Soc. **115**, 4345 (1993).

(77) G. D. Scholes and K. P. Ghiggino, J. Chem. Phys. (submitted).

(78) R. D. Harcourt and W. Roso, Can. J. Chem. **56**, 1093 (1978).

CHAPTER FOUR

Experimental Studies of Some Rigidly-Linked Bichromophores

I. Aims

The objective of this experimental work is to explore the significance of through-bond mediated EET and exciton interactions in a series of novel, rigidly-linked bichromophores. Such work was alluded to in Chapter 1, and developed theoretically in Chapter 2 (cf. Section 2.7) and Chapter 3 (Section 3.13). The impetus behind this experimental study is a need to obtain a qualitative indication of the possible significance of bridge-mediated interactions in general. A direct measurement of EET will be presented in Section III of this chapter, followed by two studies of exciton-type interactions (i.e. interactions between identical chromophores) in Sections IV and V. Each of these sections are self-contained, but will be united via the general conclusions drawn in the final section of this chapter. The experimental results have been published by the author during the course of this work.[1,2]

To obtain insight into contributions made by the various EET mechanisms, time-resolved fluorescence studies of the novel molecule A–6–N (Figure 4.1) are reported in Section III. In this molecule the naphthalene (N) donor and anthracene (A) acceptor chromophores are held in a known, well defined orientation and separation (12.2 Å) by a rigid bisnorbornyl–bicyclo[2.2.0]hexane bridge, six σ-bonds in length. This configuration is of particular interest since EET between the naphthalene ^1L_b and anthracene ^1L_a states should not occur via a direct Coulombic mechanism (in the absence of vibronic coupling) due to the orthogonality of the respective donor and acceptor transition dipoles. The contributions by other EET mechanisms can thus be investigated in some detail. The spectroscopy of A–6–N is compared to its corresponding model compounds, A–6 and N–2 (cf. Figure 4.1).

We describe three possible mechanisms for propagating intermediate-range interchromophore electronic energy transfer. In order to explain the high measured rate of naphthalene-to-anthracene EET in A–6–N it is proposed that a significant pathway involves a through-bond mediated "short-range" mechanism.

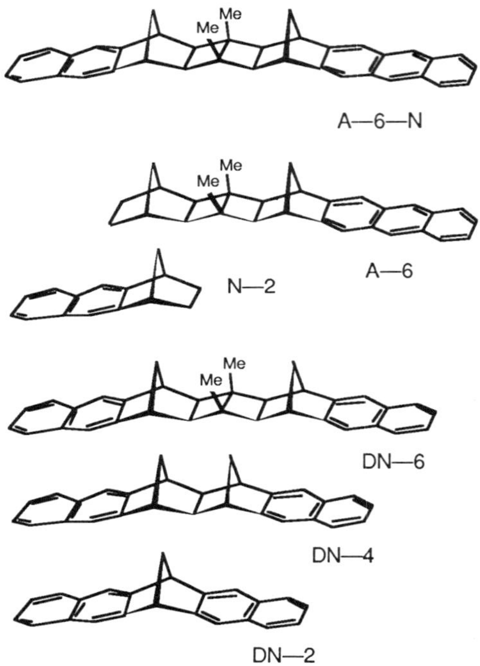

Figure 4.1. The novel bichromophoric systems studied in this chapter and their corresponding model chromophores.

The purpose of Section IV is to explore the role played by through-space and through-bond coupling on exciton interactions by examining electronic absorption spectra of the naphthalene-bridge-naphthalene systems, DN–2, DN–4 and DN–6 (Figure 4.1), in which the pair of naphthalene chromophores are held at well defined distances and orientations by a rigid polynorbornyl-type bridge of variable length (two, four or six σ bonds respectively). Electronic absorption spectra of the intense naphthalene S_3 ($^1A_g \rightarrow 2^1B_{3u}$) band are reported here, since its high oscillator strength should make it more amenable to analysis by simple exciton theory using the dipole approximation (in the absence of any through-space overlap effects). The effect of solvent dielectric constant is also discussed.

Finally, in Section V, the fluorescence emission spectra and fluorescence lifetimes of this series of dinaphthyl molecules are reported. These properties are shown to be a sensitive probe of relative electronic couplings in the weak coupling limit (i.e., where no obvious spectral perturbations are evident, cf. Chapter 2).

II. Experimental Section

Synthesis of the compounds examined in this work was undertaken by our collaborators, Professor Michael Paddon-Row and co-workers at The University of New South Wales, and has been reported previously.[3] The geometries of all compounds were determined (also by M.N.P.-R.) using MNDO geometry optimization within a C_{2v} symmetry constraint.

For fluorescence experiments, solutions of the compounds were prepared in *n*-hexane (Ajax Spectrosol) to a maximum optical density of less than 0.05 at 290 nm so as to exclude reabsorption effects and intermolecular interactions. Samples were degassed by repeated freeze-pump-thaw cycles prior to study. The concentration of solutions (*n*-hexane, Ajax Spectrosol; acetonitrile-210, Ajax Unichrom, HPLC grade) used for the electronic absorption experiments were typically $2.0 - 3.0 \times 10^{-5}$ M. At these concentrations intermolecular interactions can be excluded and maximum absorbances were in the range 0.1 to 0.8.

Spectra were recorded in 1 cm path length (fluorescence and A–6–N absorption experiments) or 1 mm path length (DN–*n* absorption experiments) quartz cuvettes at 295 K. Steady state (corrected) fluorescence emission and excitation spectra were measured using a computer-interfaced Hitachi F-4010 fluorimeter, with 1.5 nm excitation and emission slit widths and 1 nm data intervals. Absorption spectra were recorded with a Cary 5 UV/VIS/NIR absorption spectrophotometer (spectral band width 0.6 nm, data interval 0.2 nm).

The fluorescence quantum yield of N–2 (Φ_{N-2}) in *n*-hexane (refractive index, n_{N-2} = 1.3749[4]) was determined relative to 1-methyl naphthalene (MN) in cyclohexane solvent (Ajax Spectrosol, n_{MN} = 1.4262[4]), with a fluorescence quantum yield reported to be 0.25[5] using the relation

$$\Phi_{N-2} = \Phi_{MN} \frac{A_{N-2}}{A_{MN}} \frac{n_{N-2}^2}{n_{MN}^2} \frac{F_{MN}}{F_{N-2}}$$

where A_{N-2} and A_{MN} are the areas under the corrected fluorescence spectra, while F_{N-2} and F_{MN} denote the fraction of light absorbed by each solute at the excitation wavelength. The quantum yield for N–2 in *n*-hexane was thus found to be 0.246, which was required to estimate the rate of Förster-type (dipole-dipole) EET (Eqs. 2.6.11 and 2.6.13) between N and A in A–6–N. A spectral overlap integral (Eq. 2.6.12) between N–2 emission and A–6 absorption spectra was also determined.

Fluorescence decay profiles were recorded using the time correlated single photon counting (SPC) method.[6] The excitation source was a cavity-dumped picosecond dye laser (Spectra Physics model 3500) synchronously pumped by a mode locked argon ion laser (Spectra Physics model 2030). Excitation pulses (~ 4 ps FWHM) at 292 nm were obtained by frequency doubling the Rhodamine 6G dye laser output with a KDP angle tuned second harmonic generating crystal. Fluorescence photons were detected, through a polariser set at the magic angle of 54.7° to the vertically-polarized excitation light (to eliminate the possibility of molecular rotational effects distorting the fluorescence decay[7]), followed by a Jobin-Yvon H-20 monochromator, with a Hamamatsu model R1564U-01 microchannel plate photomultiplier tube as detector. The total response function of the system was ~ 85 ps FWHM. Data were collected over 512 channels of a Tracor-Northern multi-channel analyser. Fluorescence decays (10 000 to 20 000 counts in the peak channel), recorded at various observation wavelengths and over a range of time scales, were analysed using non-linear least squares iterative reconvolution software to extract the fluorescence lifetimes. Details of the time-resolved fluorescence instrumentation and data analysis procedures have been extensively described in other theses and publications from this laboratory.[8,9,10]

III. Intramolecular Electronic Energy Transfer in A-6-N

Results. The difference between A–6–N and A–6 fluorescence excitation spectra, monitoring only the anthracene emission at 405 nm, (Figure 4.2) can be attributed to naphthalene absorption. This is seen clearly by inspection of the absorption spectra of the model chromophores N–2 and A–6 compared with the bichromophore A–6–N in Figure 4.3, which indicates that the A–6–N absorption spectrum is essentially a sum spectrum of N–2 and A–6. The observation that absorption by the naphthalene chromophore leads to anthracene emission provides clear evidence for excitation energy transfer in A–6–N.

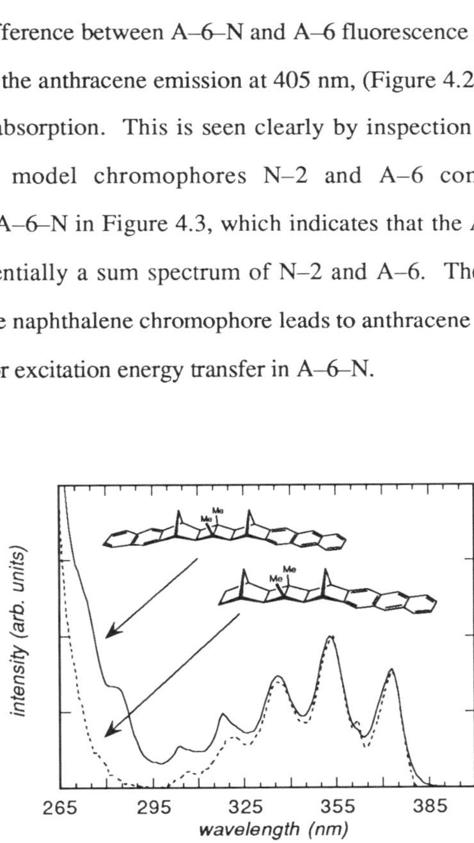

Figure 4.2. Fluorescence excitation spectra of A-6-N and A-6 in *n*-hexane solvent (295 K). The emission was monitored at 405 nm in both cases, where fluorescence is entirely due to anthracene. Differences in the spectra are attributed to absorption by the naphthalene chromophore in A–6–N followed by energy transfer to the anthracene moiety. Reprinted with permission from Ref. 2. Copyright [1993] American Chemical Society.

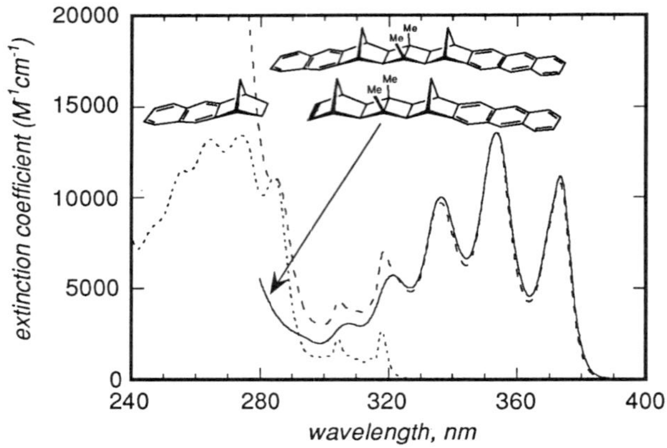

Figure 4.3. Absorption spectra of A–6–N, N–2 and A–6 in *n*-hexane (295 K). The N–2 absorption was intensity-normalized to A–6–N at 284 nm and the A–6 was intensity-normalized to A–6–N at 354 nm.

It was found from time-correlated single photon counting measurements that the N–2 and A–6 fluorescence decay profiles could be fit by single exponential decay kinetics, recovering fluorescence lifetimes of the model compounds were 63.33 ns (τ_{N-2}) for N–2 and 5.25 ns (τ_{A-6}) for A–6. For A–6–N with excitation at 292 nm, where only the naphthalene chromophore absorbs, and monitoring emission from the anthracene moiety at 400 nm, decay curves collected over different time scales could be reliably fit with a double exponential rise and decay function

$$I(t) = A_1 \exp(-t/\tau_1) + A_2 \exp(-t/\tau_2)$$

A decay collected on the shortest time scale is illustrated in Figure 4.4 and yielded values of $\tau_1 = 34 \pm 1$ ps (A_1 = -10400 counts), $\tau_2 = 5.22 \pm 0.01$ ns (A_2 = 10800 counts) with a χ^2 (goodness of fit parameter) of 1.1. Such a kinetic scheme

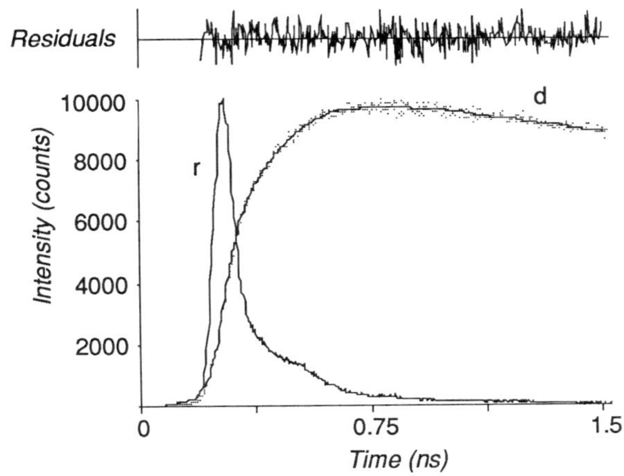

Figure 4.4. A-6-N fluorescence decay profile (d) and instrument response function (r) showing the rise and decay of anthracene emission after naphthalene excitation. The points are the experimental data and the solid line is the fitted curve. Reprinted with permission from Ref. 2. Copyright [1993] American Chemical Society.

is consistent with rapid, irreversible formation of excited state anthracene (via the excited naphthalene) and its subsequent decay, with a rate constant of $1/\tau_1$ s^{-1} – corresponding to that for the A–6 model. Hence the measured rate of EET is

$$k_{EET} = \tau_1^{-1} - \tau_{N-2}^{-1}$$

$$= 2.9 \times 10^{10}\ s^{-1}$$

The quantum efficiency, Q, of EET is close to unity,

$$Q = k_{EET}/(k_{EET} + k_{N-2})$$

$$= 0.999$$

The naphthalene-anthracene spectral overlap integral was determined via Eq. (2.6.12) - that is, without including the solvent refractive indices. It was estimated to be $4.2 \times 10^{-13}\ M^{-1}cm^3$ (cf. Figure 4.5).

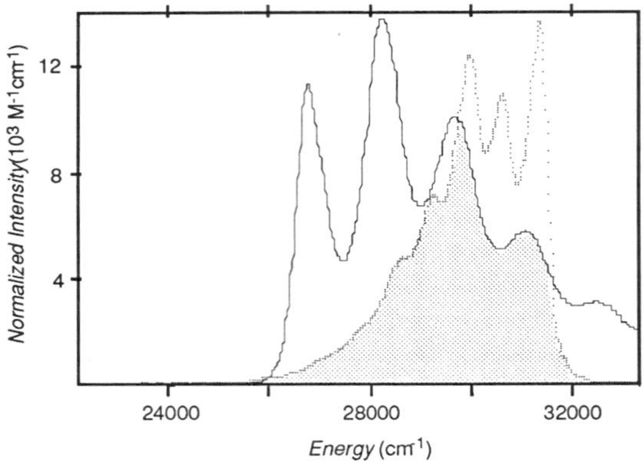

Figure 4.5. Overlayed naphthalene emission (dotted curve) and anthracene absorption (solid curve), used to determine the spectral overlap integral.

Discussion. At the excitation wavelength (292 nm), both ^1L_b and ^1L_a naphthalene states will be populated. However, internal conversion from the ^1L_a state will be fast,[11] so that EET can be assumed to proceed from a relaxed Boltzmann population of the naphthalene ^1L_b state. EET between naphthalene and anthracene is often assumed to occur between this naphthalene ^1L_b ($^1B_{3u}$) state (long-axis polarized and dipole-allowed, although very weak) and the anthracene ^1L_a ($^1B_{2u}$) state (short-axis polarized and intense); the lowest excited states of the chromophores. However, in A–6–N, the orthogonal orientation of the respective transition dipoles of the two chromophores (cf. Figure 4.6) is such that a direct Coulombic interaction between these two pure electronic states is forbidden.

It should be noted that there exists a second dipole-allowed anthracene transition lying below (in energy) the naphthalene ^1L_b band; namely the very weak, long-axis polarized, anthracene ^1L_b transition.[12,13] EET from naphthalene ^1L_b to anthracene ^1L_b electronic states has a non-zero orientation factor and so dipole-dipole and dipole-octopole interaction terms could all have some significance.

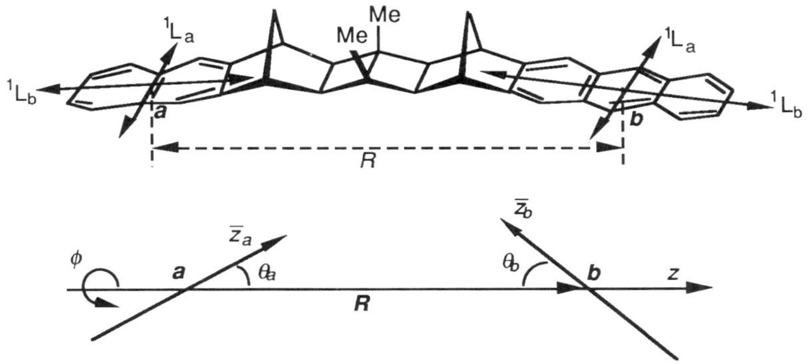

Figure 4.6. The A-6-N molecule, showing relative orientations of the relevant transition moments.

It is known[14] that most of the intensity observed in naphthalene fluorescence emission is vibronically induced. When the pure electronic transition (1^1B_{3u}, long-axis polarized) couples with quanta of ground state b_{1g} vibrations, the resultant fluorescence has substantial intensity borrowed from the ^1L_a (1^1B_{3u}, short-axis polarized) transition. Hence, although at first the Coulombic EET process between naphthalene ^1L_b and anthracene ^1L_a states appears to be forbidden in the A–6–N molecule (due to the orthogonality of donor and acceptor transition dipoles), most of the naphthalene fluorescence emission intensity will be vibronically-induced and short-axis polarized. A direct Förster-type (Coulombic) EET naphthalene-"^1L_a" to anthracene-^1L_a EET mechanism may thus obtain significance.

The dipole transition moments for the naphthalene and anthracene groups in A–6–N may be estimated from reported experimental gas phase oscillator strengths. The anthracene ^1L_b oscillator strength was taken to be 0.002 ($\bar{\nu}$(0-0) ≈ 30 500 cm^{-1}).[15] An estimate for the oscillator strength of the pure electronic component of the naphthalene ^1L_b state is 0.0002[16] (and $\bar{\nu}$(0-0) = 32 020 cm^{-1}).[17] So, transition dipole moments relevant to the long-axis polarized transitions (^1L_b) are $\mu^{naph}(^1L_b$, purely electronic) ~ 4×10^{-31} Cm and $\mu^{anth}(^1L_b)$ ~ 1×10^{-30} Cm. For short-axis polarized transitions (anthracene ^1L_a and naphthalene emission with substantial borrowed ^1L_a intensity), values, similarly estimated from experimentally determined oscillator strengths,[17,18] are μ^{naph}("^1L_a", short-axis polarized) $\approx 1.2 \times 10^{-30}$ Cm and $\mu^{anth}(^1L_a) \approx 9.2 \times 10^{-30}$ Cm. The substantially larger interaction obtained between the latter pair of transition moments implies that the major part of the direct interaction may be estimated from a Förster expression, with transition dipoles oriented along the chromophore short axes.

In this case, an experimentally measured spectral overlap integral (Chapter 2, Eq. 2.6.12), should provide the most reliable estimate of the rate of EET, for the leading dipole-dipole interaction. The MNDO optimized geometry of A–6–N gives the following values for the parameters (for long-axis polarized transitions, cf. Figure 4.6): $R = 12.24$ Å, $\theta_a = 44.29°$ and $\theta_b = 36.72°$ ($\phi = 0°$), but for short-axis polarized transitions $\theta_a = \theta_b = 90°$. The spectral overlap integral for this system was determined experimentally to be 4.2×10^{-13} M^{-1}cm^3. Using the theory from Chapter 2, Eqs. (2.6.11 - 2.6.13), the rate of Förster-type (dipole-dipole) EET, neglecting solvent, was found to be 1.2×10^9 s^{-1}. This is over an order of magnitude smaller than the experimentally-measured rate of 2.9×10^{10} s^{-1}.

Octopole transition moments have been previously estimated for both naphthalene ^1L_b and anthracene ^1L_a transitions to explain the large exciton splittings observed in pure single crystals,[15,19,20] indicating they are of the order 10^{-48} Cm^3 for both naphthalene and anthracene. Values have also been obtained for naphthalene in a recent theoretical study,[21] and for ethene in Chapter 3 of this thesis. A reliable estimate of the contribution of these higher order transition moments to the rate of EET is difficult to obtain, nonetheless it is clear that although the dipole-octopole interaction term may be significant, it will be less important than the dipole-dipole term. This implies that, in this case, contributions to the

Coulombic interaction from higher multipole interactions cannot account for the difference between the calculated Förster rate for EET and experiment.

Thus far, we have considered only the direct Coulombic interaction between two isolated chromophores. At the other limit, where there is a high density of a third body (such as solvent molecules) between donor and acceptor, the interaction is viewed as occurring in a dielectric medium. In this case one simply changes ε_0 to ε, the medium dielectric constant (cf. Förster's Equation,[22] cf Eqs. 2.2.11 and 2.2.13). However, when there are only few intervening molecules then the direct Coulombic interaction can be accompanied by three-body processes, leading to a modification in electronic coupling and a change in distance dependence.[23] This was discussed in Chapter 3, Section XIII.

The connecting bridge between the two chromophores may therefore act as the third body and mediate the EET process by the exchange of virtual excitation. Craig and Thirunamachandran[23] have derived an expression for this relayed dipole-dipole interaction, given in Chapter 3, Eq. (13.5). To evaluate the approximate contribution that the relayed Coulombic interaction makes to the overall electronic coupling the bisnorbornylbicyclo[2.2.0]hexane bridge is treated as a single relay unit, then Eq. (3.13.5) is rewritten as

$$u^{\mathrm{rel}} \approx \frac{u^{\mathrm{d\text{-}d}}(\text{donor-bridge}).u^{\mathrm{d\text{-}d}}(\text{bridge-acceptor})}{(E_{\mathrm{donor}} - E_{\mathrm{bridge}})} \tag{3.1}$$

Assuming a $\sigma-\sigma^*$ transition of the bridge has a transition moment of approximately 10^{-29} Cm oriented along the short axis, donor (and acceptor) to bridge separations are 6.3 Å, and the acceptor-bridge energy gap is estimated from spectra to be approximately -26000 cm^{-1}, the relayed Coulombic contribution to the overall electronic coupling is estimated to be -0.14 cm^{-1}, compared to 2.72 cm^{-1} for the dipole-dipole interaction. In this case, then, such an interaction interferes *destructively* with the direct Coulombic interaction, but its magnitude is only very small.

Model calculations[24] suggest that direct, through–space orbital interactions between the two chromophores in A–6–N are negligible. Consequently, intramolecular EET in A–6–N, mediated by a mechanism based solely on through–space interchromophore orbital overlap should be unimportant.

However, EET occurring via a long–range mechanism involving through-bond (TB) coupling could be significant in A–6–N.[25-27] In this case EET could be occurring between the naphthalene ^1L_b state or vibronically-induced intensity (cf Chapter 2, Section IX) and either the anthracene ^1L_b or ^1L_a states. The result would be a substantially greater rate of EET than that due to the Coulombic mechanism alone. Of particular relevance is the observation of TB mediated intramolecular singlet-singlet EET occurring via the "short-range" mechanism in molecular systems comprising bridges similar to that present in A–6–N.[28-30] The significant contribution to the electronic interaction matrix element evidently made by this TB interaction is explained by its origins: it is essentially an orbital overlap-dependent interaction, mediated over large through-space separations via the connecting bridge (cf. Chapter 3, Section XIII).

Hence, observations and analysis from the experimental investigation reported here illustrate the importance of a through-bond interaction contributing to EET in A–6–N. It is likely that such a mechanism of intramolecular electronic energy transfer will be significant in related systems, particularly with smaller interchromophore separations, provided that the geometric conformation of the bridge is favourable.[30] The effect of bridge conformation in promoting or inhibiting through-bond interactions has been extensively studied for both electron[31,32] and energy transfer.[26,27,30]

Conclusions. The results of this investigation have significant bearing on studies of electronic energy transfer in aromatic polymers, bichromophoric molecules and biological systems. It is shown for the rigidly-linked donor-acceptor system studied here that the usual dipole-dipole approximation provides a poor estimate for the rate of energy transfer; the through-bond interaction being of more significance. In the present example the dipole approximation underestimates the rate of naphthalene to anthracene energy transfer by over one order of magnitude. In general, both chromophore orientation and the geometric conformation of any linking bridge will be important in determining the mechanisms of energy transfer in multichromophoric supramolecular systems.

IV. Exciton Interactions in DN-*n*

Results. The dinaphthyl-polynorbornyl molecules examined here and in the next section are shown in Figure 4.1, where N–2 is the (two-bond) model chromophore and DN–*n* symbolizes an *n*-bond dinaphthyl compound. Three interchromophore separations R_1, R_2 and R_3, representing C1-C1', C3-C3' and C5-C5' through-space distances respectively (Figure 4.7) are collected in Table 4.1. R_2 defines the separation, R, between centres of the electric dipole transition moments. The orientation factor relevant to the dipole-dipole interaction (Eq. 4.1), $\eta = \cos\alpha + 3\cos^2\theta$,, along with the angles α and θ defining the orientation of transition dipole moments (cf. Figure 4.7) for each compound are also listed.

Table 4.1. Optimized parameters as determined by the MNDO method. R_1, R_2 and R_3 define interchromophore separations and the orientation of the long-axis polarized dipole transition moments are given by the angles α and θ, which are used to calculate the orientation factor, η.

	R_1, Å	R_2, Å	R_3, Å	α, deg	θ, deg	η
DN-2	2.45	6.50	10.59	112.8	33.6	1.695
DN-4	4.89	8.46	12.07	94.4	42.8	1.538
DN-6	7.56	11.26	15.00	98.9	40.5	1.580

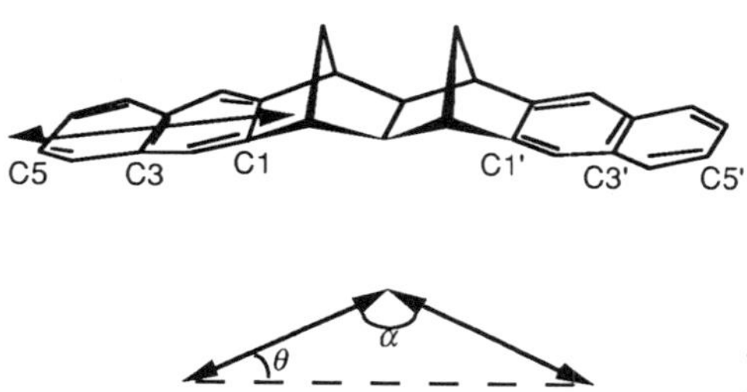

Figure 4.7. The DN-4 molecule with geometry labels and orientation of transition moments relevant to Table 4.1.

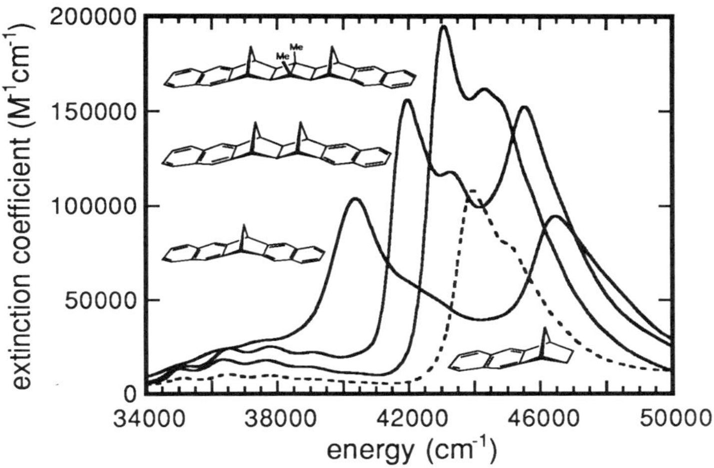

Figure 4.8. UV absorption spectra of the DN-*n* series of molecules in *n*-hexane solvent (298 K) in the region of the dipole-allowed naphthalene S_2 and S_3 electronic transitions. Reprinted with permission from Ref. 1. Copyright [1993] American Chemical Society.

Electronic absorption spectra of each molecule in *n*-hexane solvent are illustrated in Figure 4.8, which shows clearly the splitting of the intense naphthalene $^1A_g \rightarrow 2^1B_{3u}$ electronic transition, S_3, absorption band as a function of interchromophore separation. Perturbations to the S_2 vibronic band are also evident. The shoulder on the high energy side of the S_3 band for N–2, which also appears in the DN-*n* spectra, is absent in the S_3 (condensed phase) absorption band of 2,3-dimethylnaphthalene.[33] A similar shoulder was, however, observed in the gas phase absorption spectrum of naphthalene,[16] although its assignment was ambiguous. Since this shoulder does not appear to be split independently of the S_3 band in the DN–*n* spectra it is more likely to be 0-0 plus a vibrational frequency rather than a higher electronic state. The major observations are collected in Table

4.2, together with the electronic coupling in cm^{-1} calculated from the dipole approximation, that is Eq. (3.11.3a), given by (Eq. 4.1),

$$u^{Coul} \approx 1.07069 \times 10^{10} f\eta / (\overline{v}_0 R^3) \quad (4.1)$$

where η (= $\cos\alpha$ + $3\cos^2\theta$, cf Figure 4.7) is the orientation factor, f is the oscillator strength for the transition at $\overline{v}_0$ cm^{-1} and R is the centre-to-centre separation of the chromophores in Å.

The oscillator strength of the N–2 S_3 transition in *n*-hexane solvent was calculated by integrating from 42 000 cm^{-1} to 50 000 cm^{-1}, and was found to be 1.0.

Negligible solvent shift was detected for N–2 in acetonitrile relative to *n*-hexane and electronic couplings for the DN–*n* series were only up to 100 cm^{-1} larger. A comparison between DN–2 spectra in the two solvents is shown in Figure 4.9.

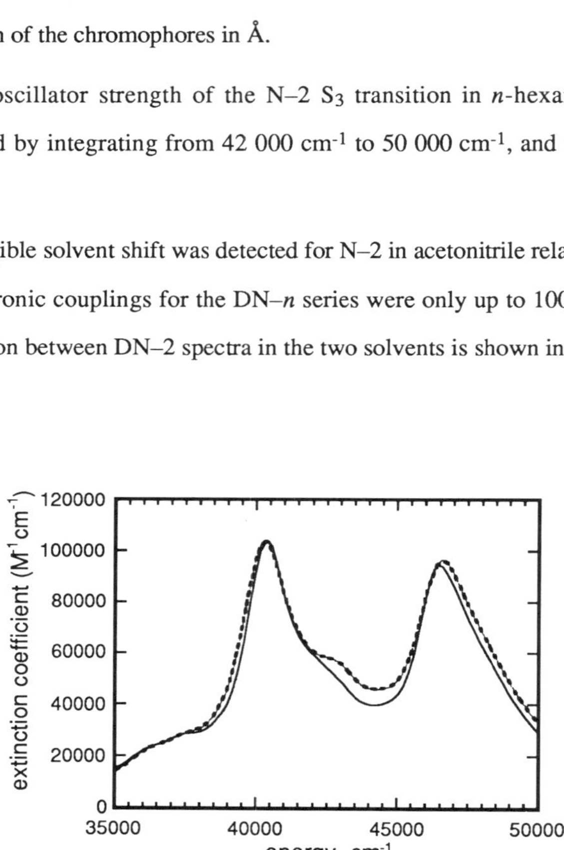

Figure 4.9. Comparison of DN-2 S_3 absorption bands in *n*-hexane (solid line) and acetonitrile (dashed line). Reprinted with permission from Ref. 1. Copyright [1993] American Chemical Society

Table 4.2. S_3 absorption band maxima ($\overline{\nu}$) and extinction coefficients (ε) in n-hexane solvent, with experimentally measured T_{14} and U (see Figure 4.10 and Chapter 3). The through-space (TS) component of T_{14} is estimated using the dipole approximation (where possible), therefore allowing the through-bond (TB) component to be estimated. T_{14}(TB) was estimated using T_{14}(exp) - T_{14}(TS).

					experiment			
	$\overline{\nu}$, cm^{-1}	ε, $M^{-1}cm^{-1}$	$\overline{\nu}$, cm^{-1}	ε, $M^{-1}cm^{-1}$	T_{14}, cm^{-1}	U, cm^{-1}	T_{14}(TS)	T_{14}(TB)
N-2	43 960	108 000						
DN-2	40 360	104 000	46 470	95 000	3060	540		
DN-4	41 980	156 000	45 540	152 000	1780	200	620	1160
DN-6	43 070	195 000	44 670	162 000	800	90	270	530

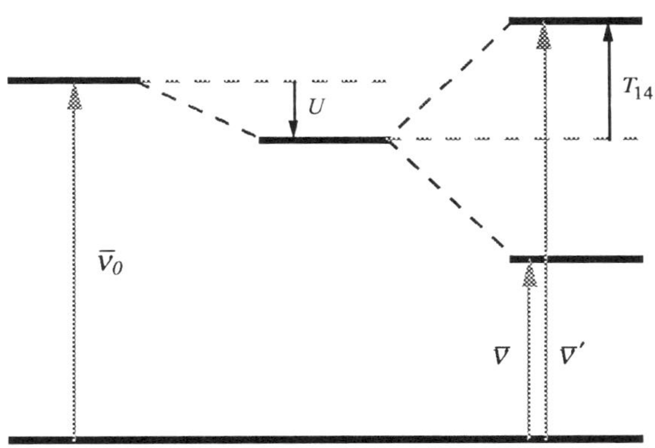

Figure 4.10. Schematic diagram illustrating contributions to the spectral perturbations observed for the DN–n bichromophores. The direct interaction T_{14} and the through-configuration interaction U are introduced in Chapter 3.

Discussion. From the data presented in Table 4.2 it is evident that the observed exciton resonance splittings (given by $2T_{14}$ in Sections 3.4 and 3.5) of the naphthalene $S_0 \rightarrow S_3$ transition in all the DN–*n* molecules are considerably larger than those values calculated using the dipole-dipole approximation. An *ab initio* (RHF method, 3-21G and 6-31+G basis set) study of the π^* splittings in ethene dimers has shown that direct through-space (TS) orbital overlap is strongly attenuated with distance, and is significant to a separation of 5 Å only.[24] An extensive series of similar calculations on polynorbornadienes found that TS interchromophore overlap was important in the two-bond diene, but insignificant for longer bridges.[24] Thus, while interactions introduced by direct TS orbital overlap might be expected to enhance the exciton resonance splitting in DN–2, it is unlikely that such interactions significantly influence the splitting of the S_3 transition in DN–4 and DN–6. For these two molecules any TS interaction should be Coulombic in origin.

It is appropriate to represent the Coulombic interaction by the multipole expansion only when the two charge distributions are non-overlapping (cf. Section 3.11). It is thus inappropriate to use Equation (4.1) for DN–2, however, it may be applicable to the DN–4 and DN–6 systems. Consideration of this was not made in a previous study of DN–2,[33] nor was the possibility of TS or through-bond interactions. It has been suggested previously that rapid convergence of the multipole expansion to the electric dipole-dipole term may not occur when the molecular size is large compared to the intermolecular separation.[34,35] Due to the high oscillator strength of the naphthalene $S_0 \rightarrow S_3$ transition, the direct electronic coupling between the chromophores should be well described as an interaction of two transition dipoles. The ratio of interchromophore separation to transition dipole length (taken to be about 5 Å) in DN–4 and DN–6 is approximately 1.7 and 2.3 respectively. Hence for the orientation of dipoles in these molecules it is expected that terms in the multipole expansion higher than the leading (i.e., dipole-dipole) term would act to alter the Coulombic interaction from that calculated using Eq. (4.1) by only between 5% and 20% (cf. Chapter 3, Section X).

Hence, if interactions mediated via the norbornyl bridge (through-bond (TB) interactions) were negligible, then the electronic couplings calculated by Eq. (4.1) for DN–4 and DN–6 should be close to those measured experimentally. The much larger spectral splittings measured than predicted by Eq. (4.1) for DN–4 and DN–6 are therefore attributed to arising from a TB mechanism.

In the presence of a TB interaction we may envisage that the "donor" and "acceptor" wavefunctions in the A-L-B systems be written as

$$\Psi_A = N_A(\phi_A + \alpha_A \phi_L) \tag{4.2a}$$

$$\Psi_B = N_B(\phi_B + \alpha_B \phi_L) \tag{4.2b}$$

which represents the mixing of the naphthalene chromophore wavefunction with the sigma orbitals of the bridge. N_A and N_B are the appropriate normalizing constants and the mixing coefficients α_A and α_B are determined variationally.[36] The degree to which the excitation is delocalized from one chromophore to another, via the bridge orbitals, is therefore dependent on how effectively the π (π^*) system of each chromophore couples to the bridge σ (σ^*), as well as the degree of coupling between each bridge unit. This dependence is encompassed by the McConnell model, Eq. (3.13.3), written below as Eq. (4.3).

$$u^{t\text{-}b} \approx \left(\frac{T^2}{-A}\right)\left(\frac{t}{-A}\right)^{m-1} \tag{4.3a}$$

$$\approx V_0 \exp\left[-\beta(m-1)\right] \tag{4.3b}$$

In the following discussion only the results for DN–4 and DN–6 will be considered since direct interchromophore orbital overlap effects are negligible in these molecules (as discussed above). An estimate for the TB interaction can be obtained by subtracting the TS interaction (assumed to be well-represented by the Coulombic component) from the observed interaction (values T_{14}(TB) in Table 4.2). To gauge the distance dependence of the TB component to the resonance interaction (in DN–4 and DN–6) the effective exponent for attenuation of the interaction, β in Eq. (4.3b), can be determined. It is found that $\beta = 0.78$ per bond for the TB component of the direct exciton resonance interaction.

These results indicate that there is a substantial contribution to the exciton interaction by a TB mechanism. It should be emphasized that this may not necessarily be the case for lower energy transitions or for less favourable bridge conformations. The question then remains: why is this TB interaction so strong?

It is likely that one of the most important factors acting to increase the chromophore-bridge electronic coupling is the high energy of the electronic transition being studied, which decreases the π^*-σ^* energy gap. This can be understood within the context of the simple McConnell model (Eq. 4.3a), from which it can be seen that the effectiveness of a bridge in propagating interactions depends both on the intrabridge coupling matrix elements, *t,* and on the energy gap, *A*, separating the chromophore π (π^*) orbitals with the localized bridge σ (σ^*) orbitals. As the π^*-σ^* energy gap, *A*, diminishes, the distance dependence gets weaker. TB interactions should therefore be more significant for this high energy electronic transition than for, say, the S_1 transition.

It has been shown that, as for the $^1E_{1u}$ electronic state of benzene,[37,38] π-σ dynamic correlation effects make an important contribution to the 2^1B_{3u} state of naphthalene as a result of its substantial ionic character.[39] This should significantly increase the mixing of the naphthalene moiety wavefunction to that of the bridge.

This should increase delocalization of the donor and acceptor wavefunctions into that of the bridge, thus enhancing the donor-bridge electronic coupling, T.

Changing the dielectric constant of the solvent may affect the exciton resonance interaction in two ways. Firstly, the oscillator strength of the transition may be altered, which would scale Eq. (4.1). Secondly, it has been suggested that the solvent dielectric constant may be introduced into expressions such as Eq. (4.1) by using the solvent optical dielectric constant,[40] n^2. However, no notable solvent effects were detected in this work, presumably since the system is in the strong coupling limit and because the primary mechanism mediating the excitation delocalization is proposed to be a through-bond interaction.

A question which should be addressed here is "how well is the solvent between such closely-separated chromophores represented as a dielectric continuum?" Given that such a description corresponds to a many-body limit,[41] and in light of recent work on water,[42] the answer is very probably "not very well at all!" Nonetheless, the question will not be addressed further in this thesis.

Conclusions. Substantially larger exciton resonance splittings are observed for the naphthalene $^1A_g \rightarrow 2^1B_{3u}$ electronic transition in the series of rigidly-linked dinaphthyl molecules, DN–2, DN–4 and DN–6, than predicted by the usual electric dipole-dipole interaction model.

Two contrasting cases are encountered. For DN–2 direct through-space orbital overlap is considered to dominate the spectroscopy, although a contribution from a through-bond interaction may also be present. A strong spectral splitting is observed and application of the usual multipole expansion is inappropriate.

For DN–4 and DN–6, however, direct through-space interchromophore overlap is very small. It is concluded that the strong delocalization of electronic excitation through both chromophores is dominated by a through-bond coupling mechanism involving the intervening σ and σ^* orbitals of the bridge, whilst the through-space dipole-dipole mechanism plays a more minor role.

The results of this study emphasize the importance of through bond interactions in enhancing energy delocalization in these linked bichromophoric systems and for the first time the efficiency of a through-bond interaction in mediating a strongly allowed molecular exciton interaction is demonstrated. It is likely that such a mechanism contributes to the interchromophore energy transfer process in other bichromophoric and multichromophoric systems.

V. Weak Coupling Interactions in DN-*n*

Results. Fluorescence emission and excitation spectra for the DN–*n* series of bichromophores in *n*-hexane solvent are shown in Figure 4.11. The significant bathochromism evident in the DN–2 emission spectrum is notable, as are the perturbations in its corresponding excitation spectrum, indicative of substantial orbital overlap-mediated interactions. The majority of the fluorescence at energy less than ca. 28 600 cm^{-1} in DN–2 was assigned to an impurity on the basis of a red-shifted excitation spectrum and a shorter fluorescence lifetime, of approximately 7 ns (compared to 25 ns for the higher energy emission). Since this impurity emission could be spectrally isolated it did not interfere with measurements of the DN–2 decay time.

Fluorescence decays (Figure 4.12) for this series of molecules (in degassed *n*-hexane solvent) were well represented by single exponential fits, as summarized in Table 4.3. There is a clear correspondence between lifetime and connecting bridge length.

Table 4.3. Fluorescence lifetime data for the DN-*n* series of molecules. Goodness of fit parameters χ^2, and DW (Durbin-Watson) are included, as is the ratio of the tabulated lifetime to that of N-2 (the model chromophore).

	τ (ns)	χ^2	DW	τ/τ_{N-2}
N-2	63.33			1
DN-2	25.03	1.2	1.9	0.395
DN-4	51.28	1.6	1.8	0.810
DN-6	60.57	1.6	1.8	0.956

Figure 4.11. Fluorescence excitation (λ_{em} = 340 nm) spectra of the DN-*n* series of molecules. Fluorescence emission (λ_{ex} = 290 nm) and excitation spectra in the naphthalene S_1 region are shown on an magnified intensity scale as the inset spectra.

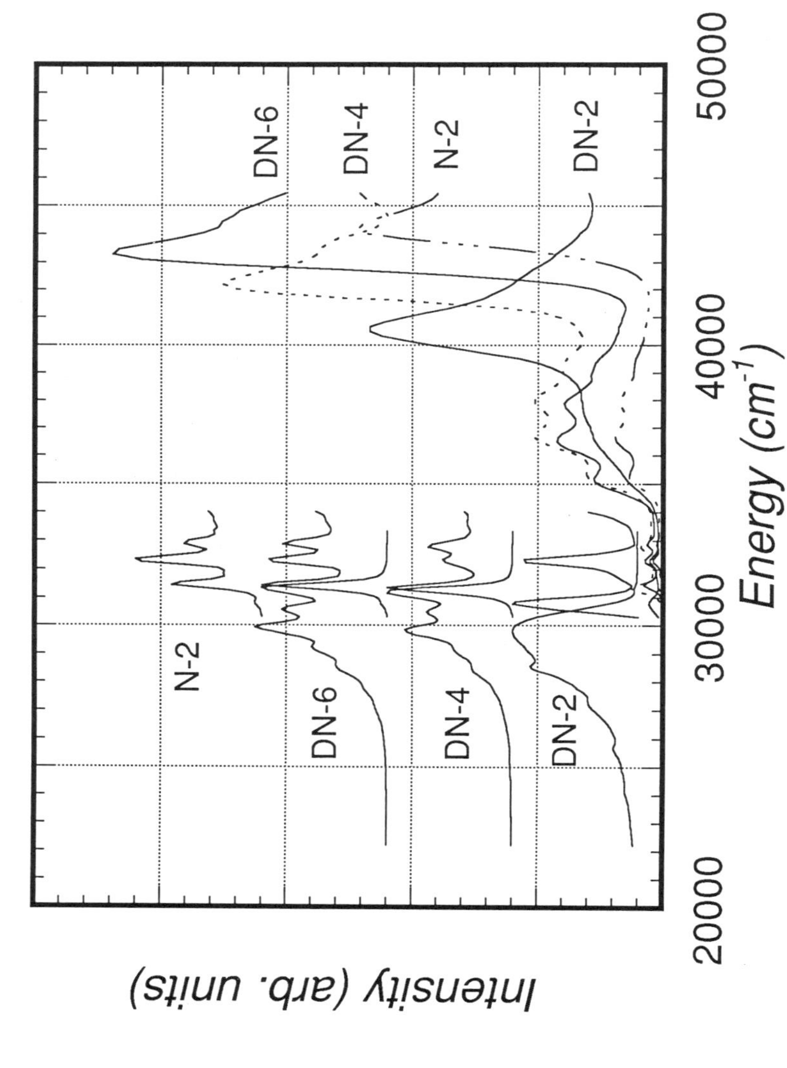

Discussion. The exciton resonance interaction, so vividly evident in the DN–*n* absorption spectra discussed in the previous section, would be expected to also affect the fluorescence emission properties (both steady-state and time-resolved) of these bichromophores.

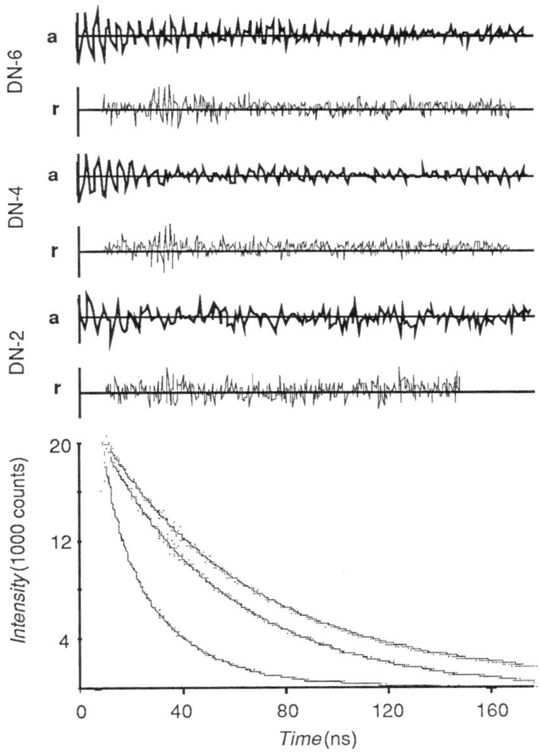

Figure 4.12. Fluorescence decay curves of (from top) DN-6, DN-4 and DN-2 in *n*-hexane, fitted to single exponential decays with lifetimes given in Table 4.3. Corresponding autocorrelation (a) and residual (r) traces are labelled accordingly. In each case the excitation wavelength was 292 nm and the observation wavelength was 330 nm.

Given an understanding of non-radiative rate theory (cf. Chapter 2), the fluorescence lifetime of the emissive state of an electronically coupled molecular pair is expected to differ from that of the lone molecule. Indeed, the non-radiative decay rate will depend on the electronic couplings in the dimer.[43] Rates of intersystem crossing may also be affected due to deviations in the singlet-triplet energy gap[44] (the singlet state will generally be split more than the triplet state). From simple considerations it can be demonstrated that the radiative rate may also be different. In extreme cases the radiative rate may be double that of the free molecule - a phenomenon known as superradiance;[45] or it may be symmetry forbidden, and hence quite small - as is the case for excimers. A treatment, from the perspective of either localized or delocalized excitation amongst the chromophores, would be possible using either the generalized master equation[46] or stochastic Liouville equation methods.[47]

One might therefore expect that fluorescence lifetime data for a single chromophore as compared to a dimer would convey information regarding the strength of the electronic coupling in the lowest excited state. Such a trend is certainly observed qualitatively in the DN-*n* lifetime data of Table 4.3. As a first step, these data are compared to the change in lifetimes predicted from consideration of radiative effects only, as shown below.

For the AB dimer, the two lowest excited states are $(\mathrm{A*B} \pm \mathrm{AB*})/\sqrt{2}$ (cf. Chapter 3, Section III). If the electronic interaction which gives rise to these states is small relative to spectral band widths, then these two states will be obscured. That is, their vibronic envelopes will overlap to a significant degree. This is the condensed phase analogue of the intermediate separation regime discussed by Craig and Thirunamachandran,[48] where the emissive state is described by $a|\mathrm{A*B}\rangle + b|\mathrm{AB*}\rangle$.

Transition moments for the dimer stationary states can be written as in Eqs. (5.1),

$$M_+ = \{\vec{\mu}(a) + \vec{\mu}(b)\}/\sqrt{2} \tag{5.1a}$$

$$M_- = \{\vec{\mu}(a) - \vec{\mu}(b)\}/\sqrt{2} \tag{5.1b}$$

The primary emission is assumed to derive from the oscillator strength due to M_+. An amount of oscillator strength deriving from M_- is mixed into this emission due to overlap of the vibronic bands. Thus,

$$f_{\text{dimer}} \propto k_{\text{rad}} \propto \left[\frac{M_+ + cM_-}{\sqrt{1+c^2}} \right]^2 \tag{5.2}$$

where the coefficient c allows mixing of some of the upper state (with transition moment M_-) into the lower (emissive) state (which has a transition moment M_+). This accounts for overlapping of these two (broadened) bands; such that if the bands are well-separated (i.e., interchromophore interactions are significant relative to line widths) then $c = 0$ and hence the dimer oscillator strength is determined by the vector sum of single chromophore transition moments given by M_+. Whereas if the bands overlap completely (i.e., interchromophore interactions are negligible in comparison with the line widths) then $c = 1$ and the "dimer" oscillator strength reduces to that of a single chromophore.

It is thus seen that, for a non-negligible interaction in these dinaphthyl systems, the radiative lifetime for long-axis polarized emission will be reduced relative to that of N–2. For DN–2 it would be reduced by a factor of 1.39 when $c = 0$ (i.e., from Eqs. 5.1a and 5.2 and the orientation of long-axis polarized transition moments given in Table 4.1). If c is greater than zero, that is the symmetric and antisymmetric state bands in the absorption spectrum overlap to some degree, this reduction in radiative lifetime will become less. Moreover, as noted in Section III, a significant degree of vibronically-induced, short-axis polarized intensity is present in the emission, and for this polarization, $M_+ = 0$.

However, when account is taken of the possible mixing of the dimer ionic configurations with the locally excited configurations (cf. Chapter 3) a further consideration emerges. It has already been noted and demonstrated (Section 3.12) that such a mixing affects the oscillator strength of the resultant state. Model calculations[49] show that for this dimer configuration (and therefore symmetry) the ionic states have high oscillator strength. Thus, when there is sufficient orbital overlap, the mixing of locally excited and ionic configurations is evidenced by a substantially increased oscillator strength for the lowest dimer excited state. This would reduce the lifetimes further and is consistent with the experimental results obtained. As mentioned earlier, any perturbations in the non-radiative rate will add to this reduction in lifetime.

Conclusions. The fluorescence lifetime data presented in this section provide a qualitative test for the theory developed in Chapter 3, as well as a further probe for the extent of through-bond-mediated interactions in the DN–*n* series of molecules. Although there are several possible factors contributing to the observed reduction of the dimer lifetimes with respect to that of the model naphthalene (N–2), it was shown that a major contribution is the mixing of ionic configurations with locally excited configurations which lead to an increase in the oscillator strength of the emissive state. This would be consistent with the through-bond interaction primarily involving a *bridge-mediated through-configuration interaction.*

VI. Overview and Conclusions

The significance of superexchange-mediated through-bond interactions were investigated experimentally using steady-state and time-resolved spectroscopic methods. These studies, presented in Sections III to V, evinced the significance of through-bond mediated interactions on both interchromophore electronic energy transfer and on exciton resonance interactions - in both the strong and weak coupling regimes.

A direct measurement and interpretation of intramolecular EET in a rigidly-linked naphthalene (donor) – anthracene (acceptor) molecule (A–6–N) with known, well-defined geometry was reported. The measured rate of EET (2.9×10^{10} s^{-1}) was found to be significantly greater than that estimated by the Förster (Coulombic dipole-dipole) model (1.2×10^{9} s^{-1}). This was attributed to the significance of a through-bond, superexchange, mechanism for promoting EET. Exciton splittings of the intense naphthalene $S_0 \rightarrow S_3$ transition were examined for a series of rigidly-linked dinaphthyl molecules, connected by two, four and six carbon-carbon σ bonds (DN–2, DN–4 and DN–6, respectively). Through-space overlap effects are negligible for DN–4 and DN–6, and it was concluded that for these systems a significant contribution by a through-bond interaction (involving the σ and σ^* orbitals of the connecting bridge) leads to the observed efficient delocalization of the excitation. The possible application of fluorescence lifetime measurements to probe the importance of interchromophore interactions was investigated for these dinaphthyl systems.

Important indications of the nature and mechanism of such TB interactions were obtained – that is, given their large magnitude, they are most likely mediated via orbital overlap-dependent interactions (as discussed in Chapter 3, Section XIII).

The results of this work have immediate, significant implications on studies of electronic energy transfer and exciton interactions in aromatic polymers, bichromophoric molecules, biological systems and other multichromophoric assemblies. It is demonstrated that care should be taken before attributing the origin of interchromophore interactions to dipole-dipole interactions only. Consideration of the role of the linking bridge in promoting electronic interactions should be made.

References

(1) G. D. Scholes, K. P. Ghiggino, A. M. Oliver and M. N. Paddon-Row, J. Am. Chem. Soc., **115**, 4345 (1993).

(2) G. D. Scholes, K. P. Ghiggino, A. M. Oliver and M. N. Paddon-Row, J. Phys. Chem., **97**, 11871 (1993).

(3) M. N. Paddon-Row and H. K. Patney, Synthesis 328 (1986).

(4) S. L. Murov, *Handbook of Photochemistry* (Marcel Dekker, 1973).

(5) I. B. Berlman, *Handbook of Fluorescence Spectra of Aromatic Molecules;* 2nd ed. (Academic Press, 1971).

(6) D. V. O'Connor and D. Phillips, *Time Correlated Single Photon Counting* (Academic Press, New York, 1984).

(7) (a) G. R. Fleming, J. M. Morris and G. W. Robinson, Chem. Phys. **17**, 91 (1976).

(b) M. Shinitzky, J. Chem. Phys., **56**, 5979 (1972).

(8) (a) T. A. Smith, Ph.D. Thesis, The University of Melbourne (1990).

(b) P. F. Skilton, Ph.D. Thesis, The University of Melbourne (1990).

(c) A. D. Scully, Ph.D. Thesis, The University of Melbourne (1989).

(9) K. P. Ghiggino, T. A. Smith and G. J. Wilson, J. Mod. Optics, **37**, 1789 (1990).

(10) K. P. Ghiggino and T. A. Smith, Prog. Reaction Kinetics, **18**, 375 (1993).

(11) J. Michl and V. Bonačić-Koutecký, *Electronic Aspects of Organic Photochemistry* (Wiley, 1990).

(12) G. D. Scholes and A. H. A. Clayton, Unpublished results (1992): *ab initio* and CNDO/S/CI calculations of the excited singlet states of anthracene.

(13) J. Michl, E. W. Thulstrup and J. E. Eggers, Ber. Bunsenges.Phys. Chem. **78**, 575 (1974).

(14) D. S. McClure, J. Chem. Phys. **22**, 1668 (1954).

(15) D. M. Friedrich, R. Mathies and A. C. Albrecht, J. Molec. Spectros. **51**, 166 (1974).

(16) D. P. Craig and S. H. Walmsley, *Excitons in Molecular Crystals* (Benjamin, New York, 1968).

(17) G. A. George and G. C. Morris, J. Mol. Spectros. **26**, 67 (1968).

(18) L. E. Lyons and G. C. Morris, J. Chem. Soc. 1551 (1959).

(19) D. P. Craig, Mol. Phys. **4**, 113 (1961).

(20) D. P. Craig and T. Thirunamachandran, Proc. Phys. Soc. **84**, 781 (1964).

(21) G. D. Scholes and K. P. Ghiggino, J. Phys. Chem. **98**, 4580 (1994).

(22) T. Förster, in *Modern Quantum Chemistry*, edited by O. Sinanoglu (Academic Press, 1965) Vol. III.

(23) D. P. Craig and T. Thirunamchandran, Chem. Phys. **135**, 37 (1989).

(24) M. N. Paddon-Row and K. D. Jordan, in *Modern Models of Bonding and Delocalization*, edited by J. F. Liebman and A. Greenberg (VCH, NY & FRG, 1988).

(25) G. L. Closs and J. R. Miller, Science **240**, 440 (1988).

(26) G. L. Closs, P. Piotrowiak, J. M. MacInnis and G. R. Fleming, J. Am. Chem. Soc. **110**, 2652 (1988).

(27) G. L. Closs, M. D. Johnson, J. R. Miller and P. Piotrowiak, J. Am. Chem. Soc. **111**, 3751 (1989).

(28) H. Oevering, J. W. Verhoeven, M. N. Paddon-Row and E. Cotsaris, Chem. Phys. Letts. **143**, 488 (1988).

(29) H. Oevering, J. W. Verhoeven, M. N. Paddon-Row and E. Cotsaris, Chem. Phys. Letts. **150**, 179 (1988).

(30) J. Kroon, A. M. Oliver, M. N. Paddon-Row and J. W. Verhoeven, J. Am. Chem. Soc. **112**, 4868 (1990).

(31) G. L. Closs, L. T. Calcaterra, N. J. Green, K. W. Penfield and J. R. Miller, J. Phys. Chem. **90**, 3673 (1986).

(32) H. Oevering, M. N. Paddon-Row, M. Heppener, A. M. Oliver, E. Cotsaris, J. W. Verhoeven and N. S. Hush, J. Am. Chem. Soc. **109**, 3258 (1987).

(33) S. J. Strickler, R. A. Cormier and J. S. Connolly, Int. J. Quant. Chem. **39**, 345 (1991).

(34) J. O. Hirschfelder, C. F. Curtis and R. B. Bird, *Molecular Theory of Gases and Liquids,* pp. 26-27, 843-846 (John Wiley and Sons, Inc., New York, 1954).

(35) J. C. Chang, J. Chem. Phys. **67**, 3901 (1977).

(36) (a) P. Bertrand, Chem. Phys. Lett. **140**, 57 (1987).

(b) P. Bertrand, Structure and Bonding **75**, 1 (1991).

(37) J. M. O. Matos, B. O. Roos and P.-A. Malmqvist, J. Chem. Phys. **86**, 1458 (1987).

(38) B. O. Roos, K. Andersson and M. P. Fülscher, Chem. Phys. Letts. **192**, 5 (1992).

(39) J. M. O. Matos and B. O. Roos, Theor. Chim. Acta **74**, 363 (1988).

(40) N. Mataga and T. Kubota, *Molecular Interactions and Electronic Spectra* (Marcel Dekker, Inc., NY, 1970).

(41) D. P. Craig and T. Thirunamachandran, Chem. Phys. **135**, 37 (1989).

(42) G. W. Robinson and S. Singh, Adv. Chem. Phys. **85**, 627 (1994).

(43) D. P. Chock, J. Jortner and S. A. Rice, J. Chem. Phys. **49**, 610 (1968).

(44) M. Kasha, H. R. Rawls and M. A. El-Bayoumi, Pure Appl. Chem. **11**, 371 (1966).

(45) R. H. Dicke, Phys. Rev. **93**, 99 (1954).

(46) V. M. Kenkre and R. S. Knox, Phys. Rev. B, **9**, 5279 (1974).

(47) T. S. Rahman, R. S. Knox and V. M. Kenkre, Chem. Phys. **44**, 197 (1979).

(48) D. P. Craig and T. Thirunamachandran, Chem. Phys. **167**, 229 (1992).

(49) (a) G. D. Scholes, Unpublished results (1993): model *ab initio* calculations of naphthalene dimers in various conformations.

(b) G. D. Scholes and A. H. A. Clayton, (1994): *ab initio* and CNDO/S/CI calculations of the excited singlet and triplet states of N-2, DN-2, DN-4 and DN-6. To be submitted for publication.

CHAPTER FIVE

Précis and Conclusions

Our experience hitherto justifies us in believing that nature is the realization of the simplest conceivable mathematical ideas. I am convinced that we can discover by means of purely mathematical constructions the concepts and the laws connecting them with each other, which furnish the key to the understanding of natural phenomena. Experience may suggest the appropriate mathematical concepts, but they most certainly cannot be deduced from it. Experience remains, of course, the sole criterion of physical utility of a mathematical construction. But the creative principle resides in mathematics. In a certain sense, therefore, I hold it true that pure thought can grasp reality, as the ancients dreamed.

Albert Einstein, *Herbert Spencer Lecture*, 1933.

To briefly recapitulate, the primary aim of the work presented in this thesis was to obtain a more detailed and thorough understanding of the mechanisms underlying excitation energy transport in complex synthetic and biological systems. The motivation for such a study being that to reliably explain or predict the photophysical response of multichromophoric assemblies, important for example, in the design of efficient light harvesting systems, the fundamentals of the EET phenomenon needed to be more thoroughly elucidated. It was evident that current theory was generally too simplistic and often misunderstood.

The thesis began (in Chapter 2) by establishing EET in the framework of radiationless transition theory. Quantum mechanical expressions for microscopic aspects of the dynamics of EET were developed to enable complex relayed EET phenomena (such as superexchange-mediated processes and energy migration) to be quantified. A more concrete understanding of coherent versus incoherent mechanisms for energy migration were obtained as its relationship to the superexchange (electronic tunnelling) mechanism was established. Via the coherent mechanism, excitation may be transferred very rapidly over several chromophores – thus significantly enhancing the efficiency of energy migration, and possibly enabling energy to be partly "directed" to a trap.

In Chapter 3 the electronic factors which promote EET were examined, elucidating previously unknown interactions important at close donor-acceptor separations. It was found that the electronic coupling contains contributions from two major types of term. Firstly, the *direct* interaction, which includes the well-known Coulombic interaction, the Dexter exchange interaction (which was shown to be of only minor significance) and the orbital overlap-dependent penetration interactions. Secondly, the *through-configuration* interaction, in which significant orbital overlap-dependent interactions are manifested by the mixing of interchromophore ionic (i.e. charge transfer) configurations with locally-excited configurations of the donor and acceptor. This was shown to be the most important orbital overlap-dependent interaction. The form and significance of higher multipole contributions to the Coulombic interaction were investigated and the distance dependence of the direct and through-configuration orbital overlap-dependent interactions were examined computationally for model ethene dimers, demonstrating quantitatively the application of the general theory. Finally, a theory for superexchange-mediated EET interactions was outlined. These results are summarized in Table 5.1 (also see Chart 3.1, p. 144).

In Chapter 4, some aspects of donor-acceptor EET and electronic interactions were investigated experimentally using steady-state and time-resolved spectroscopic methods on a series of novel rigidly-linked bichromophores. In light of the theory established in the preceding chapters it found that, even in bichromophores with well-defined and known interchromophore separations and orientation factors, there are difficulties in obtaining detailed quantitative information on the contributions by various EET mechanisms. It could be shown qualitatively, however, that superexchange interactions mediated through the connecting σ bonds of the bridge (through-bond interactions) were very significant in the systems studied.

The ideas and theory presented in this thesis have impact on current experimental analysis – especially in a qualitative manner. The implications can be important and clarifying and should not be ignored in preference to simplistic ideas. Moreover, further insight can be obtained by quantitative or semi-quantitative application of the framework given in this work, as will now be illustrated.

Many examples of supramolecular systems were provided in Chapter 1 where EET plays an important role subsequent to light absorption. These included model bichromophores (cf. Chapter 4), complex synthetic light harvesting molecules[1] and the photosynthetic antenna system.[2] An example of one such system, a synthetic light harvesting polymer, is discussed below.

Table 5.1.. Summary of the primary electronic interactions in an excited dimer. R refers to the through-space donor-acceptor separation and N refers to the number of connecting bonds (or other bridging units).

Symbol	Equations (Chapter 3)	Name & R-dependence	Physical Interpretation
u^{Coul}, $2J$	5.5a 10.14-15 11.3, 11.6-7	Coulombic Interaction $R^{-3} + R^{-4} + ...$	Multipolar coupling between the transition moments resulting in the de-excitation of the donor with a concommitant excitation of the acceptor. The interaction has long-range character; being independent of orbital overlap.
u^{short}		First-order orbital overlap-dependent interactions	Has three components: $-Z$, $-P$ and U.
$-Z$	5.5b	Dexter exchange interaction $\exp(-2\alpha R)$	Arises from antisymmetry requirements of the total (supermolecular) wavefunction. It describes a "correction" to u^{Coul} due to interpenetration of molecular orbitals. Negligible in magnitude.
$-P$	5.9	Penetration interaction $\exp(-2\alpha R)$	One electron penetration interaction (interchromophore electron-nuclear attraction) associated with direct excitation transfer via T_{14}.
U	6.8-9 6.11	Through-configuration interaction $\exp(-2\alpha R)$	A configuration superexchange interaction mediated by ionic bridging configurations.
$u^{t\text{-}b}$	13.2 - 13.5	Through-bond interaction $\exp(-2\beta N)$	A superexchange interaction wherein electronic coupling (especially that due to u^{short}) is promoted over large through-space separations via bridging bonds (often σ-type), solvent, etc. connecting donor and acceptor.
u^{ind}	12.6	Coulombic induction interaction R^{-3}	Arises from the dynamic polarization (distortion) of the charge distribution on the acceptor due to the transition moment of the donor. (Does not effect EET.)

Energy migration (EM) is a process by which the excitation localized at one chromophore in a multichromophoric system is relocated by a multistep process to another chromophore of the same type, elsewhere in the system. For a polymer a different chromophore (with a lower excitation energy) may be tagged onto the chain, in which case the EM process will be terminated when excitation is transferred to this "trap". Observation of the time-dependence of fluorescence from the trap chromophore, subsequent to initial excitation of the polymer, can provide information on the EM process.[3]

Most work to date on EM in polymers has concentrated on dipole-dipole energy transfer hops between nearest neighbour chromophores. Interchromophore separations in aromatic polymers are relatively small (ca. 4 to 8 Å), so the dipole approximation may not provide a good indication of the actual electronic coupling between chromophores. The nature of this electronic coupling at such separations has been pursued in detail in this thesis, as has the ramifications of a relayed transfer of excitation, including "superexchange" via intervening σ bonds or chromophores (through bond, TB, interactions). Superexchange via chromophores identical with the donor (on which the initial excitation is localized) is a coherent EM process.

In a recent study the rapid temporal grow-in of anthracene emission following excitation of acenaphthylene chromophores in poly(acenaphthylene) (PACE) end-tagged with anthracene was measured experimentally.[3] The rate of this grow-in could not be accounted for using a model which assumes energy migration steps between nearest-neighbour chromophores only, promoted by the dipole-dipole ("Förster") mechanism. It has been seen that at short to intermediate interchromophore separations, measured rates of EET in multichromophoric systems (such as bichromophores, aromatic polymers, or biological systems) can sometimes be substantially larger than predicted by simple dipole-dipole theory, and the rates of energy migration and trapping processes are known to be strongly dependent upon the magnitude of this electronic coupling. An application of the theory outlined in this thesis to a model PACE dyad, typical of that expected to compose the bulk-polymerized polymer, will now be presented, and some comments regarding the treatment of coherent EM hops will be made.

A "real" polymer system is complex, as it consists of many conformational-, as well as stereo-, isomers. In this example, just the *gauche*-threo-disyndiotactic PACE dyad, discussed in a recent communication,[4] will be considered. The structure of the acenaphthylene monomer was geometry-optimized with a C_{2v} symmetry constraint using the semiempirical INDO method. The dyad geometry was then optimized, keeping the internal geometry of each chromophore fixed. The optimized dyad structure provided the orientation and separation of dipole transition moments (for the lowest singlet electronic transition), as shown in Figure 5.1. An identical dyad, but without any connecting σ bonds (i.e., a pair of C_{2v} symmetry naphthalene chromophores) is also shown.

Approximate excited state calculations on both types of model dyad, (1) and (2), and "monomer" (i.e. C_{2v} naphthalene [model (1)] and acenaphthylene [model (2)] were carried out using the CI-singles method in Gaussian-90[5] (minimal, STO-3g, basis set – which has been shown to be often adequate for the quantitative determination of through-bond interactions[6]). This investigation of the PACE dyad was reported recently;[7] details of the steps in calculation of the electronic coupling are listed below. Note that when these calculations were undertaken, just the *direct* interactions (i.e., those leading to spectral splitting) were examined. The through-configuration interaction (which would contribute to the through-bond interaction) was therefore not considered.

(1)

(2)

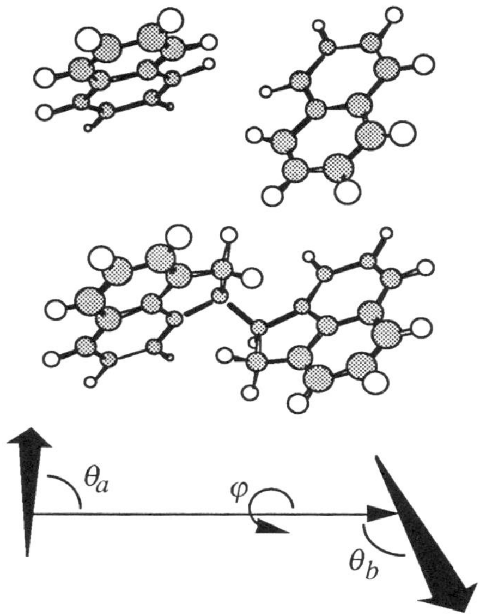

Figure 5.1. Optimized geometries of the model naphthalene dimer (1) and acenaphthylene dyad (2), together with the orientation of the dipole transition moments for the long-axis-polarized (purely electronic) acenaphthylene excited state ($\theta_a = 151°$, $\theta_b = 150°$, $\phi = 109°$ and $R = 6.42$ Å.

(i) Estimation of higher multipole and through-bond interactions.

- From the *ab initio* calculation of dyad (1) the total Coulombic interaction was determined (by assuming that the orbital overlap-dependent interactions were negligible at this separation and orientation of the chromophores) to be 4.83 cm^{-1}.

- A similar calculation of *monomer* (1) determined the dipole transition moment (*relevant to this calculation*) to be 1.48×10^{-30} Cm. Hence, from Eq. (3.11.3a) with an orientation factor (cf. Figure 5.1) of 1.43, the dipole-dipole interaction in the dyad (1) is estimated to be 5.38 cm^{-1}. Thus, by difference, the contribution to the total Coulombic interaction (primarily due to a dipole-octopole interaction) is approximately -0.55 cm^{-1}. This implies that the Coulombic interaction in the PACE dyad is well-represented by the leading dipole-dipole term.

• For the dyad (2), the total (direct) electronic coupling was calculated to be 9.64 cm^{-1}. The dipole transition moment for monomer (2) was calculated to be 1.48×10^{-31} Cm (interestingly, ten-fold smaller than for monomer (1)), so the dipole-dipole interaction (and hence, to a close approximation, the total Coulombic interaction) is estimated to be only 0.053 cm^{-1}.

• The difference between the calculated electronic coupling and the estimated Coulombic interaction in dyad (2) is attributed to a through-bond (TB) interaction (cf. the calculation of dyad (1) - with no connecting bonds). Hence the TB interaction in this PACE dyad is estimated to be 9.6 cm^{-1}.

(ii) Inclusion of vibronic coupling.

• The acenaphthylene dipole transition moment (including contributions from vibronic coupling) was determined from the reported oscillator strength[8] of 0.01 to give a predominantly short-axis polarized transition moment of 2.8×10^{-30} Cm. Using an orientation factor of 1.50, appropriate for the short-axis polarized transition moment, an "experimentally-relevant" Coulombic interaction of 20 cm^{-1} was determined (again, higher multipole contributions in this instance are assumed to be small). Note the substantial increase in the significance of the dipole-dipole interaction over that due to purely electronic transitions.

(iii) The total electronic coupling in the PACE dyad $\Rightarrow$ rate of EET.

• The total electronic coupling for the PACE dimer consists of the Coulombic part of 20 cm^{-1} (estimated in (ii)) and the TB part of 9.6 cm^{-1} (estimated in (i)). Thus the total electronic coupling is approximately 31 cm^{-1}.

• Using a spectral overlap integral[9] $J = 2.61 \times 10^{-16}\ M^{-1}cm^3$, donor fluorescence emission quantum yield (ethanol), $\Phi_A = 0.39$, and fluorescence lifetime (ethanol), $\tau_A = 46$ ns,[10] the rate of dipole-dipole EET (in ethanol solvent) is found to be 2.3 $\times\ 10^{10}\ s^{-1}$. If, for simplicity in the present example, it is assumed that Franck-Condon factors for the TB interactions are similar to those for the Coulombic mechanism (which is not necessarily a good approximation), the total rate – including TB interactions – increases by a factor of two over the dipole-dipole rate: $5.4 \times 10^{10}\ s^{-1}$.

(iv) Coherent energy transfers in the EM process.

Monte Carlo simulations based on a simple model have been carried out for anthracene-tagged PACE systems.[3] The calculations considered the possibilities of radiative and non-radiative decay processes as well as EET to a nearest-neighbour, such that the excitation undergoes a one-dimensional random walk along the polymer. The resulting simulated fluorescence rise and decay profile for the anthryl trap severely underestimated the rate of trap sensitization. It is probable, in the light of the developments in Chapter 2, that the deficiency of the model is *primarily* due to the limitation of the Pauli master equation-type approach (i.e., the assumption of nearest-neighbour incoherent transfer steps), rather than an underestimation of electronic couplings or neglect of direct next-to-nearest neighbour incoherent transfer steps.

An important modification of the simulation just described would be to employ a method involving *probability densities* (e.g. the generalized master equation[11]). Coherent EET pathways may be incorporated into a Pauli master equation-type formulation using the expressions outlined in Section 2.8 (Eqs. 2.8.24, 2.8.29-2.8.33), so as to describe the non-Markoffian EM process.[12] Information regarding the temperature-dependence of the damping of coherencies (i.e., thermalization of intermediate states) is intrinsic in the equations of Section 2.8.

Consideration of the connection of generalized master equation-based expressions with optical spectra are provided in the work of Kenkre and Knox.[11]

The example discussed above serves to illustrate the complex interplay of the various EET processes in the transport of excitation energy in multichromophoric systems. It is hoped that the studies outlined in this thesis will provide a firmer foundation for improving the understanding and optimization of such processes in other natural and synthetic molecular assemblies.

References

(1) V. Balzani and F. Scandola, *Supramolecular Photochemistry* (Ellis Horwood Ltd., Chichester, 1991).

(2) R. M. Pearlstein, in *Excitons,* edited by E. I. Rashba and M. D. Sturge (North-Holland, 1982) p. 735.

(3) T. A. Smith and K. P. Ghiggino, Prog. Reaction Kinetics **18**, 375 (1993).

(4) G.D. Scholes and K.P. Ghiggino, Chem. Phys. Lett. **188**, 140 (1992).

(5) M. J. Frisch, M. Head-Gordon, G. W. Trucks, J. B. Foresman, H. B. Schlegel, K. Ragavachari, M. Robb, J. S. Binkley, C. Gonzalez, D. J. DeFrees, D. J. Fox, R. A. Whiteside, R. Seeger, C. F. Melius, J. Baker, R. L. Martin, L. R. Kahn, J. J. P. Stewart, S. Topiol and J. A. Pople, Gaussian, Inc. (Pittsburgh PA, 1990).

(6) (a) K. D. Jordan and M. N. Paddon-Row, Chem. Rev. **92**, 395 (1992).

(b) K. D. Jordan and M. N. Paddon-Row, in *Modern Models of Bonding and Delocalization,* edited by J. F. Liebman and A. Greenberg (VCH, New York, 1988) p. 115.

(7) G. D. Scholes and K. P. Ghiggino, J. Photochem. Photobiol., A: Chem. **80**, 355 (1994).

(8) A. Bree, Mol. Phys. **6**, 563 (1963).

(9) T. A. Smith, Ph.D. Thesis, The University of Melbourne (1990).

(10) J.B. Birks, *Photophysics of Aromatic Molecules,* (John Wiley, 1970).

(11) V. M. Kenkre and R. S. Knox, Phys. Rev. B **9**, 5279 (1974).

(12) L. van Hove, Physica **23**, 441 (1957).

PART THREE

Analysis of Optical Spectra of Photosynthetic Reaction Centers Absorption, Hole-Burning and Resonance Raman Spectra

Analysis of Optical Spectra of Photosynthetic Reaction Centers Absorption, Hole-Burning and Resonance Raman Spectra

Reza Islampour[a,b] and Michi-toshi Hayashi[a,c]*

(a) Institute of Atomic and Molecular Sciences,

Academia Sinica,

P.O. Box 23-166

Taipei, Taiwan 10764 ROC

(b) Department of Chemistry,

University for Teacher Education,

49 Mofateh Ave., Tehran, Iran

and

(c) Department of Chemistry

and

Center for the Study of Early Events in Photosynthesis,

Arizona State University,

Tempe, AZ 85287-1604, USA

Contents

1. Introduction

The photosynthetic reaction center (RC) of purple non-sulfur bacteria is the core molecular assembly, located in a membrane of the bacteria, that initiates a series of electron transfer (ET) reactions subsequent to energy transfer events. One of the most important functions of photosynthesis, then, involves the conversion of solar energy into chemical energy. The bacterial photosynthetic RCs have been characterized in greater detail, both structurally and functionally, than any other transmembrane protein complexes[1-3]. The three dimensional structures of the RCs have been elucidated from the molecular level; the X-ray structure analyses of the crystallized RCs[4-8], which were isolated from *Rhodobacter* (*Rb.*) *sphaeroides* and *Rhodopseudomans* (*Rps.*) *viridis*, manifest the identity and functions of the chromophores of the RCs. The *Rb. sphaeroides* RC, for instance, which is the best characterized of the purple bacterial RCs from a spectroscopic point of view, consists of nine cofactors. Two bacteriochlorophyll *a* (Bchl) molecules together form an electronically coupled pair (P), situated near the periplasmic side of the membrane, through which the RC shows an approximately two-fold symmetry forming two branches (L and M). Below P and to each branch are two monomer Bchl molecules (B_L or B_M), two bacteriopheophytin *a* (Bphe) molecules (H_L or H_M), and two ubiquinone molecules (Q_L or Q_M). The two branches end with a nonheme iron approximately situated between the two ubiquinones.

Ultra-short time-resolved laser measurements have shed light on the primary event of the energy conversion process, and revealed that process consists of a series of ET reactions leading to the separation of charge across the photosynthetic membrane[9-29]. Light energy absorbed by the antenna of the photosynthetic membrane, is transferred, via a series of energy transfer processes, to the coupled pair P in the RC. The excitation of the special pair P initiates ET; within the next 0.3 to 4 picoseconds (ps), ET occurs from the excited singlet of P (P^*), forming the charge separated state $P^+H_L^-$ where H_L^- is a reduced Bphe H_L molecule on the L branch.

Electron transfer from H_L^- to Q_L proceeds with a time constant of about 200ps. This is followed by ET from Q_L to Q_M which takes place in hundreds of microseconds. A series of substantially slower steps complete the processing of the light energy into chemical energy. Since the state $P^+Q_L^-$ is extremely stable against charge recombination, the final ET can be considered irreversible.

The related and much investigated questions concern the mechanism of ET between P and H_L [12,16,17,30-42]: (1) ET proceeds almost exclusively along the L-branch, while M is inactive; (2) to what extent the state $P^+B_L^-$ is actually populated or involved as a virtual state allowing direct ET; (3) the ET rates exhibit an inverse temperature dependence. Early picosecond resolution transient absorption measurements on RCs gave no evidence for the population of kinetically resolvable intermediate states between P^* and $P^+H_L^-$[9,12,33,43]. As higher-resolution transient absorption experiments were performed[17,42], it discovered that one could not model the initial ET reaction as a single-exponential decay of P^* forming $P^+H_L^-$ at all wavelengths. It becomes clear that the room temperature decay kinetics were wavelength dependent.

Different groups have interpreted the wavelength-dependent kinetics differently. Kiramier and Holten have proposed that the wavelength dependence results from conformational heterogeneity in the RC population (one-step model)[42]. Zinth and his co-workers have coupled measurements of the wavelength-dependent kinetics to a spectral analysis of the decay component amplitudes to support a model involving $P^+B_L^-$ as a true intermediate (two-step model)[17], as had been proposed by Shuvalov and his co-workers[35]. In addition, Fleming and his co-workers have analyzed the bleaching kinetics of H_L in terms of the sequential two-step mechanism[11], while recent work from the same laboratory has suggested a combination of sequential and super-exchange models to interpret the spontaneous emission of P^* [44]. The experimental studies conducted thus far have led to the qualitative observation that the interpretation of the kinetics at a single or a few selected wavelengths is problematic. The amplitude of a kinetic analysis, for example, are not directly representative of the populations of intermediate states (if they exist) but are, rather, complicated by the extinction coefficients of the participating states and the proposed mechanism itself.

Proposing an excitonic-vibronic model involving ten possible electronic states and six vibrational modes, Lin and his co-workers have recently analyzed femtosecond time-resolved, three-dimensional spectra of wild-type RCs of *Rb. sphaeroides* at room temperature [13]. The excitonic-vibronic model takes into account excitonic states of the special pair P^*, *i.e.*, P^*_- and P^*_+, which are the asymmetric and symmetric linear combinations of the two Bchl monomers comprising P, and their cation states P^+_- and P^+_+ as well. The model was based on quantum chemistry calculations reported by Warshel[45], their semi-empirical MO calculations and analysis of the ground state absorption spectra of the RCs. Their theoretical analysis indicated that more than three electronic states are required in order to construct the observed femtosecond time-resolved spectra of the RCs and interpret the wavelength dependent kinetics, and that the multi-mode effect should be considered in calculation of the ET rate. Applying a similar model consisting of five possible electronic states and seven vibrational modes, the same group has quite recently analyzed femtosecond time-resolved, three-dimensional spectra of wild-type RCs of *Rb. capsulatus* at a low temperature [46].

Femtosecond measurements of RCs at low temperatures exhibit new features in their time-resolved profiles. Vos *et al.* [13] have recently performed femtosecond experiments on *R26* mutant RCs of *Rb. sphaeroides* and D_{LL} mutant RCs of *Rb. capsulatus*, reporting short-lived oscillations, *i.e.*, quantum beats in the time-resolved profiles at a temperature of 10K. Providing the effects of temperature, excitation and probing wavelength on the pump-probe stimulated emission profiles, the same laboratory[47] has also reported more detailed femtosecond experimental data on the excited state dynamics of the dimer P^* of D_{LL} mutant RCs of *Rb. capsulatus*. The femtosecond time-resolved profiles showed oscillatory component whose Fourier analysis suggests that several vibrational modes with 10 ~ 100 cm^{-1}

are involved. The D_{LL} mutant RC, lacking of the H_L , is incapable of ET on a time scale of the fs measurements. Thus, this mutant provides information on vibrational processes in the excited state of the special pair P, in particular, information on vibrational relaxation. In addition, the same group[48] have lately reported quantum beats in the time-resolved profiles of membrane-bound mutant RCs of *Rb. sphaeroides* in the near-infrared stimulated emission band of the special pair P^* at 10 K. The quantum beats appearing in the time-resolved profiles strongly indicate that the femtosecond measurements at such a low temperature observe ET accompanying non-equilibrated vibrational motions. In other words, ET takes place before vibrational relaxation and transition achieve vibrational equilibrium.

In general, in order to observe the vibrational coherence in an ET system, there should exist at least one vibrational mode with a reasonably large coupling constant (*i.e.*, Huang-Rhys factor) so that this particular mode can be coherently pumped by the pumping laser with appropriate laser pulse-duration compatible with the vibrational frequency of the particular mode. From the observed oscillation, one can determine the vibrational frequency and the observed decay rate in this case is contributed by vibrational relaxation, ET and dephasing. Due to the interference effect caused by temperature (i.e. thermal populations) and pumping too many vibrational levels, the vibrational coherence can often become unobservable. It is important to note that for the case in which both ET and vibrational relaxation can be observed in the coherence measurements, it implies that vibrational equilibrium is not established when ET takes place. In this case, the conventional ET theory[49-54] cannot be used. Taking into account vibrational relaxation, Lin's group has theoretically analyzed those coherence experimental data[46,55,56]. It is indeed demonstrated that a model consisting of one dominant mode with a relatively large coupling constant can reproduce experimentally observed transient behaviors well. Their analyses show the RC systems achieves their vibrational equilibrium within about 300 to 500 femtoseconds at a low temperature, which is, in turn, comparative to ET rate.

High resolution, frequency-domain spectroscopic measurements[31,57-72] have also revealed important vibrational information of the photosynthetic RCs. Boxer and his co-workers[57,58,62] have recently reported hole burning (HB) spectra on *Rb. sphaeroides* wild-type RCs and their mutant RCs, and Small's group [63-71] has extensively performed HB measurements on mutant and chemically altered RCs of *Rb. sphaeroides* and *Rps. viridis*: both of their HB results revealing low-frequency vibrational modes that make an important contribution to optical features such as the band-width of absorption line-shape, as well as to the rate of the ET of the RCs. In particular, Small's group[44,45] reported a frequency of 120 and 145cm^{-1} for wild-type *Rb. sphaeroides* RCs and wild-type *Rps viridis* RCs, respectively. They observe phonon-side-band in the HB spectra whose mean phonon frequency is 30cm^{-1} and one-phonon profile width is approximately 40 cm^{-1}[70,71].

High-frequency modes have been observed in resonance Raman (RR) scattering of RCs [75-80]. The near IR resonance Raman (RR) spectra of P and B bands of *Rb. sphaeroides* have recently been reported by Boxer, Mathies and their co-workers[75,76], and Bocin's group[77-80], each of measurements exhibiting rich Raman peaks in their spectra. These investigations of the vibrational modes are very important when one would like to estimate the rate of ET in RCs based on calculation methods. In particular, the coupling constants of these vibrational modes observed in their spectra need to be determined.

Quite recently, much attention has been paid to transient behaviors of high frequency modes of RCs as well. Hochstrasser and his co-workers have reported femtosecond coherent transient infrared spectra of R26 mutant RCs of *Rb. sphaeroides*, providing information on the transient dynamics of vibrational modes in 1560-1960 cm^{-1} region[81]. Zinth and his co-worker have, on the other hand, measured transient behavior of high frequency modes in 1000-1800 cm^{-1} region by applying femtosecond infrared technique to R26 mutant RCs of *Rb. sphaeroides*[82].

By applying mutagensis techniques to replace side-residues of the proteins holding the RC core complex, a number of efforts have been paid for elucidation of the mechanism of this uni-directional asymmetry of ET[83]. Modifying side-residues leads to various changes in the energy gaps, the strength of electronic couplings and the vibrational modes of RCs. Since ET rate can be expressed as a function of electronic coupling and vibrational modes, the role of vibrational modes on such uni-directionality are considered to be important as well.

Detailed quantum chemical calculations have been performed to the electronic structure of the ground state, its electronically excited state P^* and the cation P^*, providing valuable information on the high quantum efficiency, the uni-directionality of the ET and the optical properties of the RCs[84-94]. In particular, information on the electronic structure of the excitonic-coupled dimer is very important. Although the excitonic effect in the special pair P has been recognized in analyzing the optical spectra RCs[84-97], this effect has not received much attention in studying the ET processes of RCs. Computational simulations are also important for an understanding of the ET mechanism of RCs, especially for the case of room temperature. Recently, taking into account protein dynamics, Chandler and co-workers[98,99] have reported computational simulation on the ET dynamics of Rps. viridis at room temperature on the basis of molecular dynamics and semi-empirical quantum chemical estimates on the potential surfaces. They have discussed the dynamics in terms of the fluctuation of the electronic energy gap, whose correlation persists beyond 1ps, showing complex kinetics consisting non exponential behaviors due to the fluctuation.

Alden *et al*, have proposed generalized electron transfer rate constant to study the temperature and energy gap dependence of rate constant[14]. It becomes clear that multi-mode effects and high frequency modes play an important role in the temperature and energy gap dependence of the ET rate constant.

The quantum chemistry calculations depend heavily on the spectra deduced from steady-state and time-resolved spectroscopy of electronic states and on the Stark effect of the RCs. Analysis on vibrational modes of the RCs, in particular, high frequency modes, will provide these calculations with additional knowledge.

The present paper is organized as follows. In Chap. 2, we present a brief review on the theory of radiation processes. In Chap. 3, we review vibronic theory that allows us to derive the molecular representation of transition probabilities. Based on the vibronic theory, we discuss optical absorption spectra and resonance Raman scattering in Chap. 4 and Chap. 5, respectively. Molecular susceptibility theory is also discussed in Chap. 6. We present molecular representation of absorption and hole-burning spectra of a system under the influence of inhomogeneity in Chap. 7. In Sec. 8.1 of Chap. 8, we attempt to analyze resonance Raman spectra experimentally observed. In Sec. 8.2 and Sec. 8.3, based on the RR analysis, we shall demonstrate absorption and hole-burning spectra of the P_-^* band and the Q_y bands of B of *R26* mutant RCs of *Rb. sphaeroides*. Finally, in Chap. 9 we attempt to conclude our theoretical studies on the resonance Raman, absorption and hole-burning spectra conducted on bacterial photosynthetic RCs.

2. Radiative Processes

2.1. Time-dependent Perturbation Series

Quantum mechanical time-dependent perturbation theory organizes interaction of light radiation and matter rather naturally into a sequence of events of decreasing probability in which the occupation numbers of the quantized electromagnetic field change by integer amounts of increasing magnitude. These events are generally called one, two, three, $\cdots$, or, in general, multi-photon processes. In this section we shall present a systematic formulation of the theory of multi-photon processes by using the time-dependent perturbation series. First we shall briefly outline the time-dependent perturbation series and then we shall show how to apply this method to the multi-photon processes.

The evolution in time of a quantum system is described by the time-dependent equation

$$i\hbar \frac{d}{dt}|t\rangle = \hat{H}(t)|t\rangle \ , \tag{2-1}$$

where $\hat{H}(t)$ is the Hamiltonian of the system. Equation (2-1) may be formally integrated to give

$$|t\rangle = \{ \sum_{m=0}^{\infty} (-i\hbar)^m \int_{t_0}^{t} dt_1 \cdots \int_{t_0}^{t_{m-1}} dt_m \, \hat{H}(t_1) \cdots \hat{H}(t_m) \} \, |t_0\rangle$$

$$= \exp_+\{ (i\hbar)^{-1} \int_{t_0}^{t} dt' \, \hat{H}(t') \} \, |t_0\rangle$$

$$= \hat{U}(t,t_0)|\mathrm{t}_0\rangle \ , \tag{2-2}$$

where the unitary operator $\hat{U}(t,t_0)$ is known as the *evolution* or time-dependent operator, and $\exp_+[\ \cdots\ ]$ is a positive time ordering exponential which means that $\hat{H}(t)$ in the power series expansion of $\exp_+[\ \cdots\ ]$ should be ordered so that time increases from right to left. By means of a unitary transformation $\hat{W}(t)$ such that $|t\rangle_{\mathrm{w}} = \hat{W}(t)\,|t\rangle$, the time development of $|t\rangle_{\mathrm{w}}$ is given by

$$i\hbar \frac{d}{dt}|t\rangle_{\mathrm{w}} = (\, i\hbar \frac{d\hat{W}}{dt} W^{\dagger} + \hat{W}\hat{H}\hat{W}^{\dagger} \,) \, |t\rangle_{\mathrm{w}} \ , \tag{2-3}$$

and any operator $\hat{A}$ is transformed into $\hat{A}_{\mathrm{w}} = \hat{W}\hat{A}\hat{W}^{\dagger}$. When one chooses $\hat{W}(t) = U^{\dagger}(t,t_0)$, the corresponding transformation is called the *Heisenberg picture*. Let us suppose that the Hamiltonian can be written as $\hat{H} = \hat{H}_0 + \hat{H}'$, in which $\hat{H}_0$ (which is time-independent) would correspond to the Hamiltonian of an isolated system and $\hat{H}'$ (which may be time-dependent) to the interaction of this system with other systems. Applying $\hat{W}(t) = U_0^\dagger(t,t_0) = \exp\{ i\hat{H}_0(t-t_0)/\hbar \}$ to Eq. (2-3), the corresponding transformation $\hat{U}_\mathrm{I} = \hat{U}_0^\dagger \hat{U}$, called the *interaction picture*, leads to

$$i\hbar \frac{d}{dt}|t\rangle_\mathrm{I} = \hat{H}'_\mathrm{I}(t)\,|t\rangle_\mathrm{I} \ , \tag{2-4}$$

where $\hat{H}'_\mathrm{I} = \hat{U}_0^\dagger \hat{H}' \hat{U}$. Equation (2-4) may also be integrated formally in a perturbation series in power of $\hat{H}'_I$:

$$\begin{aligned}
|t\rangle_\mathrm{I} &= \{ \sum_{m=0}^{\infty} (i\hbar)^{-m} \int_{t_0}^{t} dt_1 \cdots \int_{t_0}^{t_{m-1}} dt_m\, \hat{H}'_\mathrm{I}(t_1) \cdots \hat{H}'_\mathrm{I}(t_m) \}\, |t_0\rangle_\mathrm{I} \\
&= \exp_+\{ (i\hbar)^{-1} \int_{t_0}^{t} dt'\, \hat{H}'_\mathrm{I}(t') \}\, |t_0\rangle_\mathrm{I} \\
&= \sum_{m=0}^{\infty} \hat{U}_\mathrm{I}^{(m)}\, |t_0\rangle_\mathrm{I} \\
&= \hat{U}_\mathrm{I}(t,t_0)\, |t_0\rangle_\mathrm{I} .
\end{aligned} \tag{2-5}$$

Let us consider a transition of a molecule caused by the interaction with a radiation field. We assume that the transition takes place between two states of the molecule, for example, an initial state $|t_0\rangle \equiv |i\rangle$ (with $\hat{H}_0|i\rangle = E_i|i\rangle$) to a final state $|f\rangle$. The transition probability per unit time for this transition is defined by[100]

$$W_{if} = |\, \langle f|\, \lim_{\substack{t_0 \to -\infty \\ t \to +\infty}} \frac{d}{dt} \sum_{m=0}^{\infty} \hat{U}_\mathrm{I}^{(m)}(t,t_0)\, |i\rangle\, |^2 . \tag{2-6}$$

There are several general methods for calculating the intensities of the multi-photon processes. One method is to use the time-dependent perturbation series introduced in Eq. (2-5). The Green function and the density matrix methods can also be used to derive general expressions for the intensities. Although one can obtain identical results is all cases, the former procedure is more revealing and is used here.

2.2. Interaction Between a Molecular System and the Radiation Field

In quantum mechanical description of the electromagnetic field, it is concluded that the radiation field formally behaves as a set of linear oscillators, each of unit mass and frequency ω_λ. It then follows that the state of the field is described by the ket $\left| \cdots n_{\lambda\mu} \cdots \right\rangle$ corresponding to the total energy $E = \sum_{\lambda\mu} (n_{\lambda\mu} + 1/2)\hbar\omega_\lambda$, or more commonly $E = \sum_{\lambda\mu} n_{\lambda\mu}\hbar\omega_\lambda$, since the zero of energy is adjusted to remove the $\hbar\omega_\lambda / 2$ term. The numbers, $n_{\lambda\mu}$, give the energy distribution among the frequency ω_λ and the polarization $\vec{\varepsilon}_{\lambda\mu}$ $(\mu = 1, 2)$. Identifying the radiation oscillators as particles with energy $E_\lambda = \hbar\omega_\lambda$, *i.e.* localized energies that are termed *photons*, we can refer $n_{\lambda\mu}$ as the number of photons with the frequency ω_λ and the polarization μ. Each photon carries a linear momentum in the direction of propagation $\vec{k}_\lambda$, and the total linear momentum of the field is given by $\vec{p} = \sum_{\lambda\mu} n_{\lambda\mu}\hbar\vec{k}_\lambda$.

When radiation interacts with matter, some photons may be removed from the radiation field while others may be added. By measuring the number, the direction and the frequency of photons in the incident and the emerging beams, one obtains the energy and the momentum of photons. Based on the measurements of changes in photon energy or momentum, or both, one can deduce corresponding changes in the matter being studied.

In the non-relativistic approximation, the unperturbed Hamiltonian describing a system of charged particles in the presence of an electromagnetic field is expressed as $\hat{H}_0 = \hat{H}_0(\text{system}) + \hat{H}_0(\text{radiation})$, and the interaction Hamiltonian is given by [101,102]

$$H' = \sum_{j=1} \{ (e_j / m_j c)\, \vec{A}(\vec{r}_j,t)\cdot \vec{p}_j + (e_j^2 / 2m_j c^2)\, A^2(\vec{r}_j,t) \}, \tag{2-7}$$

where $\vec{p}_j$ is the momentum of the *j*th particle conjugated to the coordinate $\vec{r}_j$, e_j and m_j , respectively, represent the (algebraic) charge and mass, and $\vec{A}(\vec{r}_j,t)$ is the vector potential of the electromagnetic field at the position $\vec{r}_j$. In the *Schrödinger picture*[100], the vector potential may be expressed as

$$\vec{A}(\vec{r}_j) = \sum_{\lambda\mu} (\hbar / 2\omega_\lambda)^{1/2} \{ \hat{a}_{\lambda\mu} \vec{A}_{\lambda\mu}(\vec{r}_j) + \hat{a}^\dagger_{\lambda\mu} \vec{A}^*_{\lambda\mu}(\vec{r}_j) \}. \tag{2-8}$$

The annihilation $\hat{a}_{\lambda\mu}$ and creation $\hat{a}^\dagger_{\lambda\mu}$ operators describe the absorption and emission of a photon with the frequency ω_λ and the polarization $\vec{\varepsilon}_{\lambda\mu}$, respectively, and they, according to the well known rules, operate on the ket state :

$$\hat{a}_{\lambda\mu} \left| \cdots n_{\lambda\mu} \cdots \right\rangle = (n_{\lambda\mu})^{1/2} \left| \cdots n_{\lambda\mu} - 1 \cdots \right\rangle \tag{2-9a}$$

and

$$\hat{a}_{\lambda\mu} \left| \cdots n_{\lambda\mu} \cdots \right\rangle = (n_{\lambda\mu})^{1/2} \left| \cdots n_{\lambda\mu} - 1 \cdots \right\rangle . \tag{2-9b}$$

By substituting Eq. (2-8) into Eq. (2-7), we may write

$$\hat{H}'_1 = \sum_j (e_j / m_j c) \sum_{\lambda\mu} (\hbar / 2\omega_\lambda)$$

$$\times \{ \hat{a}_{\lambda\mu} \vec{A}_{\lambda\mu}(\vec{r}_j) \cdot \vec{p}_j + \hat{a}^\dagger_{\lambda\mu} \vec{A}_{\lambda\mu}(\vec{r}_j) \cdot \vec{p}_j \} \tag{2-10}$$

and

$$\hat{H}'_2 = \sum_j (e_j^2 / 2m_j c^2) \sum_{\lambda\mu} \sum_{\lambda'\mu'} (\hbar^2 / 4\omega_\lambda \omega_{\lambda'})^{1/2}$$

$$\times \{ \hat{a}_{\lambda\mu} \hat{a}_{\lambda'\mu'} \vec{A}_{\lambda\mu}(\vec{r}_j) \cdot \vec{A}_{\lambda'\mu'}(\vec{r}_j) + \hat{a}_{\lambda\mu} \hat{a}^\dagger_{\lambda'\mu'} \vec{A}_{\lambda\mu}(\vec{r}_j) \cdot \vec{A}^*_{\lambda\mu}(\vec{r}_j)$$

$$+ \hat{a}_{\lambda\mu}\hat{a}^{\dagger}_{\lambda'\mu'}\vec{A}_{\lambda\mu}(\vec{r}_j)\cdot\vec{A}_{\lambda'\mu'}(\vec{r}_j) + \hat{a}^{\dagger}_{\lambda\mu}\hat{a}^{\dagger}_{\lambda'\mu'}\vec{A}^{*}_{\lambda\mu}(\vec{r}_j)\cdot\vec{A}^{*}_{\lambda'\mu'}(\vec{r}_j)\,\}\,. \quad (2\text{-}11)$$

It is obvious that, involving the creation and/or annihilation of more than two photons in the field, the $\hat{H}'_2$ in Eq. (2-11) can not contribute to the simple absorption and emission sequential processes.

We are now ready to insert the interaction Hamiltonian $\hat{H}'$ into the perturbation series to derive the transition probabilities of various multi-photon processes.

2.2.1. One-Photon Processes

Let us consider a molecular system in the state $|i\rangle$ interacting with a single mode of a quantized radiation field that contains $n_{\lambda\mu}$ photons. If the system is excited to the higher state $|f\rangle$ when the field loses a photon, the transition $|I\rangle \equiv |i;\ \cdots\ n_{\lambda\mu}\ \cdots\rangle \rightarrow |F\rangle \equiv |f;\ \cdots\ n_{\lambda\mu}-1\ \cdots\rangle$ is regarded as an *absorption* process. The transition $|I\rangle \equiv |i;\ \cdots\ n_{\lambda\mu}\ \cdots\rangle \rightarrow |f;\ \cdots\ n_{\lambda\mu}+1\ \cdots\rangle$, on the other hand, represents an *emission* process. The first term in Eq. (2-10) is associated with the absorption process involving a photon whose energy, momentum and polarization are $\hbar\omega_\lambda$, $\hbar\vec{k}_\lambda$ and $\vec{\varepsilon}_{\lambda\mu}$, respectively, while the second term in Eq. (2-10) describes the emission of a photon with the same characteristics.

Making use of Eqs. (2-10) and (2-6), to first order in $\hat{H}'$, one can find

$$W^{\text{abs}}_{if} = (\pi/\hbar c^2)(n_{\lambda\mu}/\omega_\lambda)\,|\,\langle f|\sum_j \frac{e_j}{m_j}\vec{A}_{\lambda\mu}(\vec{r}_j)\cdot\vec{p}_j\,|i\rangle\,|^2$$

$$\times\,\delta(\omega_{fi}-\omega_\lambda), \quad (2\text{-}12)$$

which represents the absorption transition probability *per unit time*, and the emission transition probability *per unit time* is expressed as

$$W_{if}^{\text{em.}} = (\pi / \hbar c^2)\{\, (n_{\lambda\mu} + 1) / \omega_\lambda \,\} \,|\, \langle f \,|\, \sum_j \frac{e_j}{m_j} \vec{A}^*_{\lambda\mu}(\vec{r}_j) \cdot \vec{p}_j | i \rangle \,|^2$$

$$\times\, \delta(\omega_{fi} + \omega_\lambda)\,. \tag{2-13}$$

It is observed that the result of the emission process can be separated into two parts: one depending on $n_{\lambda\mu}$, *i.e.*, on the number of photons present, corresponding to the transition probability per unit time of *induced* or *stimulated* emission, and the other independent of $n_{\lambda\mu}$ giving the transition probability per unit time of *spontaneous* emission. The spontaneous transitions can be considered as a process induced through fluctuation of a radiation field due to the zero point energy of the field or the electromagnetic vacuum.

In the Fourier decomposition of the vector potential (Eq. (2-8)), the *plane waves* or *spherical waves* may be chosen for the amplitudes $\vec{A}_{\lambda\mu}(\vec{r}_j)$ to describe the transition probabilities associated with the multi-pole moments of the radiating system [98]. For plane waves in the radiation cavity with volume *V*, for instance, one can easily find $\vec{A}_{\lambda\mu}(\vec{r}_j) = (4\pi c^2 / V)^{1/2} \vec{\varepsilon}_{\lambda\mu} \exp(\, i\vec{k}_\lambda \cdot \vec{r}_j)$. If the dimension of the system is small compared with the wavelength of radiation ($\lambda >> 10^{-8}\,\text{cm}$ for a molecular system), one may then approximate $\exp(\, i\vec{k}_\lambda \cdot \vec{r}_j) \approx 1$, which is called the *electric dipole* approximation.

Notice that in stimulated emission the initial state is $|I\rangle \equiv \big|\, i;\, \cdots\, n_{\lambda\mu}\, \cdots \big\rangle$ and the final state $|F\rangle \equiv \big| f;\, \cdots\, n_{\lambda\mu} + 1 \cdots \big\rangle$ is one in which a photon has been added to the same mode of radiation. This means that a photon emitted by a stimulated emission process has the same polarization, phase and direction of propagation as the incident photons. Stimulated emission is said, therefore, to be a *coherent* process[103]. For the spontaneous emission, however, the incident photons are not required for the process to occur, and the emitted photons are not constrained to any particular mode. Their phases, polarizations and directions of propagation may range over all possible values. Spontaneous emission, therefore, is said to be an *incoherent* process. As a result, the transition probability for the spontaneous emission must be summed over all modes and polarizations. Notice that the number of modes with frequencies between ω_λ and $\omega_\lambda + d\omega_\lambda$ propagating within the solid angle $d\Omega$ in the direction of $\vec{k}_\lambda$ is given by $\{ V/(2\pi c)^3 \} \omega_\lambda^2 d\omega_\lambda d\Omega$.

2.2.2. Two-photon Processes

Ordinary absorption and emission are one-photon processes, since in every transition the field loses or gains a single photon while the molecular system undergoes a corresponding transition of state to conserve the total energy (molecule plus field). If there exists no state $|f\rangle$ of a molecule such that this conservation holds, the molecules will not absorb a photon through this simple optical process, but the interaction may still take place via the higher order terms as seen in the time-dependent perturbation series in Eq. (2-5). In the case in which a system has an inversion symmetry (*e.g.*, an atom), a one-photon (dipole) transition requires initial and final states to have opposite parity. For a second-order transition, the selection rule requires two states to have the same parity. As a consequence, because of symmetry restrictions, the states that can not be populated by one-photon transitions, may become accessible by second-order transitions.

In a one-photon process, the term $\hat{H}_2'$ in Eq. (2-11) can be neglected. It can be seen that $\hat{H}_2'$ contains the terms $\hat{a}_{\lambda\mu}\hat{a}_{\lambda'\mu'}$, $\hat{a}_{\lambda\mu}\hat{a}^\dagger_{\lambda'\mu'}$, $\hat{a}^\dagger_{\lambda\mu}\hat{a}_{\lambda'\mu'}$ and $\hat{a}^\dagger_{\lambda\mu}\hat{a}^\dagger_{\lambda'\mu'}$. Corresponding to the annihilation of two photons or two-photon absorption, the term $\hat{a}_{\lambda\mu}\hat{a}_{\lambda'\mu'}$ transforms a field state $|\cdots n_{\lambda\mu} \cdots n_{\lambda'\mu'} \cdots\rangle$ into a new state $|\cdots n_{\lambda\mu} - 1 \cdots n_{\lambda'\mu'} - 1 \cdots\rangle$. The term $\hat{a}_{\lambda\mu}\hat{a}^\dagger_{\lambda'\mu'}$ transforms the same state into another state $|\cdots n_{\lambda\mu} - 1 \cdots n_{\lambda'\mu'} + 1 \cdots\rangle$, thus, this process is associated with both absorption and emission of a photon of type $\lambda\mu$ and $\lambda'\mu'$, respectively. The other two terms are also associated with two photon processes as well. In the case in which two photons are involved, as seen above, the interaction term $\hat{H}'_2$ must be considered. For consistency, the contribution from $\hat{H}'_2$ to the transition probability is to be taken to first order in $\hat{H}'$, while the term from $\hat{H}'_1$ for such processes is taken to second order in $\hat{H}'$. In the following, we shall briefly derive an expression of the transition probability per unit time for the two-photon processes including two-photon absorption, two-photon emission, and scattering.

***Two-photon absorption*:**

Here let us consider an initial state of the total system that is given by $|I\rangle \equiv |\, i;\, n_{\lambda\mu}, n_{\lambda'\mu'}\,\rangle$ whose energy is expressed as $E_I = E_i + n_{\lambda\mu}\hbar\omega_\lambda + n_{\lambda'\mu'}\hbar\omega_{\lambda'}$. A final state of the total system is given by $|F\rangle \equiv |\, f;\, n_{\lambda\mu}-1, n_{\lambda'\mu'}-1\,\rangle$ with the energy $E_F = E_f + (n_{\lambda\mu}-1)\,\hbar\omega_\lambda + (n_{\lambda'\mu'}-1)\,\hbar\omega_{\lambda'}$. The contribution from $\hat{H}'_2$ to the total matrix element of the first order can be easily calculated:

$$\langle F|\hat{U}_{\mathrm{I}}^{(1)}|I\rangle = (-\pi i/2c^2)(n_{\lambda\mu}n_{\lambda'\mu'}/\omega_\lambda\omega_{\lambda'})^{1/2}$$

$$\times \langle f|\sum_j (e_j^2/m_j)\,\vec{A}_{\lambda\mu}(\vec{r}_j)\cdot\vec{A}_{\lambda'\mu'}(\vec{r}_j)\,|i\rangle$$

$$\times\,\delta(\omega_{fi}-\omega_\lambda-\omega_{\lambda'}) \quad . \qquad \text{(2-14)}$$

The calculation of the matrix element of $\langle F|\,\hat{U}_{\mathrm{I}}^{(2)}\,|I\rangle$, however, requires intermediate states $|K\rangle$ that are connected to the initial and final states by one-photon processes. The intermediate states have two possible forms $|k;\, n_{\lambda\mu}, n_{\lambda'\mu'}-1\rangle$ with the energy $E_K = E_k + n_{\lambda\mu}\hbar\omega_\lambda + (n_{\lambda'\mu'}-1)\,\hbar\omega_{\lambda'}$ and $|k;\, n_{\lambda\mu}-1, n_{\lambda'\mu'}\rangle$ with the energy $E_K = E_k + (n_{\lambda\mu}-1)\,\hbar\omega_\lambda + n_{\lambda'\mu'}\hbar\omega_{\lambda'}$.

$$\langle F|\hat{U}_{\mathrm{I}}^{(2)}|I\rangle = (\pi i / c^2)\,(n_{\lambda\mu} n_{\lambda'\mu'} / \omega_{\lambda}\omega_{\lambda'})^{1/2}$$

$$\times \sum_{k \neq i} \{ \frac{\langle f|\sum_j (e_j / m_j)\,\vec{A}_{\lambda'\mu'}(\vec{r}_j)\cdot\vec{p}_j\,|k\rangle\langle k|\sum_{j'} (e_{j'} / m_{j'})\,\vec{A}_{\lambda\mu}(\vec{r}_{j'})\cdot\vec{p}_{j'}|i\rangle}{\hbar(\omega_{ki} - \omega_{\lambda} - i\Gamma_k)}$$

$$+ \frac{\langle f|\sum_j (e_j / m_j)\,\vec{A}_{\lambda\mu}(\vec{r}_j)\cdot\vec{p}_j\,|\mathrm{k}\rangle\langle k|\sum_{j'} (e_{j'} / m_{j'})\,\vec{A}_{\lambda'\mu'}(\vec{r}_{j'})\cdot\vec{p}_{j'}\,|i\rangle}{\hbar(\omega_{ki} - \omega_{\lambda'} - i\Gamma_k)} \}$$

$$\times \delta(\omega_{fi} - \omega_{\lambda} - \omega_{\lambda'})\ . \tag{2-15}$$

Adding the two resultant matrix elements, the transition probability per unit time for the two-photon absorption becomes

$$W_{if} = (\pi / 2c^4)(n_{\lambda\mu} n_{\lambda'\mu'} / \omega_{\lambda}\omega_{\lambda'})$$

$$\times \mid \sum_{k \neq i} \{ \frac{\langle f|\sum_j (e_j / m_j)\,\vec{A}_{\lambda'\mu'}(\vec{r}_j)\cdot\vec{p}_j|k\rangle\langle k|\sum_{j'} (e_{j'} / m_{j'})\,\vec{A}_{\lambda\mu}(\vec{r}_{j'})\cdot\vec{p}_{j'}\,|i\rangle}{\hbar(\omega_{ki} - \omega_{\lambda} - i\Gamma_k)}$$

$$+ \frac{\langle f|\sum_j (e_j / m_j)\,\vec{A}_{\lambda\mu}(\vec{r}_j)\cdot\vec{p}_j\,|k\rangle\langle k|\sum_{j'} (e_{j'} / m_{j'})\,\vec{A}_{\lambda'\mu'}(\vec{r}_{j'})\cdot\vec{p}_{j'}\,|i\rangle}{\hbar(\omega_{ki} - \omega_{\lambda'} - i\Gamma_k)} \}$$

$$- \frac{1}{2}\langle f|\sum_j (e_j^2 / m_j)\,\vec{A}_{\lambda\mu}(\vec{r}_j)\cdot\vec{A}_{\lambda'\mu'}(\vec{r}_j)\,|i\rangle\,|^2\,\delta(\omega_{fi} - \omega_{\lambda} - \omega_{\lambda'})\ . \tag{2-16}$$

To avoid singularities at the resonance conditions $\omega_{ki} \approx \omega_{\lambda'}$ or $\omega_{ki} \approx \omega_{\lambda}$, the damping instant Γ_k has been introduced. Notice that the contribution from $\hat{H}_2'$ to transition probability vanishes in the dipole approximation.

Two-photon emission:

In this case, initial, final states, respectively, are $|I\rangle \equiv \left| i;\, n_{\lambda\mu}, n_{\lambda'\mu'} \right\rangle$, $|F\rangle \equiv \left| f;\, n_{\lambda\mu}+1, n_{\lambda'\mu'}+1 \right\rangle$ with the energies $E_I = E_i + n_{\lambda\mu}\hbar\omega_\lambda + n_{\lambda'\mu'}\hbar\omega_{\lambda'}$ and $E_F = E_f + (n_{\lambda\mu}+1)\,\hbar\omega_\lambda + (n_{\lambda'\mu'}+1)\,\hbar\omega_{\lambda'}$. Two intermediate states are assumed to be $|K_1\rangle = \left| k, n_{\lambda\mu}, n_{\lambda'\mu'}+1 \right\rangle$ and $|K_2\rangle = \left| k, n_{\lambda\mu}+1, n_{\lambda'\mu'} \right\rangle$ whose energies are $E_K = E_k + n_{\lambda\mu}\hbar\omega_\lambda + (n_{\lambda'\mu'}+1)\,\hbar\omega_{\lambda'}$ and $E_K = E_k + (n_{\lambda\mu}+1)\hbar\omega_\lambda + n_{\lambda'\mu'}\,\hbar\omega_{\lambda'}$, respectively. The transition probability per unit time is calculated in a similar fashion as seen in the two-photon absorption case, and the result is given by

$$W_{if} = (\pi/2c^4)\{\, (n_{\lambda\mu}+1)(n_{\lambda'\mu'}+1)/\omega_\lambda\omega_{\lambda'} \,\}\,\Bigg|$$

$$\times \sum_{k\neq i} \Bigg\{ \frac{\langle f|\sum_j (e_j/m_j)\vec{A}^*_{\lambda'\mu'}(\vec{r}_j)\cdot\vec{p}_j\,|k\rangle\langle k|\sum_{j'} (e_{j'}/m_{j'})\,\vec{A}^*_{\lambda\mu}(\vec{r}_{j'})\cdot\vec{p}_{j'}\,|i\rangle}{\hbar(\omega_{ki}+\omega_\lambda - i\Gamma_k)}$$

$$+ \frac{\langle f|\sum_j (e_j/m_j)\,\vec{A}^*_{\lambda\mu}(\vec{r}_j)\cdot\vec{p}_j\,|k\rangle\langle \mathrm{k}|\sum_{j'} (e_{j'}/m_{j'})\,\vec{A}^*_{\lambda'\mu'}(\vec{r}_{j'})\cdot\vec{p}_{j'}\,|i\rangle}{\hbar(\omega_{ki}+\omega_{\lambda'} - i\Gamma_k)} \Bigg\}$$

$$- \frac{1}{2}\langle f|\sum_j (e_j^2/m_j)\vec{A}^*_{\lambda\mu}(\vec{r}_j)\cdot\vec{A}^*_{\lambda'\mu'}(\vec{r}_j)\,|i\rangle\,\Bigg|^2 \delta(\omega_{fi}+\omega_\lambda+\omega_{\lambda'})\,. \quad (2\text{-}17)$$

Scattering process:

The general scattering process involves a conversion of the frequency ω_λ, energy $\hbar\omega_\lambda$, momentum $\hbar\vec{k}_\lambda$ and polarization $\vec{\varepsilon}_{\lambda\mu}$ of an incident photon to the frequency $\omega_{\lambda'}$, energy $\hbar\omega_{\lambda'}$, momentum $\hbar\vec{k}_{\lambda'}$ and polarization $\vec{\varepsilon}_{\lambda'\mu'}$ of a scattered photon while the molecular system undergoes a transition from one of its eigenstates to another. This two photon process can be well described with a basis set consisting of initial, final, and intermediate states: $|I\rangle \equiv \left| i;\, n_{\lambda\mu}, n_{\lambda'\mu'} \right\rangle$,

$|F\rangle \equiv |f;\ n_{\lambda\mu}-1, n_{\lambda'\mu'}+1\rangle$, $|K\rangle \equiv |k;\ n_{\lambda\mu}, n_{\lambda'\mu'}+1\rangle$ and $|K\rangle \equiv |k;\ n_{\lambda\mu}-1, n_{\lambda'\mu'}\rangle$, respectively. The energies of these states are given in a similar fashion as already seen above. As mentioned before, the term $\hat{H}'_2$ contributes to scattering in its first order of the perturbation series and the term $\hat{H}'_1$ in its second order of the same series. The transition probability per unit time this scattering becomes

$$W_{if} = (\pi/2c^4)\{\, n_{\lambda\mu}(n_{\lambda'\mu'}+1)/\omega_\lambda\omega_{\lambda'}\,\}\,|$$

$$\times \sum_{k\neq i}\{\frac{\langle f|\sum_j (e_j/m_j)\vec{A}^*_{\lambda'\mu'}(\vec{r}_j)\cdot\vec{p}_j\,|k\rangle\langle k|\sum_{j'}(e_{j'}/m_{j'})\vec{A}_{\lambda\mu}(\vec{r}_{j'})\cdot\vec{p}_{j''}\,|i\rangle}{\hbar(\omega_\lambda-\omega_{ki}+i\Gamma_k)}$$

$$-\frac{\langle f|\sum_j (e_j/m_j)\vec{A}_{\lambda\mu}(\vec{r}_j)\cdot\vec{p}_j\,|k\rangle\langle \mathrm{k}|\sum_{j'}(e_{j'}/m_{j'})\vec{A}^*_{\lambda'\mu'}(\vec{r}_{j'})\cdot\vec{p}_{j'}\,|\mathrm{i}\rangle}{\hbar(\omega_{\lambda'}+\omega_{ki}-i\Gamma_k)}\}$$

$$+\langle f|\sum_j (e_j^2/m_j)\,\vec{A}_{\lambda\mu}(\vec{r}_j)\cdot\vec{A}^*_{\lambda'\mu'}(\vec{r}_j)\,|i\rangle\,|^2\,\delta(\omega_{fi}-\omega_\lambda+\omega_{\lambda'})\ . \qquad (2\text{-}18)$$

We now apply the dipole approximation to the scattering process. In this approximation, $\vec{A}_{\lambda\mu}(\vec{r}_j) \approx (4\pi c^2/V)^{1/2}\vec{\varepsilon}_{\lambda\mu}$ and the last term in Eq. (2-18) is the proportional to $\delta_{if}\vec{\varepsilon}_{\lambda\mu}\cdot\vec{\varepsilon}_{\lambda'\mu'}$, so that, in the dipole approximation, the term $\vec{A}\cdot\vec{A}$ interacts only with a molecular state once and does not change the molecular state, *i.e.*, interaction that is important in Rayleigh scattering. In Appendix I, we have derived an expression for $\delta_{if}\vec{\varepsilon}_{\lambda\mu}\cdot\vec{\varepsilon}_{\lambda'\mu'}$. Substituting Eq. (A-6) into Eq. (2-18) yields the transition probability per unit time for the spontaneous scattering within the electric dipole approximation:

$$W_{if} = [(2\pi)^3/\hbar^2V^2][n_{\lambda\mu}(n_{\lambda'\mu'}+1)](\omega_\lambda\omega_{\lambda'})$$

$$\times \left| \sum_{k \neq i} \left[\frac{(\vec{\varepsilon}_{\lambda\mu} \cdot \vec{M}_{ki})(\vec{\varepsilon}_{\lambda'\mu'} \cdot \vec{M}_{fk})}{\omega_\lambda - \omega_{ki} + i\Gamma_k} + \frac{(\vec{\varepsilon}_{\lambda\mu} \cdot \vec{M}_{fk})(\vec{\varepsilon}_{\lambda'\mu'} \cdot \vec{M}_{ki})}{-\omega_\lambda + \omega_{fk} + i\Gamma_k} \right] \right|^2$$

$$\times \delta(\omega_{fi} - \omega_\lambda + \omega_{\lambda'}) \quad , \qquad (2\text{-}19)$$

where $\vec{M}_{ki}$ and $\vec{M}_{fk}$ are the transition dipole moments and we have used the fact that $\omega_{fk} + \omega_{ki} = \omega_\lambda - \omega_{\lambda'}$. Multiplying Eq. (2-19) by the density of the final states of scattered photons $[V/(2\pi c)^3]\,\omega_{\lambda'}^2 d\omega_{\lambda'} d\Omega$ leads to the total number of the scattered photons propagating in a interval of solid angle $d\Omega$ around $\vec{k}_{\lambda'}$ per unit time in a frequency interval $d\omega_{\lambda'}$.

The scattering also may be expressed in term of a differential scattering cross section. The differential scattering cross section is defined by $\dfrac{\text{the number of scattered photons within the solid angle } d\Omega \text{ per unit time}}{\text{the number of incident photons crossing unit area } (n_{\lambda\mu} c/V) \text{ per unit time}}$. We then obtain the differential scattering cross section:

$$\frac{d^2\sigma}{d\omega_{\lambda'} d\Omega} = (1/\hbar^2 c^4)(n_{\lambda'\mu'} + 1)(\omega_\lambda \omega_{\lambda'}^3)$$

$$\times \left| \sum_{k \neq i} \left\{ \frac{(\vec{\varepsilon}_{\lambda\mu} \cdot \vec{M}_{ki})(\vec{\varepsilon}_{\lambda'\mu'} \cdot \vec{M}_{fk})}{\omega_\lambda - \omega_{ki} + i\Gamma_k} + \frac{(\vec{\varepsilon}_{\lambda\mu} \cdot \vec{M}_{fk})(\vec{\varepsilon}_{\lambda'\mu'} \cdot \vec{M}_{ki})}{-\omega_\lambda + \omega_{fk} + i\Gamma_k} \right\} \right|^2$$

$$\times \delta(\omega_{fi} - \omega_\lambda + \omega_{\lambda'}) \ . \qquad (2\text{-}20)$$

The factor $(n_{\lambda'\mu'} + 1)$ in Eq. (2-20) has an interpretation similar to that in one-photon emission. If $n_{\lambda'\mu'} = 0$ in the initial state $|I\rangle$, we have *spontaneous scattering* whose probability per unit time, according to Eq. (2-19), is proportional to the number of incident photons $n_{\lambda\mu}$. When $n_{\lambda'\mu'} \neq 0$, we have, in addition to spontaneous scattering,, a term proportional to the product $n_{\lambda\mu} n_{\lambda'\mu'}$. The transition probability associated with the latter requires the presence of scattered photons; that is, the scattered photons stimulate the production of other scattered photons. For this reason, the process is known as *stimulated scattering*. At ordinary intensities, stimulated scattering can be neglected, because $n_{\lambda'\mu'}$ is small.

The state with energy $E_i + \hbar\omega_\lambda$ is called a *non-resonant state* if it does not coincide with an eigen-state of the scattering system, similarly for states with the energy $E_f - \hbar\omega_\lambda$. Such states are said to be *virtual* states since they are not observable. When the frequency of the incident radiation is far removed from any resonance, the effects of damping are small and generally may be ignored since none of denominators in Eq. (2-20) are close to zero. However, when the incident frequency is in the vicinity of an absorption band, the energy denominators for this band become so small that their effects dominate the sum in Eq. (2-20). It is then no longer possible to ignore damping.

When the frequency of the incident light is inside a molecular absorption band, the scattering phenomenon is called *resonance Raman* scattering, which shares the common feature with fluorescence, that is, in both cases, the irradiation occurs within an absorption band of the molecular system. In such cases, the distinction between fluorescence and resonance Raman scattering becomes less sharp. When the incident light is pulsed, by examining the line shape of the emitted radiation, it is often possible to distinguish between the two types of phenomenon. In fluorescence, the excited state created through the absorption process exhibits a delay associated with its lifetime. Therefore, after the pulse is turned off the emitted radiation will contain an exponentially decaying tail, characteristic of the excited state. In resonance Raman scattering, the relaxation of the excited state is not observed and the scattered light mainly follows the time-dependent of the incident light; *i.e.*, emission stops when the interaction between the molecular system and the pulse of incident radiation ends.

3. Vibronic Theory

3.1. Construction of Molecular Hamiltonian

A stationary state of a polyatomic molecule is described by eigen-states and eigen-values of the complete molecular Hamiltonian $\hat{H}_0$, where $\hat{H}_0$ contains (in the non-relativistic approximation) terms depending upon the electronic and nuclear variables. The electronic transitions of molecules depend on the characteristics of the nuclear as well as electronic motions. The motions of the nuclei can be divided into three kinds: (1) the translational motion of the molecule as a whole, which is discarded because of no spectroscopic interest; (2) the rotation of the molecule; and (3) the vibrations of the nuclei within the molecule. The molecular motions are not independent of each other, and the interactions between them lead to the complexity of their dynamics and optical spectra. In addition, the existence of several sources of angular momenta and a variety of ways of coupling them together may make such complexity even puzzled. It is urgently necessary to obtain some useful information on molecular systems from analyzing such complex spectra.

Let us here consider a molecule containing N nuclei and n electrons. In this case, the molecule has 3(N+n) independent Cartesian coordinates with respect to a laboratory fixed frame (LF frame). In order to separate translational, rotational, vibrational, and electronic coordinates, the set of 3 (N+n) LF Cartesian coordinates is transformed to a set of generalized coordinates, called molecular coordinates, defined by subsets {3 translational}, {3 or 2 rotational}, {3N-6(or 5) vibrational}, and {3 n electronic} variables. The corresponding transformation in the Hamiltonian operator $\hat{H}_0$ for energy of the molecule gives an expression in which the energy of each separate motion and the interactions between each two motions are very revealing. Furthermore, the Born-Oppenhiemer approximation which approximates the nuclear and electronic motions independent, becomes more clear in the transformed Hamiltonian operator.

By choosing a molecular frame located at the center of mass of nuclei which satisfies the Eckart conditions[104,105] and factorizes the translational energy, we can write $\hat{H}_0$ in the form[106,107]

$$\hat{H}_0 = \left(-\hbar^2 \sum_s \nabla_s^2 / 2m - \hbar^2 \sum_{s,t} \vec{\nabla}_s \cdot \vec{\nabla}_t / 2m_n \right) + \left(-\frac{1}{2}\hbar^2 \sum_k \partial^2 / \partial Q_k^2 \right)$$

$$+ \frac{1}{2} \sum_{ij} \mu_{ij} (\hat{M}_i - \hat{G}_i - \hat{L}_i)(\hat{M}_j - \hat{G}_j - \hat{L}_j) + \hat{U}(Q) + \hat{V}(\vec{r}, Q) , \qquad (3\text{-}1)$$

where $\vec{r} \equiv \{ r_{is};\ i = x,y,z,\ s = 1,2,\cdots,n \}$ denotes the set of electronic variables, $Q \equiv \{ Q_k;\ k = 1,2,\cdots,3N-6 \}$ is the set of vibrational variables of the nuclei , m is the mass of an electron, and m_n is the total mass of nuclei. The first two terms in Eq. (3-1) represent the electronic and vibrational kinetic energy of the molecule, respectively. The next two terms are the contribution of rotational kinetic energy and the energy of interactions among the various types of molecular motions to the total kinetic energy of the molecule (the operators $\hat{J}_i, \hat{G}_i$, and $\hat{L}_i$ represent the i component (with respect to the molecular frame) of total angular momentum, so-called vibrational angular momentum, and electronic angular momentum, respectively, and μ_{ij} is the inverse of the effective inertia tensor). Finally, the last term is the potential energy of the entire system of electrons and nuclei.

In the treatment of the electronic states of large molecules one usually neglects details concerning the rotational motion and concentrates on the dynamics of the vibrational and the electronic motions. With this approximation, and with the additional neglect of the mass polarization term in the electronic kinetic energy (*i.e.*, $-\hbar^2 \sum_{s,t} \vec{\nabla}_s \cdot \vec{\nabla}_t / m_n$), the resulting vibronic Hamiltonian can be written as

$$\hat{H}_0(\vec{r}, Q) = -\frac{\hbar^2}{2} \sum_{s=1}^{n} \nabla_s^2 / m - \frac{\hbar^2}{2} \sum_{k=1}^{3N-6} \partial^2 / \partial Q_k^2 + V(\vec{r}, Q) . \qquad (3\text{-}2)$$

In spite of the approximations already made, the exact vibronic eigen-states can not be obtained, because the potential $V(\vec{r},Q)$ does not allow a separation of the variables. However, from early spectroscopic evidence, it emerged that the molecular energy levels could be very successfully described in terms of a specific hierarchy. Rotational levels belong to particular vibrational levels which, in turn, are associated with specific electronic states. Thus, the early experimental observations suggested already that the electronic and nuclear motions could, in fact, be treated separately. The theoretical justification for these observations were made in a well known paper of Born and Oppenhiemer[108,109]. Here we are not concerned with details of the adiabatic approximation, but with those aspects which are related to our discussions in the following sections.

The adiabatic approximation present the vibronic eigen-states as a product of an electronic eigen-state $\phi_m(\vec{r},Q)$, depending parametrically on the set of the vibrational variables Q, and a vibrational eigen-state $\chi_{mj}(Q)$. The vibronic Hamiltonian $\hat{H}_0(\vec{r},Q)$ can be expanded in the adiabatic basis set as follows

$$\hat{H}_0 = \sum_{nk,mj} |\phi_n\chi_{nk}\rangle H_{nk,mj}\langle\phi_m\chi_{mj}| \, , \tag{3-3}$$

where

$$H_{nk,mj} = \delta_{nm}\varepsilon_{mj}\langle\chi_{nk}|\chi_{mj}\rangle - \frac{\hbar^2}{2}\sum_i \int dQ\,\chi^*_{nk}\,[\,\int d^3\vec{r}\,\phi^*_n\partial^2\phi_m / \partial Q_i^2\,]\,\chi_{mj}$$

$$-\hbar^2\sum_i \int dQ\,\chi^*_{nk}\,[\,\int d^3\vec{r}\,\phi^*_n\partial\phi_m / \partial Q_i\,]\,\partial\chi_{mj} / \partial Q_i \tag{3-4}$$

If the diagonal matrix elements are equal or nearly equal $|H_{mj,mj}| \approx |H_{nk,nk}|$, that is, there is a degeneracy or near degeneracy, then there is a coupling between vibronic states through the off-diagonal matrix elements. This leads to the Jahn-Teller and the pseudo-Jahn-Teller effects[110-113].

In adiabatic approximation of Born-Huang it is assumed that $| H_{mj,mj} - H_{nk,nk} | >> | H_{nk,mj} |^2$, and therefore, the off-diagonal elements $| H_{nk,mj} |$ is neglected. In adiabatic approximation of Born-Oppenheimer it is further assumed that the terms of the type $\langle \phi_n | \partial^2 / \partial Q^2 | \phi_n \rangle$ are negligible [114-116]. This mathematical formalism is in accord with physical description that the electronic energies provides the potential energy for the nuclear motion and that the vibrational motion takes place on the one electronic potential energy surface. When the neglect of the off-diagonal elements in the adiabatic bases, is not justified, poor vibronic energies and wavefunctions are obtained. In this case, we may take $H_{nk,mj}$ as a small perturbation and approximate the (non-degenerate) vibronic states $|\psi_{mj}\rangle$ by the expansion

$$|\psi_{mj}\rangle = |\phi_m \chi_{mj}\rangle + \sum_{nk} (\varepsilon_{mj} - \varepsilon_{nk})^{-1} H_{nk,mj} |\phi_n \chi_{nk}\rangle + \cdots \quad . \tag{3-5}$$

It may be proved that the off-diagonal matrix elements can be safely approximated by

$$H_{nk,mj} = -\hbar^2 \sum_i \int dQ\, \chi^*_{nk} \,[\, \int d^3 \vec{r}\; \phi^*_n \partial\phi_m / \partial Q_i \,]\, \partial\chi_{mj} / \partial Q_i \tag{3-6}$$

3.2. Vibronic Matrix Elements

The transition matrix element (which determines the strength of an optical transition) for the vibronic transition $av'' \leftrightarrow bv'$ is given by*

$$\langle av'' | \hat{\vec{M}} | bv' \rangle = \langle \chi_{av''} | \vec{\mu}_{ab} | \chi_{bv'} \rangle$$

$$+\sum_{cv} \{ (\varepsilon_{bv'} - \varepsilon_{cv})^{-1} H_{cv,bv'} \langle \chi_{av''} | \vec{\mu}_{ac} | \chi_{cv} \rangle$$

* The lower state quantities are labeled by a double prime and the upper state quantities by a single prime.

$$+(\varepsilon_{av''}-\varepsilon_{cv})^{-1}H_{av'',cv}\langle\chi_{cv}\,|\,\vec{\mu}_{cb}|\chi_{bv'}\rangle\,\}\,+\,\cdots\quad , \tag{3-7}$$

where Eq. (3-5) has been used. Because of the parametric dependence of the electronic wavefunctions on the nuclear variables, the electronic transition moment

$$\vec{\mu}_{ab}(Q)=\int d^3\vec{r}\;\phi_a^*(\vec{r},Q'')\hat{\vec{M}}(\vec{r})\phi_b(\vec{r},Q')\quad , \tag{3-8}$$

depends on Q. Here $\hat{\vec{M}}(\vec{r})$ is the electronic dipole moment. We now introduce two additional approximations. First, the vibrational spacing is usually small compared with the energy gap between coupled electronic states, so that $\varepsilon_{bv'}-\varepsilon_{cv}\approx E_b-E_c$. Secondly, the electronic matrix element as a whole is expanded in the vicinity of the equilibrium configuration of one of the two electronic states (say, the ground electronic state)

$$\vec{\mu}_{ab}(Q)=\vec{\mu}_{ab}(0)+\sum_k\,(\partial\vec{\mu}_{ab}/\partial Q_k'')_0Q_k''+\cdots \tag{3-9}$$

and

$$H_{cv,bv'}=-\hbar^2\sum_k\,\langle\phi_c|\partial/\partial Q_k''|\phi_b\rangle_0\langle v|\partial/\partial Q_k''|v'\rangle+\cdots\quad . \tag{3-10}$$

Substituting of Eqs. (3-9) and (3-10) in Eq. (3-7) leads to the following expression for the transition matrix element

$$\langle av''|\hat{\vec{M}}|bv'\rangle=\vec{\mu}_{ab}(0)\langle v''|v'\rangle$$

$$+\sum_{kc}\,\{\,\vec{\mu}_{ac}(0)\langle\phi_c|\partial/\partial Q_k''|\phi_b\rangle_0-\vec{\mu}_{cb}(0)\langle\phi_a|\partial/\partial Q_k''|\phi_c\rangle_0\,\}$$

$$\times\,\langle v''|Q_k''|v'\rangle+\hbar^2\sum_{kc}\,\{\,(E_c-E_b)^{-1}\,\vec{\mu}_{ac}(0)\langle\phi_c|\partial/\partial Q_k''|\phi_b\rangle_0$$

$$\times (E_c - E_a)^{-1}\ \vec{\mu}_{ab}(0)\langle\phi_a|\partial/\partial Q_k''|\phi_c\rangle_0\ \}\ \langle v''|\partial/\partial Q_k''|v'\rangle\ +\ \cdots\ . \qquad (3\text{-}11)$$

The approximation embodied in

$$\langle av''|\hat{\vec{M}}|bv'\rangle \approx \vec{\mu}_{ab}(0)\langle v''|v'\rangle\ , \qquad (3\text{-}12)$$

is known in electronic spectroscopy as the Condon approximation. This leads to the usual Franck-Condon description of the intensity distribution within a band in terms of squares of vibrational overlap integrals.

The first term in Eq. (3-11) is called the allowed component. If this term vanishes (because of symmetry), the transition is called forbidden. Forbidden electronic transitions can usually gain intensity by borrowing from an allowed transition via two mechanisms. The first is based on deviations from the adiabatic approximations, i.e., on the fact that electronic and nuclear motions are not exactly separable. The second arises when one makes the additional approximation that the electronic transition moment is independent of the nuclear variables (the Condon approximation).

The conventional Herzberg-Teller (HT) theory[117] of intensity borrowing is based on deviations from the Condon approximation. The HT induced transitions moment is given by the second term of Eq. (3-11)

$$\langle av''|\hat{\vec{M}}|bv'\rangle_{HT} = \sum_k\ (\partial\vec{\mu}_{ab}/\partial Q_v'')_0\langle v''|Q_k''|v'\rangle$$

$$= \sum_{kc}\ \{\ \vec{\mu}_{ac}(0)\langle\phi_c|\partial/\partial Q_k''|\phi_b\rangle_0 - \vec{\mu}_{cb}(0)\langle\phi_a|\partial/\partial Q_k''|\phi_c\rangle_0\ \}\ \langle v''|Q_k''|v'\rangle$$

$$= \sum_{kc}\ \{\ \vec{\mu}_{ac}(0)H_{cb}^k - \vec{\mu}_{cb}(0)H_{ac}^k\ \}\ \langle v''|Q_k''|v'\rangle\ , \qquad (3\text{-}13)$$

where $H_{ab}^{k} = \langle \phi_a | \partial / \partial Q_k'' | \phi_b \rangle_0$. Notice that one may neglect one of the two eigen-state derivatives relative to the other, namely, that of the eigen-state for which the electronic energy gap is largest**.

The Born-Oppenheimer (BO) theory of intensity borrowing predicts[118,119] that the induced transition moment is given by the third term of Eq. (3-11).

$$\langle av'' | \hat{\vec{M}} | bv' \rangle_{BO} = \hbar^2 \sum_{kc} \{ (E_c - E_b)^{-1} \vec{\mu}_{ac}(0) \langle \phi_c | \partial / \partial Q_k'' | \phi_b \rangle_0$$

$$+ (E_c - E_a)^{-1} \vec{\mu}_{cb}(0) \langle \phi_a | \partial / \partial Q_k'' | \phi_c \rangle_0 \} \langle v'' | \partial / \partial Q'' | v' \rangle$$

$$= \sum_{k} \{ \vec{\mu}_{ac}(0) B_{cb}^{k} + \vec{\mu}_{cb}(0) B_{ac}^{k} \} \langle v'' | \partial / \partial Q_k'' | v' \rangle , \qquad (3\text{-}14)$$

where $B_{ac}^{k} = (E_c - E_a)^{-1} \langle \phi_a | \partial / \partial Q_k'' | \phi_c \rangle_0$. As before, we may assume that one energy denominator to be so much larger that the other, that the corresponding term may be neglected.

We have been so far concerned with the dependence of the electronic transition moment on Q . Now we consider the change in molecular dimensions upon electronic excitation.

A change in the equilibrium geometry of a molecule resulting in alterations of bond lengths and angles may change the magnitude and the direction of some or all the normal coordinates. This is easily proved. Suppose $\vec{R}$ is a vector whose components are the changes in equilibrium position from the initial to the final states, $\vec{R} = (\vec{r}_0)'' - (\vec{r}_0)'$. In terms of the mass-weighted Cartesian displacement coordinates $\vec{\rho}''$ and $\vec{\rho}'$, this becomes $\vec{\rho}' = \vec{\rho}'' + \vec{R}$ (where $\vec{R}$ is now mass-weighted). By employing the transformation from Cartesian to normal coordinates $\vec{\rho}'' = L''Q'', \vec{\rho}' = L'Q'$, we obtain

---

** Notice that $\langle \phi_a | \partial / \partial Q_k'' | \phi_b \rangle = (E_b - E_a)^{-1} \langle \phi_a | \partial \hat{H}_e / \partial Q_k'' | \phi_b \rangle$, where $\hat{H}_e$ is the electronic Hamiltonian

$$Q' = (L')^{-1} L'' Q'' + (L')^{-1} \vec{R} = SQ'' + \vec{D} \ , \qquad (3\text{-}15)$$

provided that the molecular symmetry remains invariant upon excitation. Here $S = (L')^{-1} L''$ is a constant (orthogonal) matrix and $\vec{D} = (L')^{-1} \vec{R}$ is a constant vector. In many treatments of electronic spectra, it is assumed that the normal coordinates of the two electronic states remain parallel upon electronic excitation. According to Eq. (3-15), as first recognized by Duschinsky[120], however, the normal coordinates of excited electronic states will generally be rotated with respect to those of the ground state. Thus in terms of the normal coordinates of the ground state, the vibrational potential of excited states will generally not be diagonal, resulting in a mixing of the normal coordinates of the excited state. Depending on the order of magnitude of the rotation, the Franck-Condon factors which determine the intensity distribution in electronic spectra, may change significantly and thereby cause a change in the vibrational structure of the spectra[120-127]. The explicit consideration of "mode scrambling" is thus a way to improve the absorption and luminescence vibronic intensities. It should be noticed that the displacements occur along the totally symmetric modes, and the rotations are for normal modes belonging to the same irreducible representation. For non-totally symmetric coordinates no shift in the equilibrium configuration is possible and hence $\vec{D} = \vec{0}$.

4. Calculation of Optical Absorption Spectra

The absorption line shape $I(\omega_1)$ due to absorption of the isotropic and unpolarized radiation of frequency ω_1 is defined as the Fourier transform of the time correlation function of the dipole moment $\hat{M}$ of the molecular system (see appendix II)

$$I(\omega_1) = \frac{1}{2\pi} \int_{-\infty}^{+\infty} d\tau \ \exp(-i\omega_1 \tau) < \hat{\vec{M}}(0) \cdot \hat{\vec{M}}(\tau) >$$

$$= \frac{1}{2\pi} \int_{-\infty}^{+\infty} d\tau \ \exp(i\omega_1 \tau) < \hat{\vec{M}}(\tau) \cdot \hat{\vec{M}}(0) > . \quad (4\text{-}1)$$

We consider a polyatomic molecule with two electronic states a and b, and N vibrational degrees of freedom. In the adiabatic approximation, the vibronic Hamiltonian of the molecule can be expressed in adiabatic basis set $\{ |a\alpha\rangle , |b\beta\rangle \}$

$$\hat{H} = \sum_{\alpha} |a\alpha\rangle (E_a + \hat{H}_a) \langle a\alpha| + \sum_{\beta} |b\beta\rangle (E_b + \hat{H}_b) \langle b\beta| \quad (4\text{-}2)$$

where E_a, E_b ($\hat{H}_a$, $\hat{H}_b$) are the electronic energies (the vibrational Hamiltonians) of the electronic states a and b, respectively. Here, for instance, α denotes collectively the set of vibrational quantum numbers in the electronic state a.

With Hamiltonian in Eq. (4-2) we may write Eq. (4-1) within the Condon approximation (Eq. (3-12)) as

$$I(\omega_1) = | \vec{\mu}_{ab}(0) |^2 \frac{1}{2\pi} \int_{-\infty}^{+\infty} d\tau \ \exp\{ -i\tau (\omega_{ba} - \omega_1) - \Gamma_{ba}|\tau| \} \ G(\tau) \quad (4\text{-}3)$$

$$= | \vec{\mu}_{ab}(0) |^2 \frac{1}{2\pi} \int_{-\infty}^{+\infty} d\tau \ \exp\{ i\tau (\omega_{ba} - \omega_1) - \Gamma_{ba}|\tau| \} \ G(-\tau) , \quad (4\text{-}4)$$

where

$$G(\tau) = < \exp(-i\hat{H}_b \tau / \hbar) \exp(i\hat{H}_a \tau / \hbar) > . \quad (4\text{-}5)$$

We may also write Eq. (4-3) in the form

$$I(\omega_1) = | \vec{\mu}_{ab}(0) |^2 \sum_{\alpha} \rho_\alpha I_\alpha(\omega_1) , \quad (4\text{-}6)$$

where

$$I_\alpha(\omega_1) = \frac{1}{2\pi}\int_{-\infty}^{+\infty} d\tau \ \exp\{ -i\tau(\omega_{ba} - \omega_1 - E_a/\hbar) - \Gamma_{ba}|\tau| \} \ g_{\alpha\alpha}(\tau)$$

$$= \frac{1}{\pi}\mathrm{Re}\int_0^\infty d\tau \exp\{ -i\tau(\omega_{ba} - \omega_1 - E_a/\hbar - i\Gamma_{ba}) \} \ g_{\alpha\alpha}(\tau) \quad (4\text{-}7)$$

with

$$g_{\alpha\alpha}(\tau) = \left\langle \alpha \right| \exp(-i\hat{H}_b\tau/\hbar) \left| \alpha \right\rangle . \quad (4\text{-}8)$$

Here $I(\omega_1)$ is defined as the absorption line shape from the *initially distributed vibronic states* and $I_\alpha(\omega_1)$ as the absorption line shape from *a single vibronic state*.

4.1. Absorption from a Single Vibronic State.

It can be seen that the central problem in resonance Raman scattering (see next section) and absorption spectroscopy is the calculation of $g_{\alpha'\alpha}(\tau)$, where $g_{\alpha\alpha}(\tau)$ is a particular case of $g_{\alpha'\alpha}(\tau)$. We now wish to calculate the function $g_{\alpha'\alpha}(\tau)$ for a displaced-distorted harmonic system, for convenience, first we shall treat the single-mode case and then generalize our results to a multi-mode system.

For the single-mode case we may write the function $g_{\alpha'\alpha}(\tau)$ as follows

$$g_{\alpha'\alpha}(\tau) = \sum_\beta \ \exp\{ -i\omega'\tau(\beta + 1/2) \} \ \langle\alpha'|\beta\rangle\langle\beta|\alpha\rangle$$

$$= \iint dQ'' d\overline{Q}'' \chi_\alpha(Q'') \ \chi_{\alpha'}(\overline{Q}'')$$

$$\times \sum_\beta \ \exp\{ -i\omega'\tau(\beta + 1/2) \} \ \chi_\beta(Q') \ \chi_\beta(\overline{Q}'), \quad (4\text{-}9)$$

where ω' is the vibrational frequency of the upper electronic state and χ_α and χ_β are the vibrational wavefunctions of the electronic states *a* and *b*, respectively.

Using the harmonic oscillator wavefunctions, and the Slater sum rule, letting $d = Q' - Q''$ be the displacement between the upper and the lower minimum potential energy surfaces, and changing the integral variables properly, Eq. (4-9) takes the following form[128]

$$g_{\alpha'\alpha}(\tau) = (\pi^2 2^{\alpha'+\alpha}\alpha'!\alpha!)^{-1/2}(u^2 - w^2)^{1/2}\exp\{-2S(1-u^2)\}$$

$$\times \iint dxdy\ H_{\alpha'}(ux - wy)\,H_{\alpha}(ux + wy)\exp\{-(x-s)^2 - y^2\}\ , \tag{4-10}$$

where

$$u = \{1 + a\ \tanh(\lambda/2)\}^{-1/2}\ , \tag{4-11a}$$

$$w = \{1 + a\coth(\lambda/2)\}^{-1/2}\ , \tag{4-11b}$$

and

$$s = -\sqrt{2S}\,(1-u^2)/u\ . \tag{4-11c}$$

Here $a = \gamma'/\gamma''$, $S = \frac{1}{2}\gamma'' d^2$, $\gamma = (\omega/\hbar)$, $\lambda = i\omega'\tau$, and $H_n(x)$ represent the nth order Hermite polynomial. Using the contour integral representation for the Hermite polynomial, and by making use of the Gaussian integral formula, Eq. (4-10) takes the following form

$$g_{\alpha'\alpha}(\tau) = \left(\frac{\alpha'!\alpha!}{2^{\alpha'+\alpha}}\right)^{1/2}(u^2 - w^2)^{1/2}\exp\{-2S(1-u^2)\}$$

$$\times \sum_k\ [2(u^2 - w^2)]^k\ \frac{(1-u^2-w^2)^{(\alpha+\alpha'-2k)/2}}{k!(\alpha-k)!(\alpha'-k)!}$$

$$\times H_{\alpha-k}\{ su(1-u^2-w^2)^{-1/2} \}$$

$$\times H_{\alpha'-k}\{ su(1-u^2-w^2)^{-1/2} \}. \tag{4-12}$$

Equation (4-12) is the general expression for the function $g_{\alpha'\alpha}(\tau)$ for a displaced-distorted harmonic system consisting of a single mode. Here we define $b=(1-a)/(1+a)$ and

$$su = -\frac{1}{2}\sqrt{2S}(1-b)(1-e^{-i\omega'\tau})/(1+be^{-i\omega'\tau}) \quad , \tag{4-13a}$$

$$1-u^2-w^2 = -b(1-e^{-2i\omega'\tau})/(1-b^2e^{-2i\omega'\tau}) \quad , \tag{4-13b}$$

$$(1-u^2) = \frac{1}{2}(1-b)(1-e^{-i\omega'\tau})/(1+be^{-i\omega'\tau}) \, , \tag{4-13c}$$

and

$$\xi = (1-b^2)/(1-b^2e^{-2i\omega'\tau}). \tag{4-13d}$$

In the low temperature limit, the population of the higher levels is negligible and only transitions originating from the zeroth vibrational level are considered. For this case, Eq. (4-12) reduces to the following form

$$g_{\alpha 0}(\tau) = \int_0^\infty d\tau \ \exp\{ -i\tau(\Omega_{ba} - \omega_1 - i\Gamma_{ba}) \} \exp\{ -2S(1-u^2) \}$$

$$\times \xi^{1/2}(1-u^2-w^2)^{\alpha/2} H_\alpha \{ su(1-u^2-w^2)^{-1/2} \} . \tag{4-14}$$

According to Eq. (4-5), for the single vibronic absorption line shape we have

$$G_\alpha(\tau) = \langle \alpha | e^{-i\hat{H}_b \tau/\hbar} e^{i\hat{H}_a \tau/\hbar} | \alpha \rangle = e^{iE_\alpha \tau/\hbar} g_{\alpha\alpha}(\tau) , \tag{4-15}$$

so that we need here the correlation function $g_{\alpha\alpha}(\tau)$. Setting $\alpha' = \alpha$ in Eq. (4-12) and introducing the result into Eq. (4-15) we obtain the required expression

$$G_\alpha(\tau) = e^{iE_\alpha \tau/\hbar} (\alpha!/2^\alpha) \exp\{ -2S(1-u^2) \} (u^2 - w^2)^{1/2}$$

$$\times \sum_k \{ 2(u^2 - w^2) \}^k \frac{(1-u^2-w^2)^{\alpha-k}}{k!\{(\alpha-k)!\}^2}$$

$$\times H^2_{\alpha-k} \{ su(1-u^2-w^2)^{-1/2} \} . \tag{4-16}$$

For the displaced oscillator case $b = 0, \omega' = \omega'' = \omega$, $1-u^2-w^2 = 0$, and noticing that $\lim_{\beta \to 0} \beta^{n/2} H_n(x\beta^{-1/2}) = (2x)^n$, Eq. (4-16) reduces to

$$G_\alpha(\tau) = e^{iE_\alpha \tau/\hbar} \exp\{ -S(1-e^{-i\omega\tau}) \}$$

$$\times L_\alpha \{ -S(e^{i\omega\tau/2} - e^{-i\omega\tau/2})^2 \} , \tag{4-17}$$

where $L_n(\mathrm{x})$ is the nth order Laguerre polynomial.

By making use of Eqs. (4-17), (4-15) and (4-7), we obtain the single vibronic state absorption line shape for a multi-mode system as

$$I_{\alpha}(\omega_1) = \frac{1}{\pi}\Big(\prod_j \alpha_j!/2^{\alpha_j} \Big)$$

$$\times \ \mathrm{Re}\sum_{k_1} \cdots \sum_{k_N} \int_0^{\tau} d\tau \ \exp\Big[-i\tau \Big\{ \Omega_{ba} - \omega_1 - \sum_j (\alpha_j \omega''_j - k_j \omega'_j) - i\Gamma_{ba} \Big\} \Big]$$

$$\times \ \exp\Big\{ -2\sum_j S_j(1-u_j^2) \Big\} \prod_j \frac{2^{k_j} \xi_j^{k_j+\frac{1}{2}} (1-u_j^2-w_j^2)^{\alpha_j - k_j}}{k_j!\{(\alpha_j - k_j)!\}^2}$$

$$\times \ H^2_{\alpha_j - k_j}\{ s_j u_j (1-u_j^2-w_j^2)^{-1/2} \} \quad . \tag{4-18}$$

If the absorption process occurs from the ground vibrational level of the lower electronic state, then Eq. (4-18) takes the following simpler form

$$I_0(\omega_1) = \frac{1}{\pi}\mathrm{Re}\int_0^{\tau} d\tau \ \exp\{ -i\tau (\Omega_{ba} - \omega_1 - i\Gamma_{ba}) \}$$

$$\times \ \Big(\prod_j \xi_j^{1/2}\Big) \ \exp\Big\{ -2\sum_j S_j(1-u_j^2) \Big\} \ . \tag{4-19}$$

4.2. Absorption Line Shape from Initially Distributed Vibronic States

The absorption line shape canonically averaged can be also easily derived. Using the closure relations, Eq. (4-5) may be written as

$$G(\tau) = (1 - e^{-\beta\hbar\omega''})\exp[i\omega''\tau/2]\sum_{\alpha} \exp[i\omega''\alpha\tau - \beta\hbar\omega''\alpha]g_{\alpha\alpha}(\tau) \ . \quad (4\text{-}20)$$

We can carry out the summation over α by the following method. By setting $\alpha' = \alpha$ in Eq. (4-10), by using the integral representation

$$H_n(x) = \frac{(-2i)^n}{\sqrt{\pi}} e^{x^2} \int_{-\infty}^{+\infty} d\zeta\, \zeta^n \exp(-\zeta^2 + 2ix\zeta) \quad (4\text{-}21)$$

and by carrying out the summation first and then the integrals we obtain

$$G(\tau) = \exp\{\ \xi/(1+\bar{n})^2\ \}$$

$$\times\ \{\xi/(1+\bar{n})^2\}\{\ 1-\varepsilon(1-2w^2)\ \}\{\ 1+\varepsilon(1-2w^2)\ \}^{\ 1/2}$$

$$\times\ \exp\{\ \frac{-2S(1-u^2)(1-\varepsilon)}{1+\varepsilon(1-2u^2)}\ \}\ , \quad (4\text{-}22)$$

where

$$\varepsilon = \ \{\ \bar{n}/(1+\bar{n})\ \}\ e^{i\omega''\tau}\ . \quad (4\text{-}23)$$

The absorption line shape for a multi-mode system then becomes

$$I(\omega_1) = \frac{1}{\pi} \mathrm{Re} \int_0^\infty d\tau \exp\{ -i\tau (\Omega_{ba} - \omega_1 - i\Gamma_{ba}) \}$$

$$\times [\prod_j \{ 1 - \varepsilon_j (1 - 2w_j^2) \} \{ 1 + \varepsilon_j (1 - 2u_j^2) \} \xi_j / (1 + \bar{n}_j)^2]^{1/2}$$

$$\times \exp[-2 \sum_j S_j (1 - u_j^2)(1 - \varepsilon_j) / \{ 1 + \varepsilon_j (1 - 2u_j^2) \}]. \tag{4-24}$$

The correlation function $< \exp(i\hat{H}_b \tau / \hbar) \exp(-i\hat{H}_a \tau / \hbar) >$ for the linear vibrational-electronic coupling case (that is, $\hat{V} = \hat{H}_b - \hat{H}_a \propto Q$) can be also calculated by the following method[129]. By using the operator identity

$$\exp\{ (\hat{a} + \hat{b}) \tau \} = \exp_- \{ \int_0^\tau d\tau' \, \hat{b}(\tau') \} \exp(\hat{a}\tau) , \tag{4-25}$$

we may then write

$$< \exp(i\hat{H}_b \tau / \hbar) \exp(-i\hat{H}_a \tau / \hbar) > \; = \; < \exp_- \{ (i/\hbar) \int_0^\tau d\tau' \, \hat{V}(\tau') \} >$$

$$= \sum_{n=0}^{\infty} \frac{(i/\hbar)^n}{n!} \int_0^\tau d\tau_1 \int_0^\tau d\tau_2 \cdots \int_0^\tau d\tau_n < \hat{T}_- \hat{V}(\tau_1) \cdots \hat{V}(\tau_n) > \; , \tag{4-26}$$

where $\hat{T}_-$ is the negative time ordering operator.

To calculate the thermal average $< Q_1 Q_2 \cdots >$ we may make use of the following rule which holds for harmonic systems[130]:
The thermal average of an odd number of Q's vanishes and the thermal average of an even number of Q's is equal to the sum of the thermally averaged pair products, the sum being over all pairs which preserves the order of the pair. For instance

$$< Q_i Q_j Q_k Q_\ell > \; = \; < Q_i Q_j >< Q_k Q_\ell >$$

$$+ < Q_i Q_k >< Q_j Q_\ell > + < Q_i Q_\ell >< Q_j Q_k > . \qquad (4\text{-}27)$$

Therefore, for the thermal average $< Q_1 Q_2 \cdots Q_{2n} >$, the number of pair products is $(2n-1)(2n-3)(2n-5) \cdots 1 = (2n)!/2^n n!$. Since each integrated, thermally averaged pair product makes the same contribution, Eq. (4-27) can be written as

$$< \exp(i\hat{H}_b \tau / \hbar) \exp(-i\hat{H}_a \tau / \hbar) >$$

$$= \sum_{n=0}^{\infty} \frac{(i/\hbar)^{2n}}{(2n)!} \cdot \frac{(2n)!}{2^n n!} \{ \int_0^\tau d\tau_1 \int_0^\tau d\tau_2 < T_- \hat{V}(\tau_1) \hat{V}(\tau_2) > \}^n$$

$$= \exp\{ -\hbar^2 \int_0^\tau d\tau_1 \int_0^{\tau_1} d\tau_2 < \hat{V}(\tau_1) \hat{V}(\tau_2) > \}$$

$$= \exp\{ -\hbar^2 \int_0^\tau d\tau_1 \, (\tau_1 - \tau_2) < \hat{V}(0) \hat{V}(\tau_1) > \} , \qquad (4\text{-}28)$$

where the last equality is justified by the fact that the thermal average of a pair is a function of just the difference of the operator's time arguments.

The thermal average $< \hat{V}(0) \hat{V}(\tau) >$ is given by[131]

$$< \hat{V}(0) \hat{V}(\tau) > \; = \sum_j \{ (1 + \bar{n}_j) \, e^{i\omega_j \tau} + \bar{n}_j e^{-i\omega_j \tau} \} . \qquad (4\text{-}29)$$

By substituting Eq. (4-29) into Eq. (4-28) and by carrying out the time integral, we obtain for the *displaced* oscillator case.

$$< \exp(i\hat{H}_b \tau / \hbar) \exp(-i\hat{H}_a \tau / \hbar) >$$

$$= \exp[-\sum_j S_j(1+2\bar{n}_j) + \sum_j S_j\{ (1+\bar{n}_j)e^{i\omega_j\tau} + \bar{n}_j e^{-i\omega_j\tau} \}]. \qquad (4\text{-}30)$$

5. Resonance Raman Scattering

We consider a spontaneous Raman scattering from an initial vibronic state $|a\alpha\rangle$ to final state $|a\alpha'\rangle$ via resonant states $\{ |b\beta\rangle \}$. The electric transition moment in adiabatic non-Condon approximation (within the Herzberg-Teller theory) is given by Eqs. (3-13) and (3-14)***

$$\left\langle b\beta \left| \hat{\vec{M}} \cdot \vec{\varepsilon}_1 \right| a\alpha \right\rangle = \mu^0_{ba}(1)\langle\beta|\alpha\rangle$$

$$+ \sum_{kc} \{ \mu^0_{bc}(1)H^k_{ca} - \mu^0_{ca}(1)H^k_{bc} \}\langle\beta|Q''_k|\alpha\rangle ,$$

$$\left\langle a\alpha' \left| \hat{\vec{M}} \cdot \vec{\varepsilon}_2 \right| b\beta \right\rangle = \mu^0_{ab}(2)\langle\alpha'|\beta\rangle$$

$$+ \sum_{kc} \{ \mu^0_{ac}(2)H^k_{cb} - \mu^0_{cb}(2)H^k_{ac} \}\langle\alpha'|Q''_k|\beta\rangle \qquad (5\text{-}1)$$

and

$$\left\langle b\beta \left| \hat{\vec{M}} \cdot \vec{\varepsilon}_1 \right| a\alpha \right\rangle \left\langle a\alpha' \left| \hat{\vec{M}} \cdot \vec{\varepsilon}_2 \right| b\beta \right\rangle = \mu^0_{ab}(2)\mu^0_{ba}(1)\langle\beta|\alpha\rangle \langle\alpha'|\beta\rangle$$

$$- \sum_{kc} \{ \mu^0_{ba}(1)\mu^0_{ac}(2)H^k_{cb} + \mu^0_{ba}(1)\mu^0_{cb}(2)H^k_{ac} \}\langle\beta|\alpha\rangle\langle\alpha'|Q''_k|\beta\rangle$$

$$+ \sum_{kc} \{ \mu^0_{bc}(1)\mu^0_{ab}(2)H^k_{ca} + \mu^0_{ca}(1)\mu^0_{ab}(2)H^k_{ab} \}\langle\alpha'|\beta\rangle\langle\beta|Q''_k|\alpha\rangle$$

$$+ \cdots , \qquad (5\text{-}2)$$

*** Hereafter we make replacements $\omega_\lambda \to \omega_1$, $\omega_{\lambda'} \to \omega_2$; $\vec{\varepsilon}_{\lambda\mu} \to \vec{\varepsilon}_1$ and $\vec{\varepsilon}_{\lambda'\mu'} \to \vec{\varepsilon}_2$.

where, for instance, $\vec{\mu}_{ab}(0)\cdot\vec{\varepsilon}_1 = \mu_{ab}^{(0)}{}_{ab}$.

Substituting Eq. (5-2) into Eq. (2-20) leads to the well known result[132,133]

$$\frac{d^2\sigma}{d\omega_2 d\Omega} = (1/\hbar^2 c^4)\ \omega_1\omega_2^3$$

$$\times \Big| \sum_{b\beta} \Big\{ \frac{\mu_{ab}^0(2)\mu_{ba}^0(1)}{\omega_1 - \omega_{b\beta,a\alpha} + i\Gamma_{b\beta}} + \frac{\mu_{ab}^0(1)\mu_{ba}^0(2)}{-\omega_1 + \omega_{a\alpha',b\beta} + i\Gamma_{b\beta}} \Big\} \langle\beta|\alpha\rangle\langle\alpha'|\beta\rangle$$

$$+ \sum_{b\beta}\sum_{kc} \Big\{ \frac{\mu_{ac}^0(2)H_{cb}^k\mu_{ba}^0(1)}{\omega_1 - \omega_{b\beta,a\alpha} + i\Gamma_{b\beta}} + \frac{\mu_{ac}^0(1)H_{cb}^k\mu_{ba}^0(2)}{-\omega_1 + \omega_{a\alpha',b\beta} + i\Gamma_{b\beta}} \Big\} \langle\beta|\alpha\rangle\langle\alpha'|Q_k''|\beta\rangle$$

$$- \sum_{b\beta}\sum_{kc} \Big\{ \frac{\mu_{ab}^0(2)H_{bc}^k\mu_{ca}^0(1)}{\omega_1 - \omega_{b\beta,a\alpha} + i\Gamma_{b\beta}} + \frac{\mu_{ab}^0(1)H_{bc}^k\mu_{ca}^0(2)}{-\omega_1 + \omega_{a\alpha',b\beta} + i\Gamma_{b\beta}} \Big\} \langle\alpha'|\beta\rangle\langle\beta|Q_k''|\alpha\rangle$$

$$- \sum_{b\beta}\sum_{kc} \Big\{ \frac{H_{ac}^k\mu_{cb}^0(2)\mu_{ba}^0(1)}{\omega_1 - \omega_{b\beta,a\alpha} + i\Gamma_{b\beta}} + \frac{H_{ac}^k\mu_{cb}^0(1)\mu_{ba}^0(2)}{-\omega_1 + \omega_{a\alpha',b\beta} + i\Gamma_{b\beta}} \Big\} \langle\beta|\alpha\rangle\langle\alpha'|Q_k''|\beta\rangle$$

$$+ \sum_{b\beta}\sum_{kc} \Big\{ \frac{\mu_{ab}^0(2)\mu_{bc}^0(1)H_{ca}^k}{\omega_1 - \omega_{b\beta,a\alpha} + i\Gamma_{b\beta}} + \frac{\mu_{ab}^0(2)\mu_{bc}^0(1)H_{ca}^k}{-\omega_1 + \omega_{a\alpha',b\beta} + i\Gamma_{b\beta}} \Big\} \langle\alpha'|\beta\rangle\langle\beta|Q_k''|\alpha\rangle \Big|^2$$

$$\times\ \delta(\omega_{a\alpha',a\alpha} - \omega_1 + \omega_2)\,. \tag{5-3}$$

Equation (5-3) is regarded as the exact expression for the Raman intensities. If the frequency of the incident laser line is far from resonance, i.e., $\omega_{b\beta,a\alpha} \approx \omega_{ba}$, Eq. (5-3) simplifies dramatically[134,135]. The energy denominators come outside the sum over vibrational levels, and the closure relation $\sum_{\beta} |\beta\rangle\langle\beta| = 1$ may be used.

The non-adiabatic corrections may be also obtained[136-138] by substituting Eq. (3-14) into Eq. (2-20).

The dominant intensity contributions, for resonant and near resonance excitation, are expected to arise from the zero-order term, so called A-term[132], in Eq. (5-3). Hereafter we concentrate on the properties of this term. At the resonance condition the non-resonant term can be neglected and we have

$$\frac{d^2\sigma}{d\omega_2 d\Omega} = K\,\omega_1\omega_2^3 \left| \sum_{b\beta} \frac{\langle\beta|\alpha\rangle\langle\alpha'|\beta\rangle}{\omega_1 - \omega_{b\beta,a\alpha} + i\Gamma_{b\beta}} \right|^2$$

$$\times\, \delta(\omega_{a\alpha',a\alpha} - \omega_1 + \omega_2) \tag{5-4}$$

where $K = (1/\hbar^2 c^4)|\,\mu^0_{ba}(1)\mu^0_{ab}(2)\,|^2$. This equation is conveniently rewritten by bring the denominators into exponential[129,139], by means of identity $(x - i\varepsilon)^{-1} = i\int_0^\infty dk \exp\{\, ik\,(x + i\varepsilon)\,\}$. The result is****

$$\frac{d^2\sigma}{d\omega_2 d\Omega} = K\,\omega_1\omega_2^3\, J_{a\alpha\to a\alpha'}(\omega_1)\; \delta(\omega_{a\alpha',a\alpha} - \omega_1 + \omega_2) \quad , \tag{5-5}$$

where

$$J_{a\alpha\to a\alpha'}(\omega_1) = \left| \sum_{\beta} \frac{\langle\alpha'|\beta\rangle\,\langle\beta|\alpha\rangle}{(\omega_{b\beta,a\alpha} - \omega_1) - i\Gamma_{b\beta,a\alpha}} \right|^2 = |\,\tilde{g}_{\alpha'\alpha}(\omega_1)\,|^2 \tag{5-6}$$

and

$$\tilde{g}_{\alpha'\alpha}(\omega_1) = \int_0^\infty d\tau\; \exp\{\, -i\tau\,(\omega_{ba} - \omega_1 - E_\alpha/\hbar - i\Gamma_{ba})\,\}\; g_{\alpha'\alpha}(\tau) \tag{5-7}$$

with

$$g_{\alpha'\alpha}(\tau) = \langle\alpha'|\; \exp(\, -i\hat{H}_b\tau/\hbar\,)\,|\alpha\rangle \quad . \tag{5-8}$$

In the derivation of Eq. (5-7), the vibrational quantum number dependence of the dephasing constant has been ignored (that is, Γ_{ba} is the average electronic dephasing constants.)

**** It is assumed that only one excited electronic state is important in the resonance Raman process.

If we assume an initial thermal distribution for the system, with distribution ρ_α , then it is straightforward to show that the resonance Raman spectra can be expressed as

$$\frac{d^2\sigma}{d\Omega d\omega_2} = \frac{1}{2} K\omega_1\omega_2^3 \int_{-\infty}^{+\infty} dt \int_0^\infty d\tau \int_0^\infty d\tau'$$

$$\times \exp\{ i(\omega_1 - \omega_2)t - i(\tau - \tau')(\omega_{ba} - \omega_1) - (\tau + \tau')\Gamma_{ba} - \Gamma_v|t| \}$$

$$\times G(\tau, \tau', t) \ , \tag{5-9}$$

where the correlation function $G(\tau,\tau',t)$ is defined as

$$G(\tau,\tau',t) = < e^{i\hat{H}_b\tau'/\hbar} e^{-i\hat{H}_a t/\hbar} e^{-i\hat{H}_b\tau/\hbar} e^{i\hat{H}_a(t+\tau-\tau)/\hbar} > \ . \tag{5-10}$$

It should be noted that the single-level RR scattering cross section is defined by

$$\frac{d^2\sigma_{a\alpha}}{d\Omega d\omega_2} = K\omega_1\omega_2^3 \sum_{\alpha'} \frac{\Gamma_{a\alpha',a\alpha}}{(\omega_1 - \omega_2 - \omega_{a\alpha',a\alpha})^2 + \Gamma^2_{a\alpha',a\alpha}}$$

$$\times J_{a\alpha\to a\alpha'}(\omega_1) , \tag{5-11}$$

which is useful in the time-resolved resonance Raman Process.

5.1. Resonance Raman Scattering from a Single Vibronic State

In most theoretical treatments of RRS the effect of frequency changes have been neglected on the assumption that this effect is small compared to that of equilibrium position shifts[139-144]. This phenomenon may in some cases lead to misinterpretations of experimental data, when undistorted oscillators are assumed to be involved in the RRS process. For the cases in which the linear displacements from the minimum are small, a quadratic interaction may dominate and the curvature of potential energy surfaces gives an effective vibrational - electronic interaction.

In earlier papers, the RRS excitation profiles in the low temperature limit are derived for general harmonic potential surfaces by using the wave packet technique[145,146] or Hizhnyakov and Tehver's[147] time correlator formalism[148]. In a more recent work, the Green function technique has also been used to calculate the state to state excitation profiles for a general harmonic potential surface[149].

From the previous section we can see that the function $g_{\alpha'\alpha}(\tau)$ plays the main role in the calculation of the resonance Raman excitation profiles and the resonance Raman spectra. In section 4, we derived a general expression for $g_{\alpha'\alpha}(\tau)$ for a displaced-distorted system:

$$g_{\alpha'\alpha}(\tau) = \left(\frac{\alpha'!\alpha!}{2^{\alpha+\alpha'}}\right)^{1/2}(u^2-w^2)\exp\{-2S(1-u^2)\}\sum_k \{2(u^2-w^2)\}^k$$

$$\times \frac{(1-u^2-w^2)^{(\alpha+\alpha'-2k)/2}}{k!(\alpha-k)!(\alpha'-k)!}$$

$$\times H_{\alpha-k}\{su(1-u^2-w^2)^{-1/2}\}$$

$$\times H_{\alpha'-k}\{su(1-u^2-w^2)^{-1/2}\} \qquad (5\text{-}12)$$

and in the low temperature limit

$$g_{\alpha 0}(\tau) = \int_0^\infty d\tau \exp\{ (\Omega_{ba} - \omega_1 - i\Gamma_{ba}) \}\exp\{ -2S(1-u^2) \}$$

$$\times \xi^{1/2}(1-u^2-w^2)^{\alpha/2} H_\alpha\{ su(1-u^2-w^2)^{-1/2} \} . \tag{5-13}$$

By using the definition of Hermite polynomials, by introducing Eqs. (5-12) and (5-13), and by noticing that $\lim_{\beta\to 0} \beta^{n/2} H_n(x\beta^{-1/2}) = (2x)^n$ and $H_{2n}(0) = (-1)^n (2n)!/n!$, we may perform the binomial and exponential expansions in Eq. (5-12) and carry out the integral over τ to obtain the expressions of the excitation profiles for two cases, *i.e.*, (i) a distorted $(S=0)$ and (ii) a displaced $(b=0)$ oscillator system in the low temperature limit:

$$J_{a0\to a\alpha}^{\text{distorted}}(\omega_1) = \frac{(b^2/4)^\alpha}{(2\alpha)!} \mid \sum_{k,\ell} (-1)^\ell \binom{\alpha}{\ell} \frac{(2\alpha+2k)!}{k!\,(\alpha+k)!}.$$

$$\times \frac{i(b^2/4)^k}{\omega_1 - \Omega_{ba} - 2\omega'(k+\ell) + i\Gamma_{ba}} \mid^2 \tag{5-14}$$

and

$$J_{a0\to a\alpha}^{\text{displaced}}(\omega_1) = \frac{S^\alpha}{\alpha!} \mid \sum_{k,\ell} (-1)^\ell \binom{\alpha}{\ell} \frac{1}{k!}$$

$$\times \frac{iS^k e^{-S}}{\omega_1 - \omega_{ba} - \omega(k+\ell) + i\Gamma_{ba}} \mid^2 . \tag{5-15}$$

For the displaced oscillator case Eq. (5-15) can also be written as

$$e^{iE_\alpha\tau/\hbar} g_{\alpha'\alpha}(\tau) = \left(\frac{\alpha!}{\alpha'!}\right)^{1/2} \exp\{ -S(1-e^{-i\omega\tau}) \}$$

$$\times (-1)^{\alpha+\alpha'} \{ \sqrt{S}(1-e^{-i\omega\tau}) \}^{\alpha'-\alpha}$$

$$\times L_{\alpha}^{\alpha'-\alpha} \{ -S(e^{i\omega\tau/2} - e^{-i\omega\tau/2})^2 \} , \qquad (5\text{-}16)$$

where $L_n^m(x)$ are the associated Laguerre polynomials.

In the above derivations, it has been assumed that there is only one mode involved in the resonance Raman scattering. However, the generalization of the above results to a multi-mode system is straightforward:

$$J_{a\alpha \to a\alpha'}(\omega_1) = (\prod_j \frac{\alpha_j!\alpha'_j!}{2^{\alpha_j+\alpha'_j}})$$

$$\times | \sum_{k_1} \cdots \sum_{k_N} \int_0^\infty d\tau \, \exp[-i\tau \{\Omega_{ba} - \omega_1 - \sum_j (\alpha_j \omega''_j - k_j \omega'_j) - i\Gamma_{ba} \}]$$

$$\times \exp\{ -2\sum_j S_j(1-u_j^2) \} \prod_j \frac{2^{k_j} \xi_j^{k_j+1/2} (1-u_j^2-w_j^2)^{(\alpha_j+\alpha'_j-2k_j)/2}}{(\alpha_j-k_j)!k_j!(\alpha'_j-k_j)!}$$

$$\times H_{\alpha_j-k_j} \{ s_j u_j (1-u_j^2-w_j^2)^{-1/2} \}$$

$$\times H_{\alpha'_j - k_j}\{ s_j u_j (1-u_j^2 - w_j^2)^{-1/2} \} \,|^2 . \tag{5-17}$$

For a two mode super-cooled system, for instance, in which one mode (ω_s) can be described by the displaced oscillator potential and the other (ω_d) can be expressed by the distorted oscillator potential , Eq. (5-17) yields

$$J_{a00\to apq}(\omega_1) = K(1-b^2)\frac{(b^2/4)^q}{(2q)!}\;\frac{S^p}{p!}$$

$$\times \Big| \sum_{k_1,\ell_1}\sum_{k_2,\ell_2} (-1)^{\ell_1+\ell_2}\binom{p}{\ell_1}\binom{q}{\ell_2}\frac{(2q+2k_2)\,!}{k_2\,!\,(q+k_2)\,!\,k_1!} \;.$$

$$\times \frac{iS^{k_1}(b^2/4)^{\,k_2}\,e^{-S}}{\omega_1 - \Omega_{ba} - (k_1+\ell_1)\,\omega_s - 2\omega'_d(k_2+\ell_2) + i\Gamma_{ba}}\,\Big|^2 \;. \tag{5-18}$$

We have derived some recurrence relations for both $g_{\alpha'\alpha}(\tau)$ and its half-Fourier transform $\tilde{g}_{\alpha'\alpha}(\omega_1)$ in Appendix III. These relations are very useful for calculating the excitation profiles and the spectra.

So far in our formalism for the calculation of the resonance Raman excitation profiles, it has been assumed that lifetime, i.e. homogeneous broadening (Γ_{ba}) is the only contribution to bandwidths. However, for a molecule embedded in a solvent bath, the solvent-induced broadening (phonon broadening) must also be taken into account[150,151]. One may look at the system as a super molecule, *i.e.*, molecule plus solvent. The high frequency modes will correspond to the internal modes of the molecule, whereas the low frequency modes will correspond to the external modes. Solvent-induced broadening is thus simulated by assigning non-zero displacement -

distortion parameters (S and b) to the low frequency modes (a solvent will induce broadening of a vibronic band if there is a substantial difference in the solvent-solute interaction between the electronically excited and ground state molecules). It has been shown that the solvent-solute interactions do not only broaden vibronic lines but also affect the vibrational intensity distribution in the profiles[151].

It should be noted that for the case in which the coherence contribution is negligible, the observed time-resolved RRS can be expressed in terms of the single-level RRS cross section as

$$\frac{\sigma(t)}{d\Omega d\omega_1} = \sum_{\alpha} \rho_{\alpha}(t) \frac{d\sigma_{\alpha}}{d\Omega d\omega_1} . \qquad (5\text{-}19)$$

where $\rho_{\alpha}(t)$ denotes the population of the system as a function of time t. The effect of the coherence contribution to time-resolved RRS has been discussed.

5.2 RRS from Initially Distributed Vibronic States

The relations found in the previous section are valid only as long as a single initial vibronic state $|\, a\alpha \rangle$ contributes to the RRS processes is involved. In this section we investigate the effect of temperature on the RRS cross section. Again, for simplicity in notation we shall first consider only one vibrational mode. The results obtained can, however, be easily generalized. By using the closure relations, Eq. (5-10) can be written as

$$G(\tau, \tau', t) = (1 - e^{-\beta\hbar\omega''}) \exp\{\, i\omega'' \,(\tau - \tau')/2 \,\}$$

$$\times \sum_{\alpha\alpha'} \exp\{\, -i\omega''\alpha(t - \tau + \tau') + i\omega''\alpha' t - \beta\hbar\omega''\alpha \,\}$$

$$\times\, g_{\alpha'\alpha}(\tau) g^{*}_{\alpha'\alpha}(\tau') \ . \tag{5-20}$$

We can now carry out the summations over α and α' by the same method as used in Eq. (4-20). By introducing the integral representation in Eq. (4-21) into Eq. (5-20) and by carrying out the integrals over x and y, we obtain

$$g_{\alpha'\alpha}(\tau) = (\pi 2^{a+a'} \alpha'!\alpha!)^{-1/2} (-2i)^{\alpha+\alpha'}$$

$$\times\, [\, (u^2 - w^2) / \{ (1-u^2)(1-w^2) \}]^{1/2} \exp\{ -2S(1-u^2)/u^2 \}$$

$$\times \int_{-\infty}^{+\infty} d\zeta_1\, \zeta_1^{\alpha} e^{-\zeta_1^2} \int_{-\infty}^{+\infty} d\zeta_2\, \zeta_2^{\alpha'} e^{-\zeta_2^2}$$

$$\times\, \exp[\, \frac{\{s + iu(\zeta_1 + \zeta_2)\}^2}{(1-2u^2)} - \frac{w^2(\zeta_1 - \zeta_2)^2}{(1-2w^2)}\,] . \tag{5-21}$$

Similarly, we have

$$g^{*}_{\alpha'\alpha}(\tau') = (\pi 2^{a+a'} \alpha'!\alpha!)^{-1/2} (-2i)^{\alpha+\alpha'}$$

$$\times\, [\, (u'^2 - w'^2) / \{ (1-u'^2)(1-w'^2) \}]^{1/2} \exp\{ -2S(1-u'^2)/u'^2 \}$$

$$\times \int_{-\infty}^{+\infty} d\zeta_3\, \zeta_3^{\alpha} e^{-\zeta_3^2} \int_{-\infty}^{+\infty} d\zeta_4\, \zeta_4^{\alpha'} e^{-\zeta_4^2}$$

$$\times\, \exp[\, \frac{\{s' + iu'(\zeta_3 + \zeta_4)\}^2}{(1-2u'^2)} - \frac{w'^2(\zeta_3 - \zeta_4)^2}{(1-2w'^2)}\,] , \tag{5-22}$$

where $u' = u^*(\tau')$, $w' = w^*(\tau')$, and $s' = s^*(\tau')$. By substituting Eqs. (5-21) and (5-22) into Eq. (5-20) and by carrying out the summations first and then integral over $\zeta_1 - \zeta_4$, we obtain

$$G(\tau,\tau',t) = (1+\bar{n})^{-1}\exp\{\ -i\,(\omega'-\omega'')(\tau-\tau')/2\ \}$$

$$\times\ \{\ \xi\xi'/f(\tau,\tau',t)\ \}^{1/2}\exp\{\ h(\tau,\tau',t)/f(\tau,\tau',t)\ \}\ , \qquad (5\text{-}23)$$

where $\xi' = \xi^*(\tau')$,

$$f(\tau,\tau',t) = 1-(\varepsilon_1^2+\varepsilon_2^2)(1-u^2-w^2)(1-u'^2-w'^2)$$

$$+\ \varepsilon_1^2\varepsilon_2^2(1-2u^2)(1-2u'^2)(1-2w^2)(1-2w'^2)$$

$$-\ 2\varepsilon_1\varepsilon_2(u^2-w^2)(u'^2-w'^2) \qquad (5\text{-}24)$$

and

$$h(\tau,\tau',t) = 2S\,[\ -\ (1-u^2)\ -(1-u'^2)-\{\ w^2(1-u^2)(1-u'^2-w'^2)$$

$$+\ w'^2(1-u'^2)(1-u^2-w^2)\ \}(\varepsilon_1^2+\varepsilon_2^2)$$

$$+2\{\ (1-u^2)(u'^2-w'^2)(1-w^2)$$

$$+\ (1-u'^2)(u^2-w^2)(1-w'^2)\ \}\ \varepsilon_1\varepsilon_2$$

$$+(1-2w^2)(1-2w'^2)\{\ (1-u^2)(1-2u'^2)$$

$$+\ (1-u'^2)(1-2u^2)\ \}\ \varepsilon_1^2\varepsilon_2^2\ +\ 2(1-u^2)(1-u'^2)$$

$$\times\ \{\ 1-\ (1-2w^2)(1-2w'^2)\varepsilon_1\varepsilon_2\ \}(\varepsilon_1+\varepsilon_2)\ ]\ , \qquad (5\text{-}25)$$

with

$$u^2-w^2 = (1-b^2)\,e^{i\omega'\tau}/(e^{2i\omega'\tau}-b^2)\quad , \qquad (5\text{-}26a)$$

$$1 - w^2 = \frac{1}{2}(1-b)(1+e^{-i\omega'\tau}) / (1-be^{-i\omega'\tau}) \ , \tag{5-26b}$$

$$1 - 2u^2 = (-b - e^{-i\omega'\tau}) / (1 + be^{-i\omega'\tau}) \ , \tag{5-26c}$$

$$1 - 2w^2 = (-b + e^{-i\omega'\tau}) / (1 - be^{-i\omega'\tau}) \ , \tag{5-26d}$$

$$\varepsilon_1 = \exp(\, i\omega'' t\,) \ , \tag{5-26e}$$

$$\varepsilon_2 = \{\, \bar{n} / (1+\bar{n})\,\} \exp\{\, i\omega''(\tau - \tau')\,\} / \varepsilon_1 \tag{5-26f}$$

and

$$\bar{n} = (e^{\beta\hbar\omega''} - 1)^{-1} \ . \tag{5-26g}$$

In the low temperature limit, $\bar{n} \to 0$, $\varepsilon_2 \to 0$, Eq. (5-23) takes a simple form

$$G(\tau,\tau',t) = \exp\{\, -i\,(\omega' - \omega'')(\tau - \tau')/2\,\}$$

$$\times \left[\, \xi\xi' / \{\, 1 - \varepsilon_1^2 (1-u^2-w^2)(1-u'^2-w'^2)\} \,\right]^{1/2}$$

$$\times \exp\left[\, \frac{-2S\{\,(1-u^2)+(1-u'^2)\,\}}{1-\varepsilon_1^2(1-u^2-w^2)(1-u'^2-w'^2)} \,\right]$$

$$\times \exp\left\{\, \frac{-2S\, w^2 (1-u^2)(1-u'^2-w'^2)\,\varepsilon_1^2}{1-\varepsilon_1^2(1-u^2-w^2)(1-u'^2-w'^2)} \,\right\}$$

$$\times \ \exp\{ \frac{-2Sw'^2(1-u'^2)(1-u^2-w^2)\,\varepsilon_1^2}{1-\varepsilon_1^2(1-u^2-w^2)(1-u'^2-w'^2)} \} \quad . \tag{5-27}$$

For the displaced oscillator, $a=1$, $b=0$, $\omega'=\omega''=\omega$, functions $f(\tau,\tau',t)$ and $h(\tau,\tau',t)$ reduces to

$$f(\tau,\tau',t) = (1+\bar{n})^{-2}$$

$$\begin{aligned} h(\tau,\tau',t)\,/\,f(\tau,\tau',t) = &-S(1+\bar{n})\ \{\ (1-e^{-i\omega\tau}) + (1-e^{i\omega\tau'}) \\ &-(1-e^{-i\omega\tau})(1-e^{i\omega\tau'})e^{i\omega t}\ \} \\ &-S\bar{n}\ \{\ (1-e^{i\omega\tau}) + (1-e^{-i\omega\tau'}) \\ &-(1-e^{i\omega\tau})(1-e^{-i\omega\tau'})\ e^{-i\omega t}\ \}\ . \end{aligned} \tag{5-28}$$

By substituting Eq. (5-21) into Eq. (5-9), the RRS cross section takes the following form

$$\begin{aligned} \frac{d^2\sigma(\omega_1,\,\omega_2)}{d\Omega d\omega_2} = \frac{1}{2}K\omega_1\omega_2^3 \int_{-\infty}^{+\infty}dt \int_0^\infty d\tau \int_0^\infty d\tau' \\ \times\ \{\ \xi\xi'/(1+\bar{n})^2 f(\tau,\tau',t)\ \}^{1/2} \\ \times\ \exp\{\ i\,(\omega_1-\omega_2)\,t - i\,(\tau-\tau')(\Omega_{ba}-\omega_1) \\ -(\tau+\tau')\Gamma_{ba} - \Gamma_v|t|\ \} \\ \times\ \exp\{\ h(\tau,\tau',t)\,/\,f(\tau,\tau',t)\ \}\ . \end{aligned} \tag{5-29}$$

The multi-mode generalization of Eq. (5-29) can be easily derived

$$\frac{d^2\sigma(\omega_1, \omega_2)}{d\Omega d\omega_2} = \frac{1}{2} K\omega_1\omega_2^3 \int_{-\infty}^{+\infty} dt \int_0^{\infty} d\tau \int_0^{\infty} d\tau'$$

$$\times \{ \prod_{j=1}^{N} \xi_j \xi'_j / (1+\bar{n}_j)^2 f_j(\tau, \tau', t) \}^{1/2}$$

$$\times \exp\{ i(\omega_1 - \omega_2) t - i(\tau - \tau')(\Omega_{ba} - \omega_1)$$

$$- (\tau + \tau')\Gamma_{ba} - \Gamma_v |t| \}$$

$$\times \exp\{ \sum_j h_j(\tau, \tau', t) / f_j(\tau, \tau', t) \} . \qquad (5\text{-}30)$$

The correlation function $< e^{i\hat{H}_b\tau'/\hbar} e^{i\hat{H}_a t/\hbar} e^{-i\hat{H}_b\tau/\hbar} e^{-i\hat{H}_a(t-\tau+\tau')/\hbar} >$ for the linear vibrational-electronic coupling case (That is, $\hat{V} = \hat{H}_b - \hat{H}_a \propto Q$) can be also obtained by the same method as used for the derivation of Eq. (4-30). By rewriting Eq. (5-10) as

$$G(\tau, \tau', t) =< (e^{i\hat{H}_b\tau'/\hbar} e^{-i\hat{H}_a\tau'/\hbar}) e^{i\hat{H}_a(\tau'-\tau+t)/\hbar}$$

$$\times (e^{i\hat{H}_a\tau/\hbar} e^{-i\hat{H}_b\tau/\hbar}) \, e^{-i\hat{H}_a(\tau'-\tau+t)/\hbar} > \qquad (5\text{-}31)$$

and by making use of the operator identity in Eq. (4-25) then we have

$$G(\tau, \tau', t) =< \exp_-\{ \frac{i}{\hbar} \int_0^{\tau'} d\tau'_1 \, \hat{V}(\tau'_1) \} \exp_+\{ \frac{-i}{\hbar} \int_0^{\tau} d\tau_1 \hat{V}(\tau_1 + \xi) \} >$$

$$= \sum_{n=0}^{\infty} \sum_{m=0}^{\infty} \frac{(i/\hbar)^n}{n!} \frac{(-i/\hbar)^m}{m!} \int_0^{\tau'} d\tau'_1 \cdots \int_0^{\tau'} d\tau'_n \int_0^{\tau} d\tau_1 \cdots \int_0^{\tau} d\tau_m$$

$$\times < \hat{T}_- \hat{T}_+ \hat{V}(t'_1) \cdots \hat{V}(t'_n)\, \hat{V}(t_1 + \xi) \cdots \hat{V}(t_m + \xi) > \qquad (5\text{-}32)$$

where $\xi = \tau' - \tau + t$ and $\hat{T}_-$ ($\hat{T}_+$) is the negative (positive) time ordering operator. Now we need to calculate the thermal averages like $< Q(\tau'_1) \cdots Q(\tau'_n) Q(\tau_1 + \xi) \cdots Q(\tau_m + \xi) >$. To that end we make use of the rule in Eq. (4-27). For fixed values of n and m we compute the contributions from all the pairing that involve ℓ thermally averaged mixed pair products and then we sum over ℓ. The number of ways of choosing ℓ of the n primed Q operators and ℓ of the m unprimed Q operators, without regard to order in which they are chosen is $\binom{n}{\ell}\binom{m}{\ell}$. For each choice, the number of mixed pair products is $\ell!$. As a result, the total number of mixed pair products becomes $\binom{n}{\ell}\binom{m}{\ell} \ell!$. The number of primed and unprimed pair products are $[(n-\ell)!/\, 2^{(n-\ell)!/2} \{(n-\ell)/2\}!]$ and $[(m-\ell)!/\, 2^{(m-\ell)!/2} \{(m-\ell)/2\}!]$, respectively. Introducing these numbers in Eq. (5-32), by noticing that both $(n-\ell)$ and $(m-\ell)$ must be even (m', m, and ℓ must be either all even or all odd), and by having in mind that each integrated pair product has the same value, we obtain[129]

$$G(\tau,\tau',t) = \exp\{ (i/\hbar)^2 \int_0^{\tau'} d\tau'_1 (\tau' - \tau'_1) < \hat{V}(0)\hat{V}(\tau'_1) >$$

$$+ (i/\hbar)^2 \int_0^{\tau} d\tau_1 (\tau - \tau_1) < \hat{V}(\tau_1)\, \hat{V}(0) >$$

$$- (i/\hbar)^2 \int_0^{\tau'} d\tau'_1 \int_0^{\tau} d\tau_1 < \hat{V}(0)\hat{V}(\tau_1 - \tau'_1 + \xi) > \} \qquad (5\text{-}33)$$

By substituting Eq. (4-29) into Eq. (5-33) and by carrying out the time integral over τ_1 and τ'_1, we finally obtain Eq. (5-29).

6. Molecular Susceptibility Theory

6.1. General Consideration

Molecular hole burning experiments involve burning and probing laser fields, their interactions with a molecular ensemble and relaxation processes. It is convenient to derive expressions for both absorption and hole burning spectra by utilizing the density matrix formalism[152-157]. For this purpose, we consider a total system consisting of a molecular ensemble that interacts with burning and probing laser fields and heat-bath modes. First we assume that each molecule in the ensemble is in a homogeneous environment. After projecting out the variables of the heat-bath modes, we find the Stochastic Liouville equations:

$$\frac{\partial}{\partial t}\hat{\rho}(t) = -i\,(\,\hat{L}_0 - i\hat{\Gamma} + \hat{L}_{\mathrm{I}}^{t}\,)\hat{\rho}(t)\,, \tag{6-1}$$

where $\hat{L}_0$ denotes the Liouville operator for a molecule in the ensemble, $\hat{\Gamma}$ represents the damping operator and $\hat{L}_{\mathrm{I}}^{t}$ is the Liouville operator for the interaction between the molecule and the radiation fields. In the dipole approximation, the interaction is given by

$$\hat{V}^{t} = -\hat{\vec{\mu}}\cdot\vec{E}(t)\,, \tag{6-2}$$

where $\hat{\vec{\mu}}$ denotes the electric dipole moment operator and $\vec{E}(t)$ represents the laser fields which consists of the burning field $\vec{E}_{\mathrm{B}}(t)$ and the probing field $\vec{E}_{\mathrm{P}}(t)$, *i.e.*,

$$\vec{E}(t) = \vec{E}_{\mathrm{B}}(t) + \vec{E}_{\mathrm{P}}(t)\,, \tag{6-3}$$

where, for example,

$$\vec{E}_{\mathrm{B}}(t) = \vec{A}_{\mathrm{B}}^{(-)}(\omega_{\mathrm{B}})e^{-it\omega_{\mathrm{B}}} + \vec{A}_{\mathrm{B}}^{(+)}(-\omega_{\mathrm{B}})e^{it\omega_{\mathrm{B}}} . \tag{6-4}$$

$\vec{E}_{\mathrm{P}}(t)$ is given by replacing the subscripts B in Eq. (6-4) with P. By defining

$$\hat{\rho}(t) = \exp[-it\hat{L}_{\mathrm{eff}}]\hat{\sigma}(t) , \tag{6-5}$$

where $i\hat{L}_{\mathrm{eff}} = i\hat{L}_0 + \hat{\Gamma}$ and by applying Eq. (6-5) to Eq. (6-1), one can find

$$\frac{\partial}{\partial t}\hat{\sigma}(t) = -\tilde{L}_{\mathrm{I}}^{t}(t)\sigma(t) , \tag{6-6}$$

where

$$\tilde{L}_{\mathrm{I}}^{t} = \exp(it\hat{L}_{\mathrm{eff}})\hat{L}_{\mathrm{I}}^{t}\exp(-it\hat{L}_{\mathrm{eff}}) . \tag{6-7}$$

A formal solution of Eq. (6-6) is given by

$$\hat{\sigma}(t) = \hat{U}(t,t_i)\hat{\sigma}(t_i) = \hat{U}(t,t_i)\hat{\rho}(t_i) , \tag{6-8}$$

where

$$\hat{U}(t,t_i) = 1 + \sum_{a=1}^{\infty} (-i)^a \int_{t_i}^{t} d\tau_a \int_{t_i}^{\tau_a} d\tau_{a-1} \cdots \int_{t_i}^{\tau_2} d\tau_1$$

$$\times \tilde{L}_{\mathrm{I}}^{\tau_a}(\tau_a)\tilde{L}_{\mathrm{I}}^{\tau_{a-1}}(\tau_{a-1})\cdots\tilde{L}_{\mathrm{I}}^{\tau_1}(\tau_1) . \tag{6-9}$$

For the first and third order solutions, one find

$$\hat{U}^{(1)}(t,t_i) = (-i)\int_{t_i}^{t} d\tau_1 \tilde{L}_{\mathrm{I}}^{\tau_1}(\tau_1) \tag{6-10}$$

and

$$\hat{U}^{(3)}(t,t_i) = (-i)^3 \int_{t_i}^{t} d\tau_3 \int_{t_i}^{\tau_3} d\tau_2 \int_{t_i}^{\tau_2} d\tau_1 \tilde{L}_{\mathrm{I}}^{\tau_3}(\tau_3)\tilde{L}_{\mathrm{I}}^{\tau_2}(\tau_2)\tilde{L}_{\mathrm{I}}^{\tau_1}(\tau_1), \tag{6-11}$$

respectively. An electric polarization can be obtained by using the relation

$$\vec{P}(t) = \text{Tr}\{ \hat{\vec{\mu}} \exp(-it\hat{L}_{\text{eff}})\hat{\sigma}(t) \} . \tag{6-12}$$

In absorption or hole burning measurements, one usually observe the rate of energy absorption Q at which a molecular ensemble absorbs the energy of the probing laser field:

$$Q = -\vec{P}(t) \cdot \frac{d}{dt} \vec{E}_{\text{P}}(t) . \tag{6-13}$$

For the linear absorption and the hole burning, one needs to derive expressions for $\vec{P}^{(1)}$ and $\vec{P}^{(3)}$, respectively.

6.2. Three-level System

We first consider a molecule which consists of three vibronic levels *g*, *n* and *m*. We assume that an optical transition due to the burning laser field occurs between the vibronic states *m* and *g*, in addition to another optical transition due to the probing laser field taking place between the vibronic states *n* and *g*. From Eqs. (6-5) and (6-12), one can find

$$\vec{P}(t) = \vec{\mu}_{gn} \, \rho_{ng}(t) + c.c.$$

$$= \vec{\mu}_{gn} \exp(\, -it\omega'_{ng} \,) \, \sigma_{ng}(t) + c.c., \tag{6-14}$$

where $i\omega'_{ng} = i\omega_{ng} + \Gamma_{ng}$.

For the first order electric polarization $\vec{P}^{(1)}(t)$, we obtain $\sigma_{ng}^{(1)}(t)$,

$$\sigma_{ng}^{(1)}(t) = -i\int_{t_i}^{t} d\tau_1 [\widetilde{L}_{\mathrm{I}}^{\tau 1}(\tau_1)\,\hat{\sigma}_i]_{ng} = -i\int_{t_i}^{t} d\tau_1 [\widetilde{L}_{\mathrm{I}}^{\tau 1}(\tau_1)]_{ng}^{gg}\,(\hat{\sigma}_i)_{gg}\,. \qquad (6\text{-}15)$$

Here it is assumed that the system is in the g-state at $t = t_i$. By using Eq. (6-10), we derive

$$\sigma_{ng}^{(1)}(t) = -i(\hat{\sigma}_i)_{gg} \int_{t_i}^{t} d\tau_1 \exp(i\tau_1 \omega'_{ng})\, V_{ng}^{\tau 1}, \qquad (6\text{-}16)$$

where

$$V_{ng}^{\tau 1} = -\vec{\mu}_{ng} \cdot \{\vec{A}_{\mathrm{P}}^{(-)}(\omega_{\mathrm{P}})e^{-it\omega_{\mathrm{P}}} + \vec{A}_{\mathrm{P}}^{(+)}(-\omega_{\mathrm{P}})e^{+it\omega_{\mathrm{P}}}\}\,. \qquad (6\text{-}17)$$

Applying the rotating wave approximation (RWA) to Eq. (6-17) and assuming that $\Gamma_{ng} t_i << -1$, one can find

$$\sigma_{ng}^{(1)}(t) = \frac{(\hat{\sigma}_i)_{gg}}{\hbar} \{\vec{\mu}_{ng} \cdot \vec{A}_{\mathrm{P}}^{(-)}\} \frac{\exp\{\, it(\omega'_{ng} - \omega_{\mathrm{P}})\,\}}{\omega'_{ng} - \omega_{\mathrm{P}}}\,. \qquad (6\text{-}18)$$

Substituting Eq. (6-18) into Eq. (6-14) leads to

$$\vec{P}^{(1)}(t) = \frac{(\hat{\sigma}_i)_{gg}}{\hbar} \frac{\vec{\mu}_{gn} \otimes \vec{\mu}_{ng}}{\omega'_{ng} - \omega_{\mathrm{P}}} \vec{E}_{\mathrm{P}}^{(-)}(t) + c.c., \qquad (6\text{-}19)$$

where $\vec{E}_{\mathrm{P}}^{(-)}(t) = \vec{A}_{\mathrm{P}}^{(-)}(\omega_{\mathrm{P}})e^{-it\omega_{\mathrm{P}}}$. One can then find the complex, linear susceptibility $\vec{\vec{\chi}}^{(1)}(\omega_{\mathrm{P}})$:

$$\vec{\vec{\chi}}^{(1)}(\omega_{\mathrm{P}}) = \frac{(\hat{\sigma}_i)_{gg}}{\hbar} \frac{\vec{\mu}_{gn} \otimes \vec{\mu}_{ng}}{\omega_{ng} - \omega_{\mathrm{P}} - i\Gamma_{ng}} \qquad (6\text{-}20)$$

or

$$\mathrm{Im}\{\, \vec{\vec{\chi}}^{(1)}(\omega_{\mathrm{P}})\,\} = \frac{(\hat{\sigma}_i)_{gg}}{\hbar} (\vec{\mu}_{gn} \otimes \vec{\mu}_{ng}) \frac{\Gamma_{ng}}{(\omega_{ng} - \omega_{\mathrm{P}})^2 + \Gamma_{ng}^2}\,. \qquad (6\text{-}21)$$

Defining $\widetilde{V}^{\tau}_{nm}(\tau) = \exp\{i\tau\omega'_{nm}\}V^{\tau}_{nm}$, for the third order nonlinear polarization $\vec{P}^{(1)}(t)$, we calculate

$$\sigma^{(3)}_{ng}(t) = (-i)^3 \int_{t_i}^{t} d\tau_3 \int_{t_i}^{\tau_3} d\tau_2 \int_{t_i}^{\tau_2} d\tau_1 [\, \widetilde{L}^{\tau_3}_{\mathrm{I}}(\tau_3)\widetilde{L}^{\tau_2}_{\mathrm{I}}(\tau_2)\widetilde{L}^{\tau_1}_{\mathrm{I}}(\tau_1)\hat{\sigma}_i\,]_{ng}$$

$$= (-i)^3 \int_{t_i}^{t} d\tau_3 \int_{t_i}^{\tau_3} d\tau_2 \int_{t_i}^{\tau_2} d\tau_1 \sum_{k=g}^{n,m}\sum_{\ell=g}^{n,m}$$

$$\times\, e^{i\tau_3\omega'_{ng}} [\hat{L}^{\tau_3}_{\mathrm{I}}]^{k\ell}_{ng} e^{-i\tau_3\omega'_{k\ell}} [\, \widetilde{L}^{t_2}_{\mathrm{I}}(\tau_2)\widetilde{L}^{t_1}_{\mathrm{I}}(\tau_1)\hat{\sigma}_i\,]_{k\ell}$$

$$= \sigma^{(3)}_{ng}(t)_1 + \sigma^{(3)}_{ng}(t)_2 , \tag{6-22}$$

where

$$\sigma^{(3)}_{ng}(t)_1 = \frac{(-i)^3}{\hbar} \int_{t_i}^{t} d\tau_3 \int_{t_i}^{\tau_3} d\tau_2 \int_{t_i}^{\tau_2} d\tau_1$$

$$\times \sum_{k=g}^{n,m} \widetilde{V}^{\tau_3}_{nk}(\tau_3)[\widetilde{L}^{\tau_2}_{\mathrm{I}}(\tau_2)\widetilde{L}^{\tau_1}_{\mathrm{I}}(\tau_1)\hat{\sigma}_i\,]_{kg} . \tag{6-23}$$

The expression for $\sigma^{(3)}_{ng}(t)_2$ can be derived in a similar fashion and is given in appendix.

Noticing that there are no optical transitions between the vibronic states n and m, one can find,

$$\sigma^{(3)}_{ng}(t)_1 = \frac{(-i)^3}{\hbar} \int_{t_i}^{t} d\tau_3 \int_{t_i}^{\tau_3} d\tau_2 \int_{t_i}^{\tau_2} d\tau_1 ,$$

$$\times\, \widetilde{V}^{\tau_3}_{ng}(\tau_3)[\widetilde{L}^{\tau_2}_{\mathrm{I}}(\tau_2)\widetilde{L}^{\tau_1}_{\mathrm{I}}(\tau_1)\hat{\sigma}_i\,]_{gg} , \tag{6-24}$$

where

$$[\widetilde{L}^{\tau_2}_{\mathrm{I}}(\tau_2)\widetilde{L}^{\tau_1}_{\mathrm{I}}(\tau_1)\hat{\sigma}_i\,]_{gg} = \frac{(\,\hat{\sigma}_i\,)_{gg}}{\hbar} \sum_{k=n}^{m} \{\, \widetilde{V}^{\tau_2}_{gk}(\tau_2)\widetilde{V}^{\tau_1}_{kg}(\tau_1)$$

$$+ \widetilde{V}_{gk}^{\tau_1}(\tau_1)\widetilde{V}_{kg}^{\tau_2}(\tau_2) \} . \tag{6-25}$$

In the RWA, we find

$$\sigma_{ng}^{(3)}(t)_1 = (\frac{i}{\hbar})^3 (\hat{\sigma}_i)_{gg} | \vec{\mu}_{mg} \cdot \vec{A}_{\rm B}^{(-)}(\omega_{\rm B}) |^2$$

$$\times \{ \vec{\mu}_{ng} \cdot \vec{A}_{\rm P}^{(-)}(\omega_{\rm P}) \} \int_{t_i}^{t} d\tau_3 \int_{t_i}^{\tau_3} d\tau_2 \int_{t_i}^{\tau_2} d\tau_1$$

$$\times e^{i\tau_3(\omega'_{ng} - \omega_{\rm P})} \{ e^{i\tau_2(\omega'_{gm} + \omega_{\rm B})} e^{i\tau_1(\omega'_{mg} - \omega_{\rm B})}$$

$$+ e^{i\tau_1(\omega'_{gm} + \omega_{\rm B})} e^{i\tau_2(\omega'_{mg} - \omega_{\rm B})} \} . \tag{6-26}$$

Assuming that $\Gamma_{mg} t_i$, $\Gamma_{ng} t_i << -1$ in the above integration, one can obtain

$$\sigma_{ng}^{(3)}(t)_1 = (\frac{i}{\hbar})^3 (\hat{\sigma}_i)_{gg} | \vec{\mu}_{mg} \cdot \vec{A}_{\rm B}^{(-)}(\omega_{\rm B}) |^2$$

$$\times \{ \vec{\mu}_{ng} \cdot \vec{A}_{\rm P}^{(-)}(\omega_{\rm P}) \} \frac{2\Gamma_{mg}}{(\omega_{mg} - \omega_{\rm B})^2 + \Gamma_{mg}^2}$$

$$\times \{ (t - t_i) \frac{e^{it(\omega'_{ng} - \omega_{\rm P})}}{i(\omega'_{ng} - \omega_{\rm P})} - \frac{e^{it(\omega'_{ng} - \omega_{\rm P})}}{i^2 (\omega'_{ng} - \omega_{\rm P})^2} \} . \tag{6-27}$$

In most cases, $\sigma_{ng}^{(3)}(t)_1$ dominates (see appendix IV) and we approximately obtain

$$\sigma_{ng}^{(3)}(t)_1 = -\frac{(\hat{\sigma}_i)_{gg}}{\hbar^3} | \vec{\mu}_{mg} \cdot \vec{A}_{\rm B}^{(-)}(\omega_{\rm B}) |^2 \{ \vec{\mu}_{ng} \cdot \vec{E}_{\rm P}^{(-)}(t) \} \frac{e^{it\omega'_{ng}}}{\omega'_{ng} - \omega_{\rm P}}$$

$$\times \frac{2(t - t_i)\Gamma_{mg}}{(\omega_{mg} - \omega_{\rm B})^2 + \Gamma_{mg}^2} . \tag{6-28}$$

We finally find the third-order nonlinear susceptibility $\overleftrightarrow{\chi}^{(3)}(\omega_{\mathrm{P}})_1$, *i.e.*,

$$\overleftrightarrow{\chi}^{(3)}(\omega_{\mathrm{P}})_1 = -\frac{(\hat{\sigma}_i)_{gg}}{\hbar^3}\frac{\vec{\mu}_{gn}\otimes\vec{\mu}_{ng}}{\omega_{ng}-\omega_{\mathrm{P}}-i\Gamma_{ng}}$$

$$\times \left| \vec{\mu}_{mg}\cdot\vec{A}_{\mathrm{B}}^{(-)}(\omega_{\mathrm{B}}) \right|^2 \frac{2(t-t_i)\Gamma_{mg}}{(\omega_{mg}-\omega_{\mathrm{B}})^2+\Gamma_{mg}^2}. \tag{6-29}$$

6.3. Multi-level Case

For the general case we replace the three states, g, m and n by the three vibronic manifolds. Notice that

$$\vec{P}(t) = Tr\{\,\hat{\vec{\mu}}\,\hat{\rho}(t)\,\} = \sum_{\{n\}}\sum_{\{g\}} \vec{\mu}_{ng}\rho_{ng}(t) + c.c.. \tag{6-30}$$

and that for $\vec{P}^{(1)}(t)$, we calculate $\sigma_{ng}^{(1)}(t)$,

$$\sigma_{ng}^{(1)}(t) = -i\int_{t_i}^{t} d\tau_1 [\,\widetilde{L}_{\mathrm{I}}^{\tau 1}(\tau_1)\,\hat{\sigma}_i\,]_{ng} = -i\int_{t_i}^{t} d\tau_1 \sum_{\{g'\}} [\,\widetilde{L}_{\mathrm{I}}^{\tau 1}(\tau_1)\,]_{ng}^{g'g'}(\,\hat{\sigma}_i\,)_{g'g'}$$

$$= -\frac{i}{\hbar}(\,\hat{\sigma}_i\,)_{gg}\int_{t_i}^{t} d\tau_1 e^{i\tau\omega'_{ng}}\,V_{ng}^{\tau}. \tag{6-31}$$

In the RWA, Eq. (6-31) becomes

$$\sigma_{ng}^{(1)}(t) = \frac{(\,\hat{\sigma}_i\,)_{gg}}{\hbar}\{\,\vec{\mu}_{ng}\cdot\vec{A}_{\mathrm{P}}^{(-)}(\omega_{\mathrm{P}})\,\}\frac{e^{it(\omega'_{ng}-\omega_{\mathrm{P}})}}{\omega'_{ng}-\omega_{\mathrm{P}}}. \tag{6-32}$$

It follows that

$$P^{(1)}(t) = \frac{1}{\hbar} \sum_{\{n\}} \sum_{\{g\}} (\hat{\sigma}_i)_{gg} \frac{\vec{\mu}_{gn} \otimes \vec{\mu}_{ng}}{\omega_{ng} - \omega_{\mathrm{P}} - i\Gamma_{ng}} \cdot \vec{E}_{\mathrm{P}}^{(-)}(t) + c.c. \tag{6-33}$$

or

$$P^{(1)}(t) = \ddot{\chi}^{(1)}(\omega_{\mathrm{P}}) \cdot \vec{E}_{\mathrm{P}}^{(-)}(t) + c.c. \tag{6-34}$$

and that

$$\ddot{\chi}^{(1)}(\omega_P) = \sum_{\{n\}} \sum_{\{g\}} \frac{(\hat{\sigma}_i)_{gg}}{\hbar} \frac{\vec{\mu}_{gn} \otimes \vec{\mu}_{ng}}{\omega_{ng} - \omega_P - i\Gamma_{ng}} \tag{6-35}$$

or

$$\mathrm{Im}\{\ddot{\chi}^{(1)}(\omega_{\mathrm{P}})\} = \sum_{\{n\}} \sum_{\{g\}} \frac{(\hat{\sigma}_i)_{gg}}{\hbar} (\vec{\mu}_{gn} \otimes \vec{\mu}_{ng}) \frac{\Gamma_{ng}}{(\omega_{ng} - \omega_{\mathrm{P}})^2 + \Gamma_{ng}^2}. \tag{6-36}$$

Next we consider $\vec{P}^{(3)}(t)$. Notice that

$$\sigma_{ng}^{(3)}(t) = \sigma_{ng}^{(3)}(t)_1 + \sigma_{ng}^{(3)}(t)_2, \tag{6-37}$$

where

$$\sigma_{ng}^{(3)}(t)_1 = \frac{(-i)^3}{\hbar} \int_{t_i}^{t} d\tau_3 \int_{t_i}^{\tau_3} d\tau_2 \int_{t_i}^{\tau_2} d\tau_1$$

$$\times \sum_{\{k\}} \widetilde{V}_{nk}^{\tau_3}(\tau_3)[\widetilde{L}_{\mathrm{I}}^{\tau_2}(\tau_2)\widetilde{L}_{\mathrm{I}}^{\tau_1}(\tau_1)\hat{\sigma}_i]_{kg} \tag{6-38}$$

and

$$\sigma_{ng}^{(3)}(t)_2 = -\frac{(-i)^3}{\hbar} \int_{t_i}^{t} d\tau_3 \int_{t_i}^{\tau_3} d\tau_2 \int_{t_i}^{\tau_2} d\tau_1$$

$$\times \sum_{\{k\}} [\widetilde{L}_{\mathrm{I}}^{\tau_2}(\tau_2)\widetilde{L}_{\mathrm{I}}^{\tau_1}(\tau_1)\hat{\sigma}_i]_{nk} \widetilde{V}_{kg}^{\tau_3}(\tau_3). \tag{6-39}$$

We shall treat only Eq. (6-38) since the contribution of Eq. (6-39) to $\vec{P}^{(3)}(t)$ is negligible (see appendix IV). Equation (6-38) can be separated as

$$\sigma_{ng}^{(3)}(t)_1 = \sigma_{ng}^{(3)}(t)_{11} + \sigma_{ng}^{(3)}(t)_{12}, \tag{6-40}$$

where

$$\sigma_{ng}^{(3)}(t)_{11} = \frac{(-i)^3}{\hbar} \int_{t_i}^{t} d\tau_3 \int_{t_i}^{\tau_3} d\tau_2 \int_{t_i}^{\tau_2} d\tau_1$$

$$\times \tilde{V}_{ng}^{\tau_3}(\tau_3)[\, \tilde{L}_{\mathrm{I}}^{\tau_2}(\tau_2)\tilde{L}_{\mathrm{I}}^{\tau_1}(\tau_1)\, \hat{\sigma}_i\,]_{gg} \tag{6-41}$$

and

$$\sigma_{ng}^{(3)}(t)_{12} = \frac{(-i)^3}{\hbar} \int_{t_i}^{t} d\tau_3 \int_{t_i}^{\tau_3} d\tau_2 \int_{t_i}^{\tau_2} d\tau_1$$

$$\times \sum_{\{g'\} \neq g} \tilde{V}_{ng'}^{\tau_1}(\tau_1)[\, \tilde{L}_{\mathrm{I}}^{\tau_2}(\tau_2)\tilde{L}_{\mathrm{I}}^{\tau_1}(\tau_1)\, \hat{\sigma}_i\,]_{g'g}\,. \tag{6-42}$$

Equation (6-42) can be simplified as

$$\sigma_{ng}^{(3)}(t)_{11} = (\frac{-i}{\hbar})^3 \int_{t_i}^{t} d\tau_3 \int_{t_i}^{\tau_3} d\tau_2 \int_{t_i}^{\tau_2} d\tau_1 (\, \hat{\sigma}_i\,)_{gg} \tilde{V}_{ng}^{\tau_3}(\tau_3)$$

$$\times \sum_{\{k\}=\{n\},\{m\}} \{\, \tilde{V}_{gk}^{(\tau_2)}(\tau_2)\tilde{V}_{kg}^{(\tau_1)}(\tau_1) + \tilde{V}_{gk}^{(\tau_1)}(\tau_1)\tilde{V}_{kg}^{(\tau_2)}(\tau_2)\, \}\,. \tag{6-43}$$

For hole-burning experiments, we have

$$\sigma_{ng}^{(3)}(t)_{11} = (\frac{-i}{\hbar})^3 \int_{t_i}^{t} d\tau_3 \int_{t_i}^{\tau_3} d\tau_2 \int_{t_i}^{\tau_2} d\tau_1 (\hat{\sigma}_i)_{gg} \tilde{V}_{ng}^{\tau_3}(\tau_3)$$

$$\times \sum_{\{m\}} \{\, \tilde{V}_{gk}^{(\tau_2)}(\tau_2)\tilde{V}_{kg}^{(\tau_1)}(\tau_1) + \tilde{V}_{gk}^{(\tau_1)}(\tau_1)\tilde{V}_{kg}^{(\tau_2)}(\tau_2)\, \}\,. \tag{6-44}$$

In the RWA, Eq. (6-44) becomes

$$\sigma_{ng}^{(3)}(t)_{11} = (\frac{i}{\hbar})^3 \{\, \vec{\mu}_{ng} \cdot \vec{A}_{\mathrm{P}}^{(-)}(\omega_{\mathrm{P}})\, \}$$

$$\times \{ \frac{(t-t_i)e^{it(\omega'_{ng}-\omega_{\rm P})}}{i(\omega'_{ng}-\omega_{\rm P})} - \frac{e^{it(\omega'_{ng}-\omega_{\rm P})}}{i^2(\omega'_{ng}-\omega_{\rm P})^2} \}$$

$$\times \sum_{\{m\}} \left| \vec{\mu}_{mg} \cdot \vec{A}_{\rm B}^{(-)}(\omega_{\rm B}) \right|^2 \frac{2\Gamma_{mg}}{(\omega'_{ng}-\omega_{\rm B})^2+\Gamma_{mg}^2} (\hat{\sigma}_i)_{gg}. \qquad (6\text{-}45)$$

In the Appendix we shall show that only $\sigma_{ng}^{(3)}(t)_{11}$ makes a dominant contribution to $\vec{P}^{(3)}(t)$. From the relation

$$P^{(3)}(t) = \vec{\vec{\chi}}^{(3)}(\omega_{\rm P}) \cdot \vec{E}_{\rm P}^{(-)}(t) + c.c., \qquad (6\text{-}46)$$

we can approximately obtain

$$\vec{\vec{\chi}}^{(3)}(\omega_{\rm p}) = -\frac{2}{\hbar^3}(t-t_i)\sum_{\{m\}}\sum_{\{n\}}\sum_{\{g\}} \frac{\vec{\mu}_{ng}\otimes\vec{\mu}_{ng}}{\omega_{ng}-\omega_{\rm P}-i\Gamma_{ng}}$$

$$\times \left| \vec{\mu}_{mg} \cdot \vec{A}_{\rm B}^{(-)}(\omega_{\rm B}) \right|^2 \frac{\Gamma_{mg}}{(\omega'_{ng}-\omega_{\rm B})^2+\Gamma_{mg}^2} (\hat{\sigma}_i)_{gg} \qquad (6\text{-}47)$$

or

$$\mathrm{Im}\{ \vec{\vec{\chi}}^{(3)}(\omega_{\rm P}) \} = -\frac{2}{\hbar^3}(t-t_i)\sum_{\{m\}}\sum_{\{n\}}\sum_{\{g\}} (\vec{\mu}_{ng}\otimes\vec{\mu}_{ng}) \frac{\Gamma_{ng}}{(\omega_{ng}-\omega_{\rm P})^2+\Gamma_{ng}^2}$$

$$\times \left| \vec{\mu}_{mg} \cdot \vec{A}_{\rm B}^{(-)}(\omega_{\rm B}) \right|^2 \frac{\Gamma_{mg}}{(\omega'_{ng}-\omega_{\rm B})^2+\Gamma_{mg}^2} (\hat{\sigma}_i)_{gg}. \qquad (6\text{-}48)$$

It follows that

$$\vec{\vec{\chi}}(\omega_{\rm p}) = \vec{\vec{\chi}}^{(1)}(\omega_{\rm p}) + \vec{\vec{\chi}}^{(3)}(\omega_{\rm p}). \qquad (6\text{-}49)$$

More accurately, $\vec{\vec{\chi}}^{(3)}(\omega_{\mathrm{P}})$ should be given by

$$\vec{\vec{\chi}}^{(3)}(\omega_{\mathrm{p}}) = -\frac{2}{\hbar^3} \sum_{\{m\}} \sum_{\{n\}} \sum_{\{g\}} (\vec{\mu}_{ng} \otimes \vec{\mu}_{ng}) \{ \frac{t - t_i}{\omega'_{ng} - \omega_{\mathrm{P}}} - \frac{1}{i(\omega'_{ng} - \omega_{\mathrm{P}})^2} \}$$

$$\times \left| \vec{\mu}_{mg} \cdot \vec{A}_{\mathrm{B}}^{(-)}(\omega_{\mathrm{B}}) \right|^2 \frac{\Gamma_{mg}}{(\omega'_{ng} - \omega_{\mathrm{B}})^2 + \Gamma_{mg}^2} (\hat{\sigma}_i)_{gg} . \qquad (6\text{-}50)$$

It follows that

$$\mathrm{Im}\{ \vec{\vec{\chi}}^{(3)}(\omega_p) \} = -\frac{2}{\hbar^3} \sum_{\{m\}} \sum_{\{n\}} \sum_{\{g\}} (\vec{\mu}_{ng} \otimes \vec{\mu}_{ng})$$

$$\times \{ \frac{(t - t_i)\Gamma_{ng}}{(\omega_{ng} - \omega_{\mathrm{P}})^2 + \Gamma_{ng}^2} - \frac{(\omega_{ng} - \omega_{\mathrm{P}})^2 - \Gamma_{ng}^2}{\{ (\omega_{ng} - \omega_{\mathrm{P}})^2 + \Gamma_{ng}^2 \}^2} \}$$

$$\times \left| \vec{\mu}_{mg} \cdot \vec{A}_{\mathrm{B}}^{(-)}(\omega_{\mathrm{B}}) \right|^2$$

$$\times \frac{\Gamma_{mg}}{(\omega'_{ng} - \omega_{\mathrm{B}})^2 + \Gamma_{mg}^2} (\hat{\sigma}_i)_{gg} . \qquad (6\text{-}51)$$

7. Molecular Absorption and Hole Burning Spectra

We have thus far derived the expression for the first and third order susceptibilities of the molecular ensemble in terms of vibronic manifold {g}, {n} and {m} of the molecules in the absence of inhomogeneity. In this section, in order to take into account the effects of inhomogeneity on the optical spectra, we assume that the molecular variables associated with those vibronic manifolds are homogeneously broadened over the molecules. For each molecule in the ensemble, we may use the adiabatic approximation (see Chapter 3) to clearly define the molecular states, *i.e.*,

$$|g\rangle \rightarrow \Phi_{e_g}(r,\{Q\})\Theta_{e_g u}(\{Q\}) = \Phi_{e_g}(r,\{Q\}) \prod_{j=1}^{N} X_{e_g u_j}(Q_j) \tag{7-1}$$

and

$$|n\rangle \rightarrow \Phi_{e_n}(r,\{Q\})\Theta_{e_n v}(\{Q\}) = \Phi_{e_n}(r,\{Q\}) \prod_{j=1}^{N} X_{e_n v_j}(Q'_j), \tag{7-2}$$

where N is the number of vibrational modes and $\Phi_{e_g}(r,\{Q\})$ and $X_{e_g u_j}(Q_j)$ denote the electronic wave function in the ground state and its vibrational wave function in the vibrational state u_j of *j*th mode, respectively.

7.1. Effects of Inhomogeneity on Molecular Absorption Spectra

Applying the displaced harmonic oscillator potential and the step ladder model[156,157] to the Eq. (4-8), we obtain

$$\mathrm{Im}\{\, \vec{\chi}^{(1)}(\omega_P, T)\,\} = \frac{(\,\vec{\mu}_{e_g e_n} \otimes \vec{\mu}_{e_n e_g}\,)}{\hbar} \int_{-\infty}^{\infty} d\tau$$

$$\times \exp\{\, i\tau(\omega_{e_n e_g} - \omega_P) - \gamma_{e_n e_g}|\tau|\,\} G_N(\tau, T), \tag{7-3}$$

where $\vec{\mu}_{e_g e_n}$ denotes the electric transition moment, $\omega_{e_g e_n}$ represents the electronic transition frequency and $\gamma_{e_g e_n}$ is the electronic dephasing constant. The correlation function of nuclear motions $G_N(\tau, T)$ in Eq. (7-3) is given by

$$G_N(\tau, T) = \prod_{j=1}^{N} g_j(\tau, T)$$

$$= \exp[\; - \sum_{j=1}^{N} S_j\{\, 2\bar{n}_j(T) + 1$$

$$- \{ \bar{n}_j(T) + 1 \} e^{i\tau\omega_j - \gamma_j|\tau|/2} - \bar{n}_j(T) e^{-i\tau\omega_j - \gamma_j|\tau|/2} \}], \qquad (7\text{-}4)$$

where ω_j denotes the vibrational frequency of the jth mode, γ_j is the rate constant of the vibrational transition from $v_j = 1$ to $v_j = 0$ in the electronically excited state, S_j $(= \frac{\omega_j}{2\hbar}\Delta_j^2$; Δ_j being the displacement between the potentials of the ground and excited states.) is the coupling constant (or the Huang-Rhys factor) and $\bar{n}_j(T) = \{\exp(\hbar\omega_j / k_B T) - 1\}^{-1}$ is the thermal population of the jth vibrational mode. We have assumed in the above derivation that the vibrational relaxation time scale in the electronically excited state much shorter than that in the electronic ground state. Since we are interested in the effects of inhomogeneity, we have also assumed the hole burning experiments are performed at such low temperatures that we can omit the vibrational pure dephasing constants.

In particular, in the case in which only the jth mode is involved, we find

$$\begin{aligned} \mathrm{Im}\{ \vec{\chi}^{(1)}(\omega_P, 0) \} &= \frac{(\vec{\mu}_{e_g e_n} \otimes \vec{\mu}_{e_n e_g})}{2\hbar} \\ &\times \int_{-\infty}^{\infty} d\tau \exp(i\tau\Delta_P - \gamma_{e_n e_g}|\tau|) g_j(\tau, 0) \\ &= \frac{(\vec{\mu}_{e_g e_n} \otimes \vec{\mu}_{e_n e_g})}{2\hbar} \exp(-S_j) \\ &\times \sum_{v_j=0}^{\infty} \frac{\gamma_{e_n e_g} + \gamma_j / 2}{(\gamma_{e_n e_g} + v_j \gamma_j / 2)^2 + (\Delta_P + v_j \omega_j)^2}, \end{aligned} \qquad (7\text{-}5)$$

where $\Delta_P = \omega_{e_n e_g} - \omega_P$.

Now we incorporate the effects of inhomogeneity on Eq. (7-3). Although there might be various possibilities in considering the inhomogeneity in the molecular variables such as $\omega_{e_n e_g}$, ω_j , S_j etc., we shall assume that the electronic transition frequency is distributed with (see appendix V)

$$g(\omega_{e_n e_g}) = \sqrt{\frac{1}{\pi D^2}} \exp\{ - (\frac{\omega_{e_n e_g} - \overline{\omega}_{e_n e_g}}{D})^2 \}, \tag{7-6}$$

where $\overline{\omega}_{e_n e_g}$ and d denote the average of the electronic transition frequency and its dispersion, respectively. Averaging over the electronic transition frequency in Eq. (7-3) with Eq. (7-6), one can find that

$$< \mathrm{Im}\{ \vec{\vec{\chi}}^{(1)}(\omega_{\mathrm{P}},0) \} >_{\mathrm{AV}} = \frac{(\vec{\mu}_{e_g e_n} \otimes \vec{\mu}_{e_n e_g})}{2\hbar}$$

$$\times \int_{-\infty}^{\infty} d\tau \exp(i\tau\overline{\Delta}_P - \gamma_{e_n e_g}|\tau| - \frac{D^2\tau^2}{4}) G_N(\tau, T) \tag{7-7}$$

and that at T = 0, in which $\frac{\hbar\omega_j}{k_B T} >> 1$ holds for all the vibrational modes, Eq. (7-7) reduces to

$$< \mathrm{Im}\{ \vec{\vec{\chi}}^{(1)}(\omega_{\mathrm{P}},0) \} >_{\mathrm{AV}} = \frac{(\vec{\mu}_{e_g e_n} \otimes \vec{\mu}_{e_n e_g})}{2\hbar}$$

$$\times \int_{-\infty}^{\infty} d\tau \exp(i\tau\overline{\Delta}_{\mathrm{P}} - \gamma_{e_n e_g}|\tau| - \frac{D^2\tau^2}{4}) G_N(\tau,0). \tag{7-8}$$

where $\overline{\Delta}_P = \overline{\omega}_{e_n e_g} - \omega_{\mathrm{p}}$. Equation (7-8) indicates that the dispersion of the electronic transition frequency leads to the Gaussian decay component in the Fourier integral.

7.2. Molecular Hole Burning Spectra

The microscopic treatment of hole-burning experiment should be compared with the phenomenological one. In other word, for hole-burning experiments, we have to consider $\vec{\vec{\chi}}^{(1)}(\omega_P)_1 + \vec{\vec{\chi}}^{(3)}(\omega_P)_1$. Applying the adiabatic basis (Eqs. (7-1) and (7-2)) to Eq. (6-48), we find the imaginary part of the third order susceptibility at T = 0:

$$\mathrm{Im}\{\,\vec{\vec{\chi}}^{(3)}(\omega_P,0)\,\} = -\frac{2\Delta t}{\hbar^3}(\,\vec{\mu}_{eg_{e_n}} \otimes \vec{\mu}_{e_n e_g}\,)|\,\vec{\mu}_{eg_{e_n}} \cdot \vec{A}_B^{(-)}(\omega_B)\,|^2$$

$$\times \int_{-\infty}^{\infty} d\tau \exp(\,i\tau\Delta_P - \gamma_{e_n e_g}|\tau|\,)G_N(\tau,0)$$

$$\times \int_{-\infty}^{\infty} d\tau' \exp(\,i\tau'\Delta_B - \gamma_{e_n e_g}|\tau'|\,)G_N(\tau',0)\,. \qquad (7\text{-}9)$$

Here it has assumed that {n} and {m} belong to the same excited electronic state. Taking an average over the electronic transition frequency in Eq. (7-9) with Eq. (7-6) leads to

$$<\mathrm{Im}\{\,\vec{\vec{\chi}}^{(3)}(\omega_P,0)\,\}>_{AV} = -\frac{2\Delta t}{\hbar^3}(\,\vec{\mu}_{eg_{e_n}} \otimes \vec{\mu}_{e_n e_g}\,)|\,\vec{\mu}_{eg_{e_n}} \cdot \vec{A}_B^{(-)}(\omega_B)\,|^2$$

$$\times \int_{-\infty}^{\infty} d\tau \exp(\,i\tau\overline{\Delta}_P - \gamma_{e_n e_g}|\tau|\,)G_N(\tau,0)$$

$$\times \int_{-\infty}^{\infty} d\tau' \exp(\,-\frac{D^2(\tau+\tau')^2}{4} + i\tau'\overline{\Delta}_B - \gamma_{e_n e_g}|\tau'|\,)G_N(\tau',0)\,. \qquad (7\text{-}10)$$

Equation (7-10) is the general expression for the molecular hole burning spectra at such low temperatures that $\frac{\hbar\omega_j}{k_B T} >> 1$ holds for all the vibrational modes.

Next we consider the photon side-band observed in hole burning spectra. For this purpose we assume that only one mode, for instance the *j*th mode, is involved. In this case, Eq. (7-10) becomes

$$< \mathrm{Im}\{ \ddot{\chi}^{(3)}(\omega_{\mathrm{P}},0) \} >_{\mathrm{AV}} = -\ddot{C} \int_{-\infty}^{\infty} d\tau \exp(i\tau\overline{\Delta}_{\mathrm{P}} - \gamma_{e_n e_g}|\tau|) g_j(\tau,0)$$

$$\times \int_{-\infty}^{\infty} d\tau' \exp(-\frac{D^2(\tau+\tau')^2}{4} + i\tau'\overline{\Delta}_{\mathrm{B}} - \gamma_{e_n e_g}|\tau'|) g_j(\tau',0), \qquad (7\text{-}11)$$

where

$$\ddot{C} = \frac{\Delta t}{2\hbar^3} (\vec{\mu}_{e_g e_n} \otimes \vec{\mu}_{e_n e_g}) | \vec{\mu}_{e_g e_n} \cdot \vec{A}_{\mathrm{B}}^{(-)}(\omega_{\mathrm{B}}) |^2 \qquad (7\text{-}12a)$$

and

$$g_j(\tau,0) = \exp[-S_j \{ 1 + e^{i\tau\omega_j - \gamma_j|\tau|/2} \}]. \qquad (7\text{-}12b)$$

Changing the variables, $\tau = \frac{x+y}{2}$ and $\tau' = \frac{x-y}{2}$, we obtain

$$< \mathrm{Im}\{ \ddot{\chi}^{(3)}(\omega_{\mathrm{P}},0) \} >_{\mathrm{AV}} = -\frac{\ddot{C}e^{-2S_j}}{2} \int_{-\infty}^{\infty} dx \int_{-\infty}^{\infty} dy$$

$$\times \exp\{ -\frac{D^2x^2}{4} - \frac{\gamma_{e_n e_g}}{2} (| x+y | + | x-y |) \}$$

$$\times \exp\{ i\tau (\overline{\omega}_{e_n e_g} - \frac{\omega_{\mathrm{B}}+\omega_{\mathrm{P}}}{2}) x + i \frac{\omega_B + \omega_P}{2} y \}$$

$$\times \sum_{v_j=0}^{\infty} \sum_{v'_j=0}^{\infty} \frac{S^{v_j+v'_j}}{v_j! v'_j!} \exp\{ i v_j \omega_j \frac{x+y}{2} - v_j \gamma_j \frac{| x+y |}{4} \}$$

$$\times \exp\{ i v'_j \omega_j \frac{x-y}{2} - v'_j \gamma_j \frac{| x-y |}{4} \}. \qquad (7\text{-}13)$$

In the case in which the Gaussian decay due to the dispersion of the electronic transition frequency is much faster than both the effective electronic dephasing and the vibrational relaxation, i.e. $D >> \gamma_{e_n e_g}, \gamma_j$, the integration in Eq. (7-13) reduces to

$$< \mathrm{Im}\{ \ddot{\chi}^{(3)}(\omega_{\mathrm{P}},0) \} >_{\mathrm{AV}} = -2\ddot{C}\sqrt{\frac{4\pi}{D^2}}e^{-2S_j}$$

$$\times \sum_{v_j=0}^{\infty}\sum_{v'_j=0}^{\infty} \frac{S^{v_j+v'_j}}{v_j!v'_j!}\exp[-\{ \frac{\overline{\Delta}_{\mathrm{P}}+\overline{\Delta}_{\mathrm{B}}+(v_j+v'_j)\omega_j}{2D} \}^2]$$

$$\times \frac{2\gamma_{e_n e_g}+(v_j+v'_j)\gamma_j/2}{\{ 2\gamma_{e_n e_g}+(v_j+v'_j)\gamma_j/2 \}^2+\{ \omega_{\mathrm{B}}-\omega_{\mathrm{P}}+\omega_j(v_j-v'_j) \}^2} . \qquad (7\text{-}14)$$

For small S_j values, we find

$$< \mathrm{Im}\{ \ddot{\chi}^{(3)}(\omega_{\mathrm{P}},0) \} >_{\mathrm{AV}} = -2\ddot{C}\sqrt{\frac{4\pi}{D^2}}e^{-2S_j}$$

$$\times [\exp\{ -(\frac{\overline{\Delta}_P+\overline{\Delta}_B}{2D})^2 \} \frac{2\gamma_{e_n e_g}}{(2\gamma_{e_n e_g})^2+(\omega_{\mathrm{B}}-\omega_{\mathrm{P}})^2}$$

$$\times S_j \exp\{-(\frac{\overline{\Delta}_{\mathrm{P}}+\overline{\Delta}_{\mathrm{B}}+\omega_j}{2D})^2 \}$$

$$\times \{ \frac{2\gamma_{e_n e_g}+\gamma_j/2}{(2\gamma_{e_n e_g}+\gamma_j/2)^2+(\omega_{\mathrm{B}}-\omega_{\mathrm{P}}-\omega_j)^2}$$

$$+ \frac{2\gamma_{e_n e_g}+\gamma_j/2}{(2\gamma_{e_n e_g}+\gamma_j/2)^2+(\omega_{\mathrm{B}}-\omega_{\mathrm{P}}+\omega_j)^2} \} + \cdots]. \qquad (7\text{-}15)$$

In particular, at $\omega_B = \overline{\omega}_{e_n e_g}$, Eq. (7-15) reduces to

$$< \mathrm{Im}\{ \ddot{\chi}^{(3)}(\omega_p, 0) \} >_{AV} = -2\ddot{C}\sqrt{\frac{4\pi}{D^2}} e^{-2S_j}$$

$$\times [\exp\{ -(\frac{\overline{\Delta}_P}{2D})^2 \} \frac{2\gamma_{e_n e_g}}{(2\gamma_{e_n e_g})^2 + \overline{\Delta}_P^2}]$$

$$+ S_j \exp\{ -(\frac{\overline{\Delta}_P + \omega_j}{2D})^2 \}\{ \frac{2\gamma_{e_n e_g} + \gamma_j / 2}{(2\gamma_{e_n e_g} + \gamma_j / 2)^2 + (\overline{\Delta}_P - \omega_j)^2}$$

$$+ \frac{2\gamma_{e_n e_g} + \gamma_j / 2}{(2\gamma_{e_n e_g} + \gamma_j / 2)^2 + (\overline{\Delta}_P + \omega_j)^2} \} + \cdots]. \tag{7-16}$$

Equation (7-16) implies that one can see a peak at $\overline{\Delta}_P = 0$, that is $\omega_P = \overline{\omega}_{e_n e_g} = \omega_B$ and the peak is associated with the first term on the left hand side of Eq. (7-16). The peak is due to the electronic 0-0 transition (zero phonon transition) and its band-width is originated entirely from the electronic dephasing process. In addition, in hole burning spectra, one can see first two components of a phonon side-bad, corresponding to the second and third terms in Eq. (7-17), located in lower and higher energy regions from the 0-0 peak, respectively. It should be noted that, by making comparison between the term with $v_j = 0$ in Eq. (7-5) and the first term on the left hand side of Eq. (7-15), one can see the band width of the 0-0 peak of hole burning spectra is two times broader than that appeared in absorption spectra.

For the multi-mode case, one can easily find

$$< \mathrm{Im}\{ \ddot{\chi}^{(3)}(\omega_P, 0) \} >_{AV} = -2\ddot{C}\sqrt{\frac{4\pi}{D^2}} \exp(-\sum_{j=1}^{N} 2S_j)$$

$$\times \prod_{j=1}^{N} \left(\sum_{v_j=0}^{\infty} \sum_{v'_j=0}^{\infty} \frac{S^{v_j+v'_j}}{v_j! v'_j!} \right) \exp\left[-\left\{ \frac{\overline{\Delta}_P + \overline{\Delta}_B + (v_j + v'_j)\omega_j}{2D} \right\}^2 \right]$$

$$\times \frac{2\gamma_{e_n e_g} + \sum_{j=1}^{N} (v_j + v'_j)\gamma_j / 2}{\{2\gamma_{e_n e_g} + \sum_{j=1}^{N} (v_j + v'_j)\gamma_j / 2\}^2 + \{\omega_B - \omega_P + \sum_{j=1}^{N} \omega_j (v_j - v'_j)\}^2}. \tag{7-17}$$

8. Optical Spectra of Photosynthetic Reaction Centers

8.1. Resonance Raman Spectra

The lowest-energy, near infrared absorption bands of the RCs result from the Q_y electronic transitions of the special pair P, two monomeric Bchl B molecules and two Bphe H molecules. The RCs isolated from R26 mutant *Rb. sphaeroides*, for example, have the near infrared absorption spectrum at room temperature consisting of three distinctive bands at about 860, 811 and 760nm, which are, respectively, assigned to the absorption transition in P, B and H[63,160,161]. However, each of those bands does not exhibit any vibrational progressions in the absorption spectrum.

Recently, reporting the Raman shifts and relative intensities, Boxer and Marthies and their co-workers have observed resonance Raman (RR) spectra of the special pair P and those of the monomeric Bchl B of *R26* mutant RCs of *Rb. sphaeroides*[75,76]. The RR measurements were performed at a temperature of 278K with excitations at 850 and 800nm that are, respectively, resonant with the lowest excitonic state P_-^*, an anti-symmetric linear combination of the two *Bchl* monomers comprising P, and the Q_y transition of B. We assume that 850nm excitation is associated with excitation of the lowest excitonic state of the dimer P. Here we summarize in Table I the Raman shifts and relative intensities experimentally

Table I.

Comparison of Raman Shifts and Intensites for Observed and Calculated Spectra of RRS of B Band of *Rb. sphaeroides R-26* mutant RCs.

Raman shift[(a)] $\Delta\nu_{obs}(cm^{-1})$	Assignment	$(\bar{\nu}_{obs} - \bar{\nu}_{cal})$ (cm^{-1})	Intensity[(a)] (obs.)	Intensity (cal.)
177	$\bar{\nu}_1$	0	1.00	1.00
212	$\bar{\nu}_2$	0	0.94	0.52
332	$\bar{\nu}_3$	0	0.45	0.36
353	$2\bar{\nu}_1$	1	0.36	0.74
385	$\bar{\nu}_1 + \bar{\nu}_2$	4	0.50	0.50
569	$2\bar{\nu}_1 + \bar{\nu}_2$	3	0.28	1.34
621	$2\bar{\nu}_2 + \bar{\nu}_1$	20	0.12	0.18
682	$2\bar{\nu}_1 + \bar{\nu}_3$	4	0.19	0.32
731	$\bar{\nu}_1 + \bar{\nu}_2 + \bar{\nu}_3$	10	0.70	0.10
761	$2\bar{\nu}_2 + \bar{\nu}_3$	5	0.10	0.07
891	$2\bar{\nu}_1 + \bar{\nu}_2 + \bar{\nu}_3$	7	0.43	0.07

(a) The observed data were taken from ref. [76].

observed in a frequency region of $0-1000\text{cm}^{-1}$ of the RR spectrum of B. The RR spectra both of P_-^* and of B exhibit Raman bands at 187, 204, 332, 564, 684, 730 899 and 1163cm^{-1}, while unique modes appear in a low-frequency region of the spectrum of P_-^* at 34, 71, 95, 128 and 484cm^{-1}.

8.1.1 Lowest Excitonic Band P_-^*

To study the RR spectrum of the lowest excitonic band of P_-^*, we utilize our Eq. (5-30) that is derived for a multi-mode system at low temperatures. With the assumption that the equilibrium nuclear geometry of P_-^* is only displaced along the low-frequency modes upon electronic excitation, we shall simulate, as an example for the application of theory, RR spectra of P_-^* for several cases. Applying the Fast Fourier transform method, we perform simulation for a system at a temperature of 0K, in the low temperature limit, that consists of four modes whose frequencies and coupling constants, respectively, are assumed to be $\omega_1 = 34\text{cm}^{-1}$, $\omega_2 = 95\text{cm}^{-1}$, $\omega_3 = 128\text{cm}^{-1}$ and $\omega_4 = 204\text{cm}^{-1}$ and $S_1 =$ (a) 2.7; (b) 2.2; (c) 1.5; (d) 0.0, $S_2 =$ 0.7, $S_3 =$ 0.15 and $S_4 =$ 0.02. These four modes were found in a low-frequency side of the of the spectrum of P_-^* as mentioned above. Table II summarizes the parameters used in the calculation. Figure 1 shows RR spectra simulated as a function of S_1. The panels (a), (b), (c) and (d) in Fig. 1, respectively, display the RR spectra calculated for, $S_1 =$ 2.7, 2.2, 1.5 and 0. In the panels (a) and (b), we find two very weak bands at 68 and 190cm^{-1}. The former band corresponds to the first overtone of the mode of 34cm^{-1}, while the later one is the first overtone of the mode of 95cm^{-1}. Figure 1 indicates that the coupling of the lower-frequency mode has a significant effect on the intensity of other modes. It should be noted here that our calculation is performed by including the four prominent modes observed in the low-frequency region of the RR spectra. The coupling constants for the first two modes are chosen such that their overtones should be obviously visible in the RR spectra.

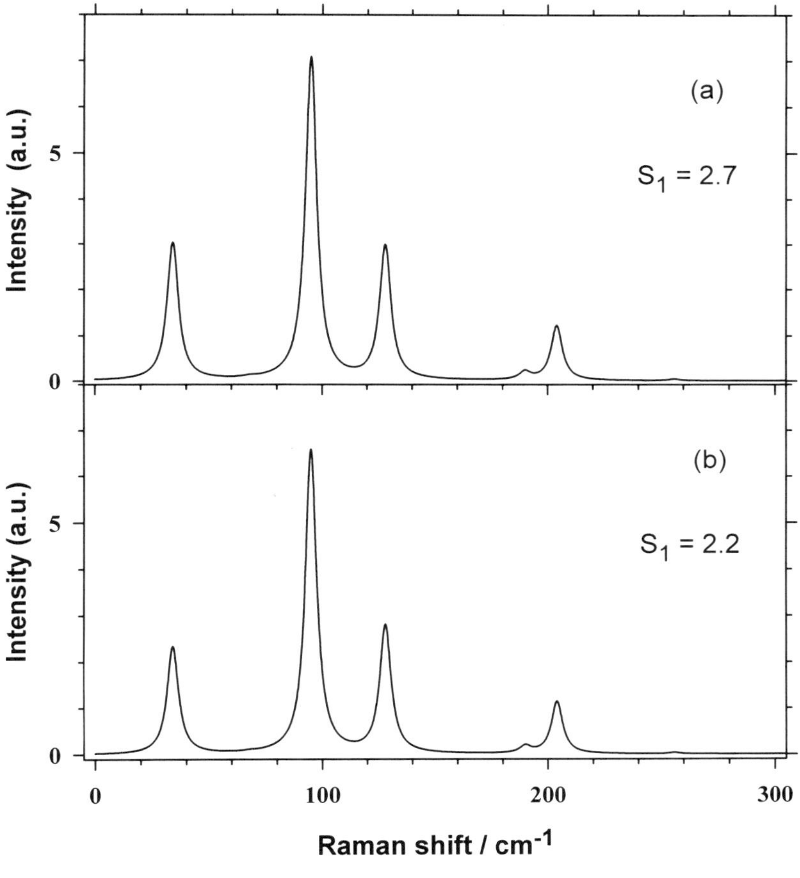

Fig. 1. Calculated resonance Raman spectra of a four-mode model for a low frequency region of the P_-^* band of *Rb. sphaeroides R26* RCs with 34, 95, 128 and 204cm^{-1}. The spectra are depicted as a function of S_1: (a) 2.7, (b) 2.2, (c) 1.5 and (d) 0.0. The rest of the parameters are listed in Table II.

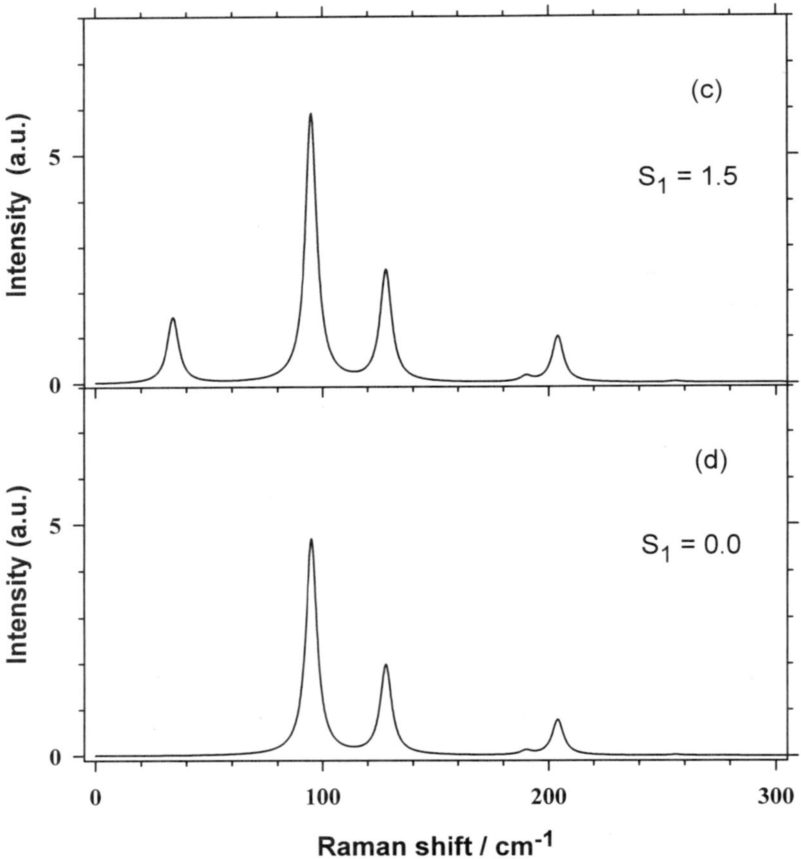

Fig. 1 (continued)

Table II.
Vibrational Frequencies and Corresponding S Values for $P \rightarrow P_-^*$ of R-26 RCs of *Rb. sphaeroides*

$P \rightarrow P_-^*$	Parameters for four vibrational modes			
	j=1	2	3	4
ω_j / cm^{-1}	34	95	128	204
S_j	2.5	0.70	0.015	0.020

Table III.
Vibrational Frequencies and Corresponding S Values for $B \rightarrow B^*$ of R-26 RCs of *Rb. sphaeroides*.

$B \rightarrow B^*$	Parameters for three vibrational modes		
	j=1	2	3
ω_j / cm^{-1}	177	212	332
S_j	0.5	0.04	0.04

8.1.2. Q_y Band of B

To interpret the Raman shifts and intensities summarized in Table I, according to our derivation in previous Chapters, we shall derive an expression of RR scattering intensity for a three-mode system that undergoes, for instance, the fundamental transition $\alpha_1\alpha_2\alpha_3 \to (\alpha_1+1)\alpha_2\alpha_3$. Recalling Eq. (5-16), we obtain an expression for the fundamental transition $\alpha \to \alpha+1$:

$$e^{i\alpha\omega\tau} g_{(\alpha+1)\alpha}(\tau) = -\sqrt{S}e^{-S} \sum_{k=0}^{\infty} S^k / k! \sum_{m=0}^{\alpha} \frac{[\alpha!(\alpha+1)!]^{1/2} S^m}{(\alpha-m)!m!(m+1)!}$$

$$\times \sum_{n=0}^{2m+1} (-1)^n \binom{2m+1}{n} e^{-i(k-m+n)\omega\tau}. \tag{8-1}$$

For the Rayleigh transition $\alpha \to \alpha$, we find

$$e^{i\alpha\omega\tau} g_{\alpha\alpha}(\tau) = e^{-S} \sum_{k=0}^{\infty} S^k / k! \sum_{m=0}^{\alpha} \frac{\alpha! S^m}{(\alpha-m)!(m!)^2}$$

$$\times \sum_{n=0}^{2m} (-1)^n \binom{2m}{n} e^{-i(k-m+n)\omega\tau}. \tag{8-2}$$

Substituting Eqs (8-1) and (8-2) into Eq. (5-7) yields

$$\tilde{g}_{(\alpha_1+1)\alpha_2\alpha_3,\alpha_1\alpha_2\alpha_3}(\omega_1) = -\sqrt{S}\exp(-S_1 - S_2 - S_{13})$$

$$\times \sum_{k_1=0}^{\infty} S_1^{k_1} / k_1! \sum_{m_1=0}^{\alpha_1} \frac{[\,\alpha_1!\,(\alpha_1+1)!\,]^{1/2} S_1{}^{m_1}}{(\alpha_1-m_1)!m_1!(m_1+1)!} \sum_{n_1=0}^{2m_1+1} \binom{2m_1+1}{n_1}$$

$$\times \sum_{k_2=0}^{\infty} S_2^{k_2} / k_2! \sum_{m_2=0}^{\alpha_2} \frac{\alpha_2! S_2{}^{m_2}}{(\alpha_2 - m_2)!(m_2!)^2} \sum_{n_2=0}^{2m_2} \binom{2m_2}{n_2}$$

$$\times \sum_{k_3=0}^{\infty} S_3^{k_3} / k_3! \sum_{m_3=0}^{\alpha_3} \frac{\alpha_3! S_3{}^{m_3}}{(\alpha_3 - m3_2)!(m_3!)^2} \sum_{n_3=0}^{2m_3} \binom{2m_3}{n_3}$$

$$\times \frac{i}{\omega_1 - \omega_{ba} - \sum_{j=1}^{3} (k_j - m_j + n_j)\omega_j + i\Gamma_{ba}} . \tag{8-3}$$

If molecules are initially distributed, with Boltzmann distribution $\rho_{\alpha_1\alpha_2\alpha_3}(T)$, then the intensity of the fundamental transition $\alpha_1\alpha_2\alpha_3 \rightarrow (\alpha_1 + 1)\alpha_2\alpha_3$ is given by

$$\frac{d^2\sigma}{d\Omega d\omega_s} = K\omega_l\omega_s^3 \sum_{\alpha_1,\alpha_2,\alpha_3} \rho_{\alpha_1\alpha_2\alpha_3}(T) J_{a(\alpha_1+1)\alpha_2\alpha_3 \rightarrow a\alpha_1\alpha_2\alpha_3}(\omega_\ell)$$

$$\times \frac{\Gamma_v}{(\omega_1 - \omega_\ell - \omega_S)^2 + \Gamma_v^2} , \tag{8-4}$$

where ω_S represents the frequency of a scattered photon, ω_ℓ stands for the frequency of the incident laser field, Γ_v is the vibrational dephasing constant. Here we have replaced the delta-function in Eq. (5-5) with the Lorentzian line shape. Expressions for RR intensities of overtone or combination bands can be also derived in a similar fashion as well.

We are now in a position to obtain a more quantitative picture of the magnitude of the vibrational-electronic coupling for the various vibrational modes. For this purpose, employing three dominant vibrational modes observed in the RR spectrum of the Q_y band of B, we shall simulate a RR spectrum of the B band. Figure 2 shows the RR spectrum at 278K calculated with Eq. (8-4) and similar expressions for overtone and combination bands. The coupling constants used in the simulation are listed in Table III. Assignment and calculated intensities for

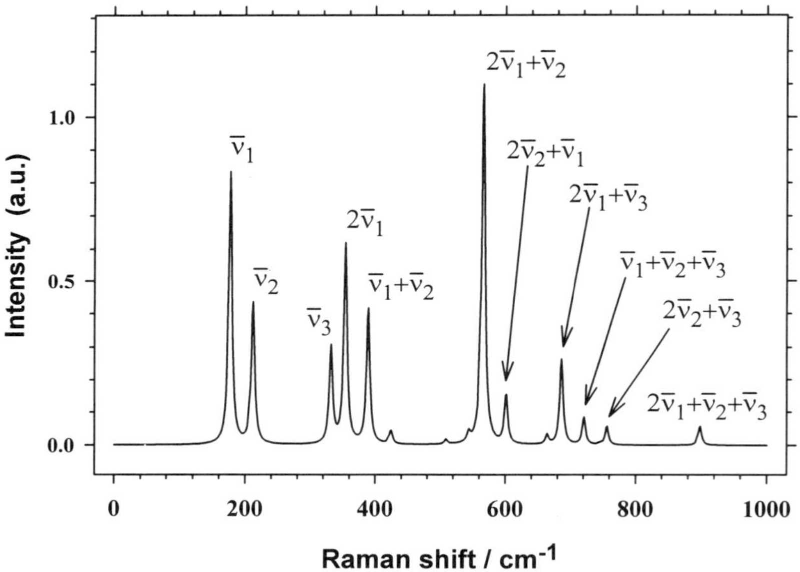

Fig. 2. Simulation of the resonance Raman spectrum of the Q_y band of B of *Rb. sphaeroides R26* RCs. The simulation is performed for a three-mode system with 177, 212 and 332 cm^{-1} The temperature is set to be 278K. The rest of the parameters used in the calculation are listed in Table III. The ratios of the overtone intensities to the fundermental ones depend strongly on the magnitude of the coupling constants.

fundamental, combination and overtone bands are also given in Table I. In this table, $\overline{\nu}_j$ corresponds to ω_j used in Table III and $\Delta\overline{\nu}_{obs} - \Delta\overline{\nu}_{cal}$ indicates the difference between observed and calculated Raman shifts are associated only with the combination and overtone bands. The coupling constants for 212 and 332cm^{-1} modes were constrained such that their overtones should be weakly visible in the simulated RR spectrum. We should note that the ratios of the overtone intensities to the fundamental ones are quite sensitive to the magnitude of the coupling constants. The simulation for the RR spectrum of B indicates that the lower frequency mode, *i.e.*, the 177cm^{-1} mode has a significant coupling with Q_y state. The electronic energy gap ω_{ba} was assumed to be 850nm. Our simulation, based on the parameters in Table III, shows a good agreement with the experimental data[76], although the calculated intensities tend to be rather smaller and larger at 891 and 569cm^{-1}, respectively. The RR peak at 731cm^{-1} in the experimental data could be a single mode which overlaps with the combination $\overline{\nu}_1 + \overline{\nu}_2 + \overline{\nu}_3$. It should be noted that, in this RR simulation, the effects of solvent-solute interactions (which not only broad vibronic lines but also affect the vibrational intensity), inhomogeneities and frequency distortions are not taken into account.

8.2. Absorption Spectra

For comparison, with Eq. (7-8), we also simulate absorption spectra of the P_-^* and B bands. We employ the molecular parameter from the analysis of the RR spectrum of the B band as seen above. The parameters for the simulation are listed in Table IV and Table V. Figure 3 shows the calculated absorption spectrum of P_-^* and B band region at T=278K. In Fig. 3, the solid broken and broken-dotted lines, respectively, indicate the calculated absorption spectrum of the P_-^* and B bands region, that of the B band component and that of the P band component.

We should note here that the Stokes shift of P_-^* with our parameters leads to 157.5cm^{-1}. There are slight discrepancies between this shift and that reported by

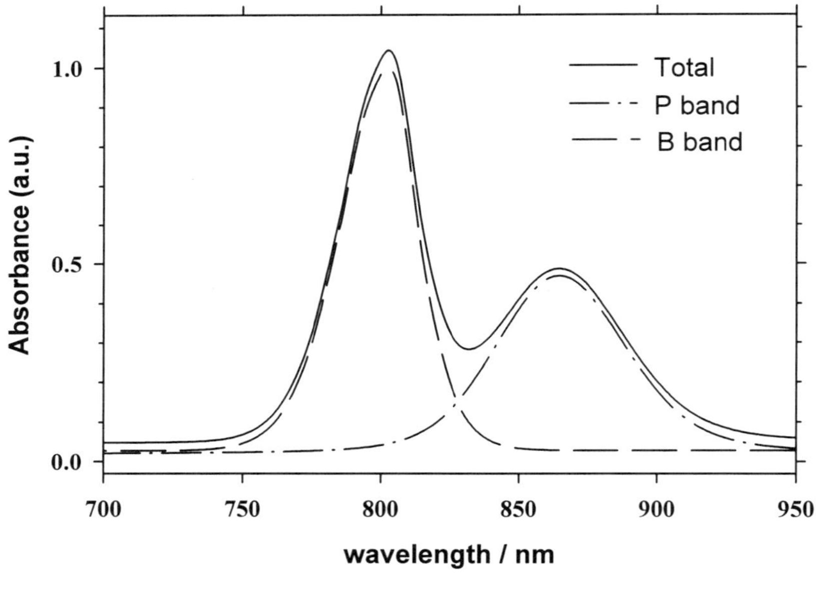

Fig. 3. Simulation of the absorption spectrum of the P_{-}^{*} and B bands of *Rb. sphaeroides R26* RCs at 278K. The parameters for the simulation are listed in Table IV and Table V. The solid, broken and broken-dotted lines indicate the calculated absorption spectrum of the P_{-}^{*} and B bands region, that of B band component and the of the P_{-}^{*} band component, respectively.

Table IV.
Parameters of Simulation for Absorption Spectrum of $P \rightarrow P_-^*$ of R-26 RCs of *Rb. sphaeroides*

$P \rightarrow P_-^*$	Parameters for four vibrational modes			
	j=1	2	3	4
ω_j / cm^{-1}	34	95	128	204
S_j	2.5	0.70	0.015	0.020
γ_j / cm^{-1}	5.0	10.0	7.42	4.65
$\sum_{j=1}^{4} S_j \omega_j$ / cm^{-1}	157.5			
Parameters for electronic states				
Γ_{ba} / cm^{-1}	1.00	D / cm^{-1}	140	

Table V.

Parameters of Simulation for Absorption Spectrum of $B \to B^*$ of R-26 RCs of *Rb. sphaeroides*

$B \to B^*$	Parameters for six vibrational modes		
	j=1	2	3
ω_j / cm^{-1}	177	212	332
S_j	0.5	0.04	0.04
γ_j / cm^{-1}	10.0	8.33	5.33
$\sum_{j=1}^{3} S_j\omega_j$ / cm^{-1}	110.26		
Parameters for electronic states			
Γ_{ba} / cm^{-1}	1.00	D / cm^{-1}	170

Small's group[66,67,69]. Their analysis was based on HB data observed by themselves. We will discuss this difference in the next section.

8.3. Hole Burning Spectra

Although the analysis of the RR spectra of the P_{-}^{*} and B of *Rb. sphaeroides R26* mutant RCs as seen above suggest that the coupling constants of the low frequency modes are large enough to participate in the optical transitions, the absorption spectra, for example, even at low temperatures, don't exhibit vibrational progressions due to those modes. There are two possible mechanisms of how optical transitions of molecules show no progressions in their spectra at low temperatures: (1) homogeneous broadening processes such as electronic dephasing or vibrational relaxation and (2) inhomogeneous broadening processes due to the existence of molecules under various environments in the ensemble so that the electronic structure of each molecule is slightly different from others. It is reasonable to consider that the above two mechanisms take part in the optical absorption process of the P_{-}^{*} and B of *Rb. sphaeroides R26* mutant RCs at low temperatures. To interpret the absorption spectra, for a better understanding of to what some extent inhomogeneity affects the energy structure of RCs, we shall demonstrate HB spectra of the P_{-}^{*} and B of *Rb. sphaeroides R26* mutant RCs. For this purpose, on the basis of our analysis of the RR and absorption spectra, we use the expression of multi-mode HB spectra, Eq. (7-16), derived in Chap. 7.

We first simulate HB spectra of P_{-}^{*} with a four-mode system consisting 34, 95, 128 and 204cm^{-1}, for various coupling constants of the 34cm^{-1} mode. Figures 4, 5 and 6 show the calculated HB spectra for $S_1 = 1.5$, 2.2 and 2.5, respectively. The rest of parameters used in the calculations are employed from Table IV. Each figure consists of 5 panels indicated by (a), (b), (c), (d) and (e) that, respectively, display the HB spectra burned at $\overline{\omega}_{ba} - \omega_B = -100, -50, 0, 50$ and 100cm^{-1}. In Fig. 4, one can see a sharp and narrow peak at the frequency in which the probing field is resonant to

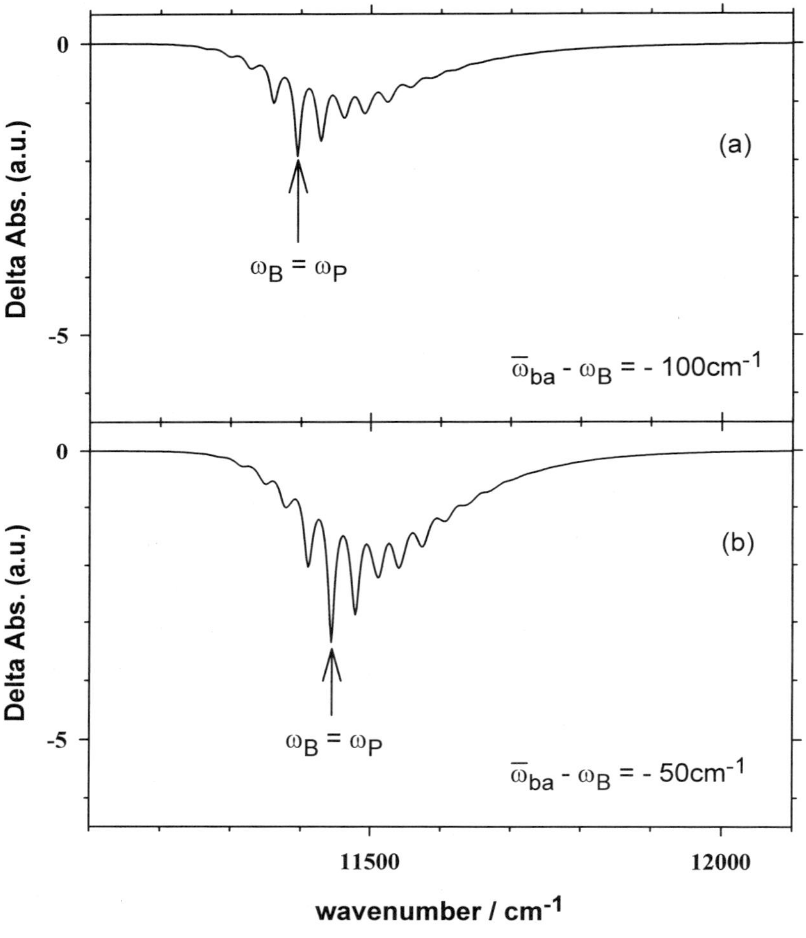

Fig. 4. Simulation of Hole-burning spectra of the P_-^* band of *Rb. sphaeroides R26* RCs at 0K. The calculations are performed for a four-mode system with the parameters listed in Table IV except for $S_1 = 1.5$, at burning frequencies $\bar{\omega}_{ba} - \omega_B =$ (a) -100, (b) -50, (c) 0, (d) 50 and (e) 100 cm^{-1}. The arrow in each panel indicates the frequency at which the probing field is resonant to the burning laser, leading to a narrow, distinctive peak in the HB spectrum of each case.

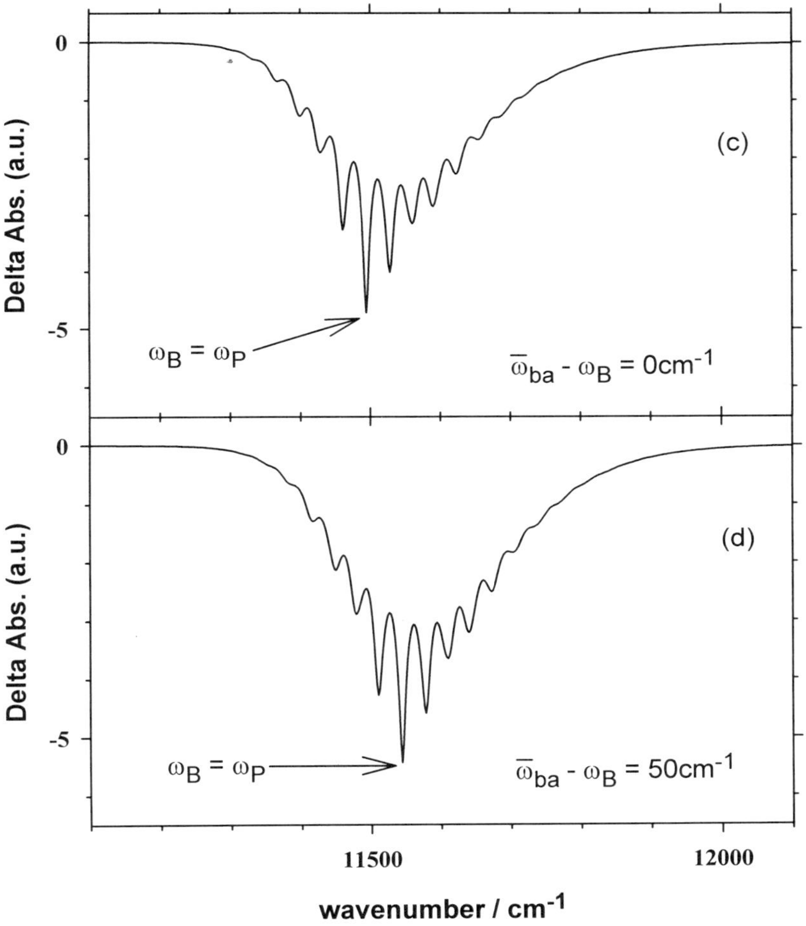

Fig. 4 (continued)

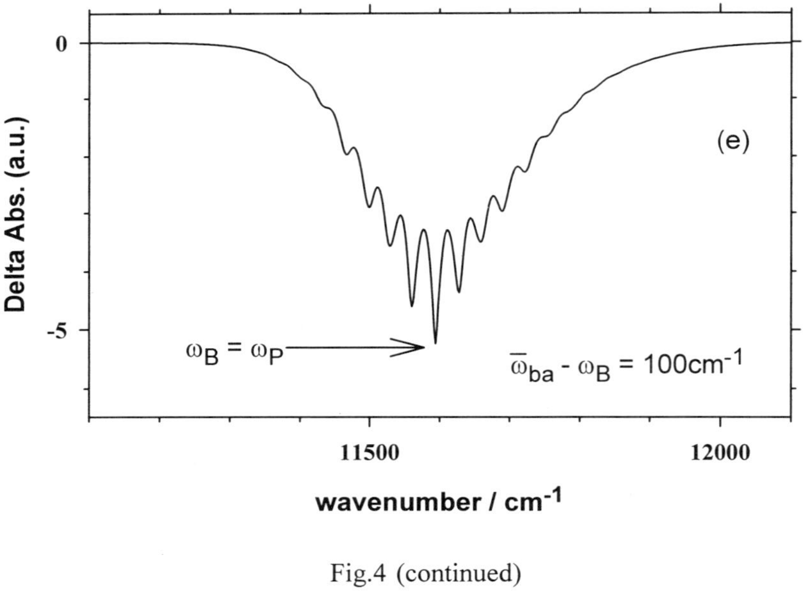

Fig.4 (continued)

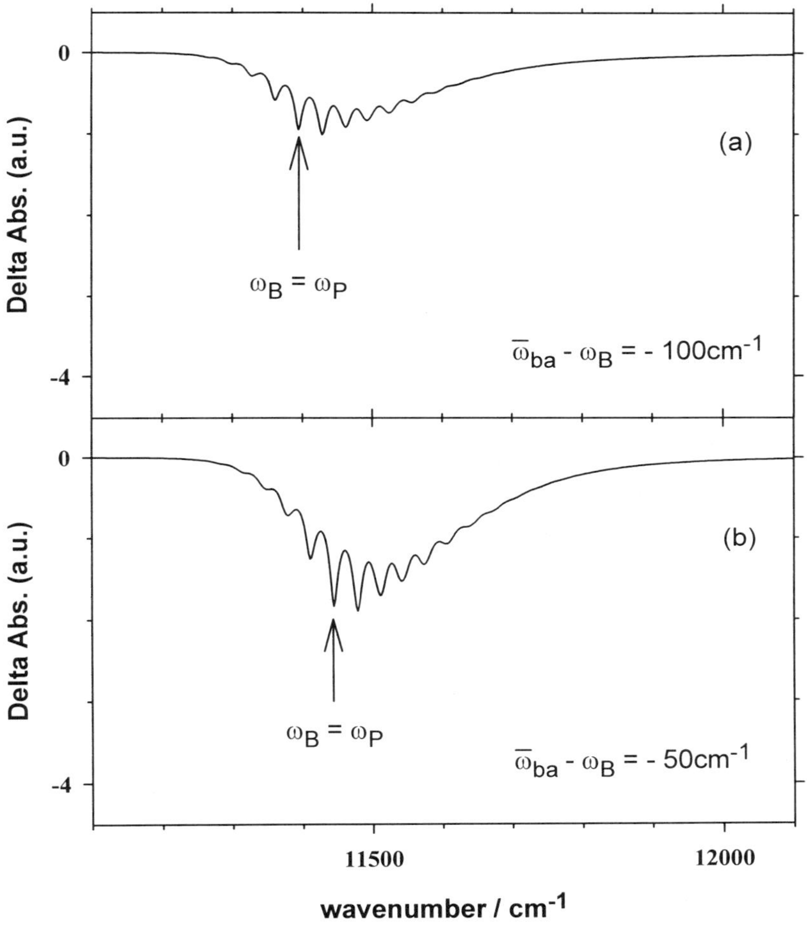

Fig. 5. Simulation of Hole-burning spectra of the P_-^* band of *Rb. sphaeroides R26* RCs at 0K. The parameters are the same as used in Fig. 4, except for $S_1 = 2.2$. Each HB spectrum shows two distinctive components: (1) a broad hole resulting mainly from homogeneous broadening processes and the large coupling constant of the $34\,cm^{-1}$ mode and (2) vibrational progressions mainly due to the $34\,cm^{-1}$ mode.

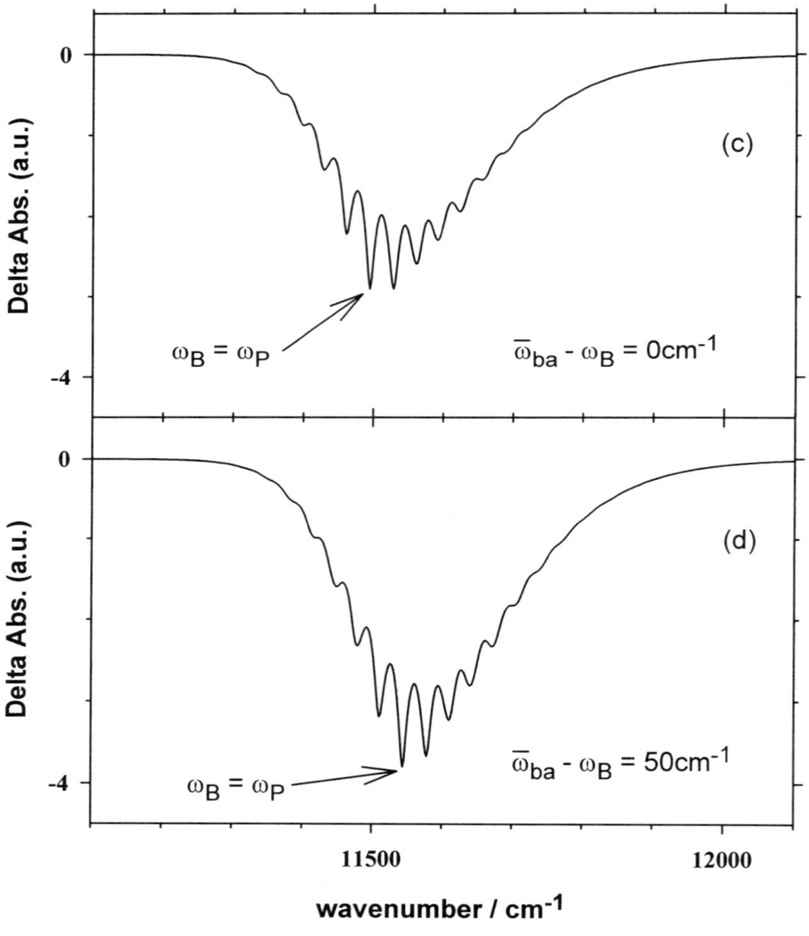

Fig. 5 (continued)

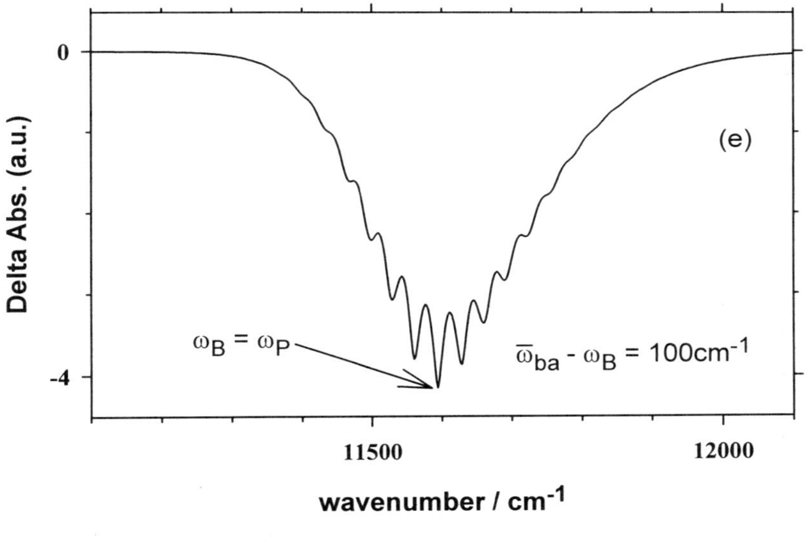

Fig. 5 (continued)

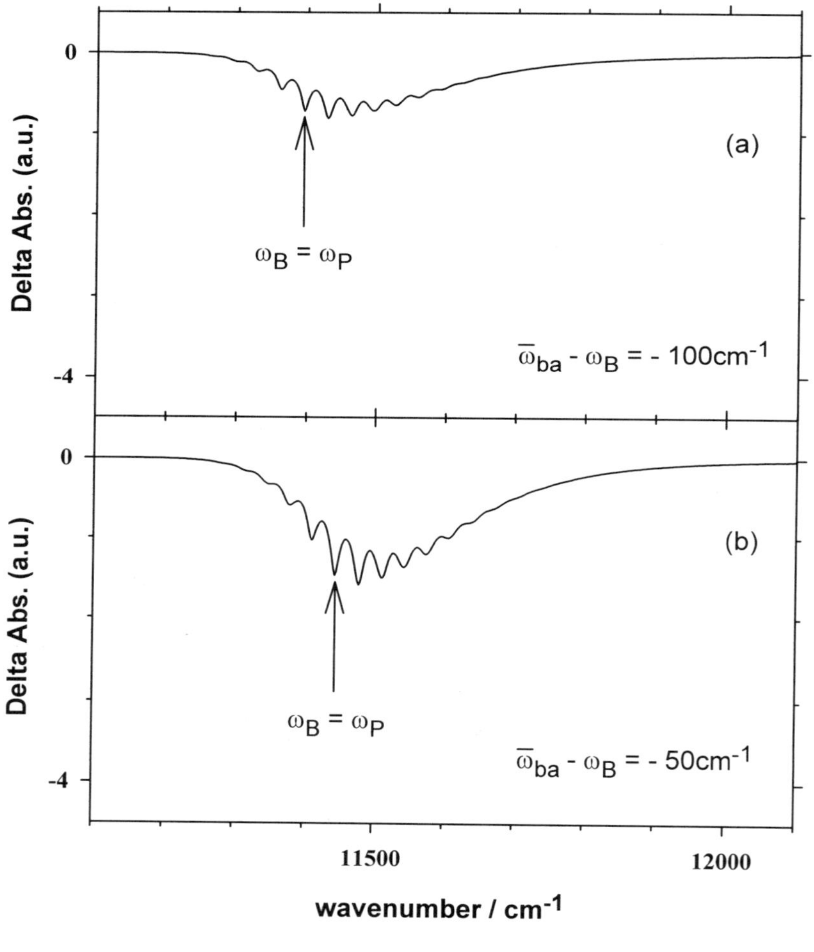

Fig. 6. Simulation of Hole-burning spectra of the P_-^* band of *Rb. sphaeroides R26* RCs at 0K. The parameters are the same as used in Fig. 4, except for $S_1 = 2.5$.

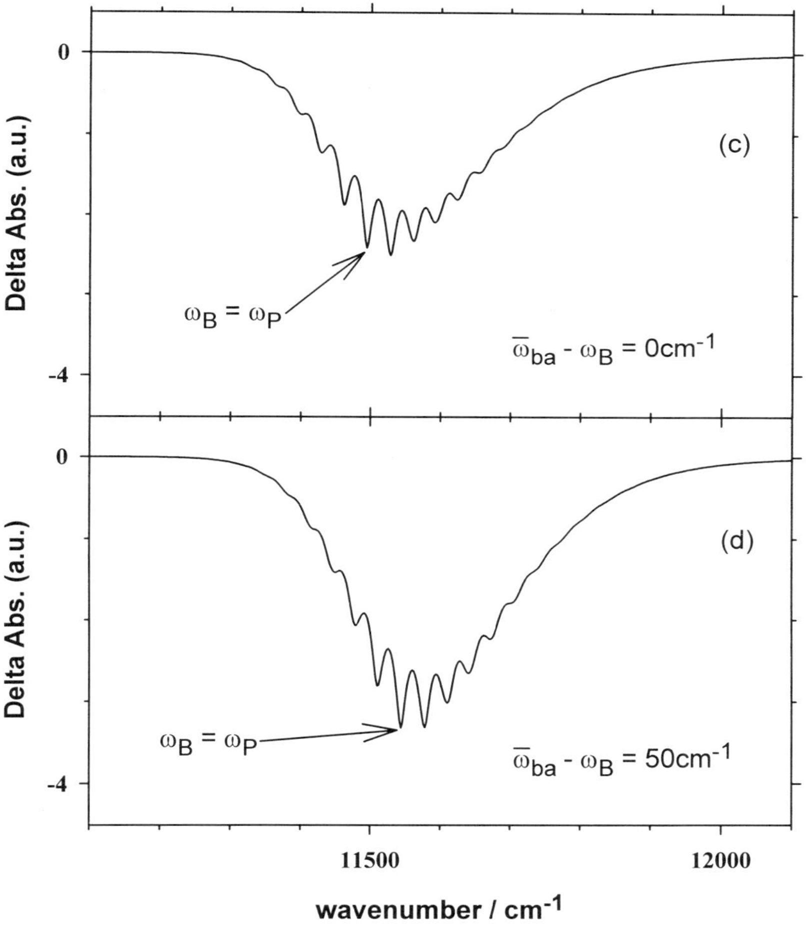

Fig. 6 (continued)

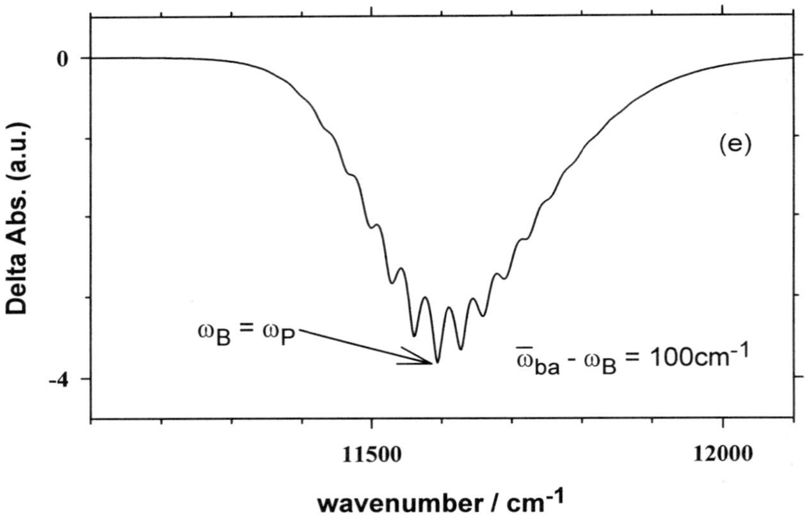

Fig. 6 (continued)

the burning laser. In addition, one can clearly see two distinctive components in the HB spectra; one is vibrational progressions mainly due to the low frequency mode and the other is a hole with a broad width resulting both from the vibrational relaxation and from the relatively large coupling constant of the 35cm^{-1} mode. Figures 5 and 6, on the other hand, exhibit broader peaks at $\omega_B = \omega_P$ compared with that as seen in Fig. 4. The broad holes, as well as the vibrational progressions, become broader as the coupling constant of the 34cm^{-1} mode increases.

It is also very much informative to demonstrate HB spectra of the Q_y band of B of *Rb. sphaeroides R26* mutant RCs. Figure 7 shows calculated HB spectra for various burning frequencies. The panels (a), (b), (c), (d) and (e), respectively, display HB spectra burned at $\overline{\omega}_{ba} - \omega_B = -100$, -50, 0, 50 and 100cm^{-1} at 0K. The calculations are performed for the three-mode system with the parameters used in the simulation of the absorption spectrum. One can see in Fig. 7 that the HB spectrum of each panel shows a sharp, distinctive peak at $\omega_B = \omega_P$, from which the spectrum exhibits side bands in the low and high frequency regions. This is mainly due to the fact that from our analysis of the RR spectrum of B shows there is no low frequency modes such as the 34cm^{-1} mode as seen in the P_-^* band.

To interpret hole-burning and absorption spectra of the primary donor state, Small has suggested that two displaced modes of frequencies 30 and 115cm^{-1} with corresponding coupling constants $S =$ 2.2 and 1.5, respectively, can fit the spectra [68,69,71] of *R26* mutant RCs of *Rb. sphaeroides*. Their analysis leads to 170 cm^{-1} of inhomogeneous broadening and 240cm^{-1} of $\sum_{j=1}^{2} S_j\omega_j$ that is associated with the Stokes shift of the system. The same group has also reported similar modes found both in protonated and in deuterated *R26* mutant RCs of *Rb. sphaeroides*, revealing 30cm^{-1} (1.8) and 120cm^{-1} (1.5) for the protonated RCs and 30cm^{-1} (2.0) and 115cm^{-1} (1.5) for the deuterated RCs and the corresponding inhomogeneous broadening are 150 and 130cm^{-1}, respectively[71].

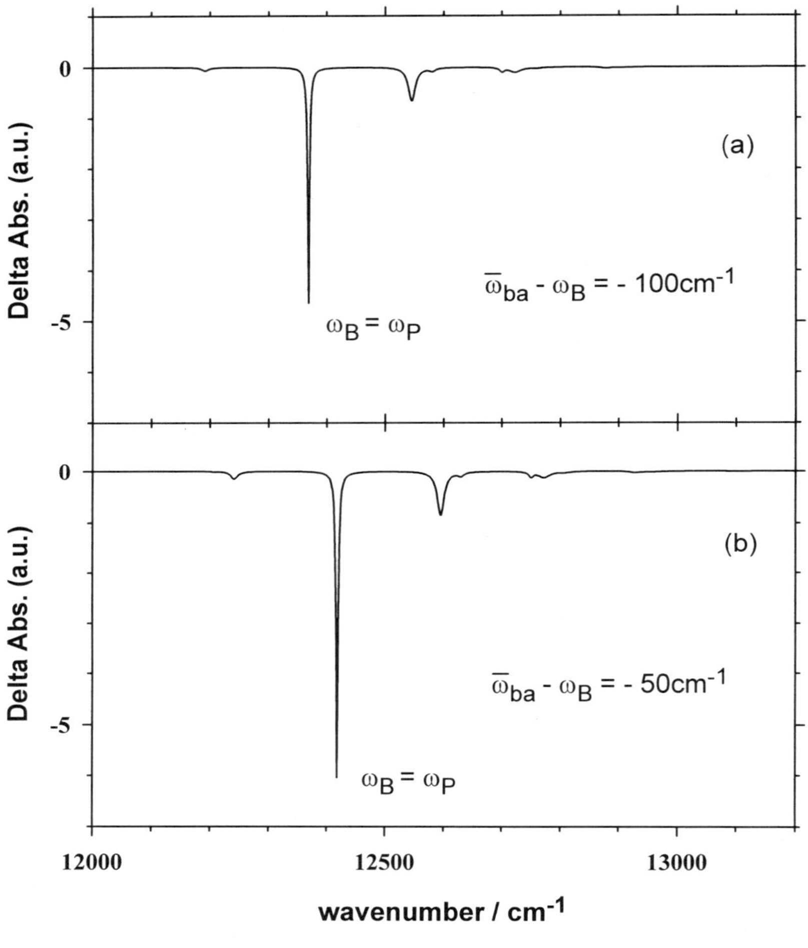

Fig. 7. Simulation of Hole-burning spectra of the Q_y band of *B* of *Rb. sphaeroides R26* RCs at 0K. The calculations are performed for the three-mode system used in Fig. 3. The paramters are listed in Table V. Without the low frequency mode as seen in the P_-^* band and the small coupling constants, Each HB spectrum shows a sharp, distinctive peak at $\omega_P = \omega_B$, clearly exhibiting side bands in the low and high frequency sides.

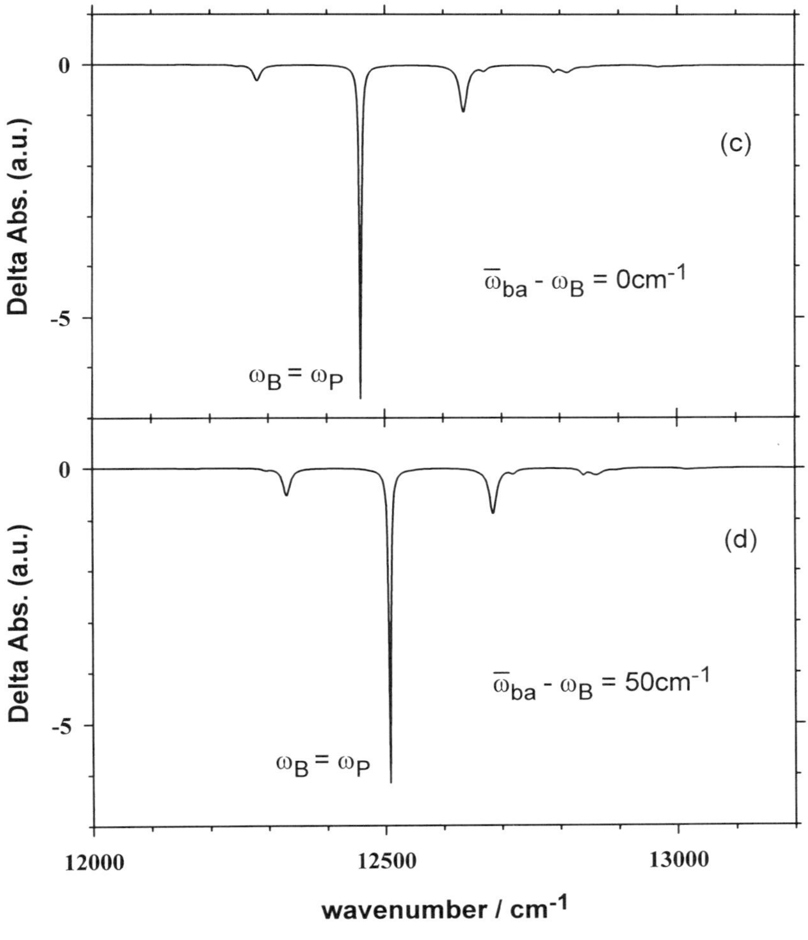

Fig. 7 (continued)

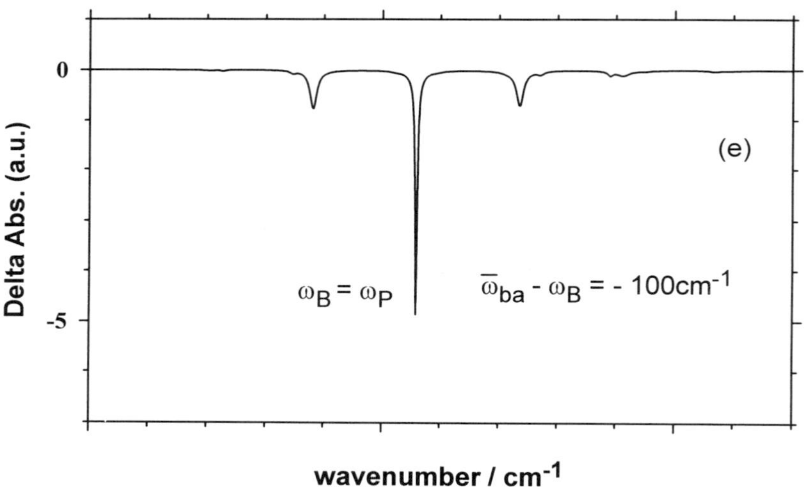

Fig. 7 (continued)

Boxer has also shown that a three-mode model can fit the absorption and hole-burning data of wild-type RCs of *Rb. sphaeroides* by employing modes of frequency ranges $20-50\text{cm}^{-1}$, $100-140\text{cm}^{-1}$ and $150-500\text{cm}^{-1}$ whose coupling constants ranging $1.7-2.6$, $0.5-1.2$ and $0.1-0.7$, respectively[62].

Based on our theoretical analysis of the RR spectrum of *R26* mutant RCs of *Rb. sphaeroides* reported by Mathies's and Boxer's groups[75,76], we find 157.4cm^{-1} of $\sum_{j=1}^{4} S_j\omega_j$ as already seen above. This discrepancy is probably due to the different properties of the two spectroscopic methods, such as symmetry restriction. Although one can find similar vibrational frequencies in both experimental data, they are different as vibrational modes. A more detailed analysis should be needed to assign those frequencies found in experimental data to molecular vibrational modes and motions. We should note here that there are also slight discrepancies concerning the coupling constants of the vibrational modes deduced from our analysis of the RR spectra and those from Small's analysis of the HB spectra. As far as ET rate is concerned, the discrepancy makes no important contribution to the ET rate constant since the energy gap of relevant states is much larger compared to the one quantum energies of those low frequency modes.

For comparison, we shall also demonstrate an absorption spectrum of one of the RCs under inhomogeneity. Figures 8 shows how a HB spectrum is associated with an absorption spectrum under the influence of inhomogeneity. The panels (a), (b) and (c) in Fig. 8 demonstrate a calculated P_-^* band absorption spectrum of one of the RCs, calculated P_-^* band absorption spectrum of the molecular ensemble used in the panel (b) of Fig. 8 and a simulated P_-^* band HB spectrum of the RCs taken from the panel (c) of Fig. 6. The 0-0 transition frequency ω_{ba} of the RC chosen here is assumed to be 11494.2cm^{-1} (870nm) that is exactly same as the averaged transition frequency $\overline{\omega}_{ba}$ used in the panel (b) in Fig. 8. Figure 8 clearly shows that the tails of the absorption and HB spectra in the high frequency regions from their distinctive peaks well follow the shape of the single RC absorption spectrum. This feature is

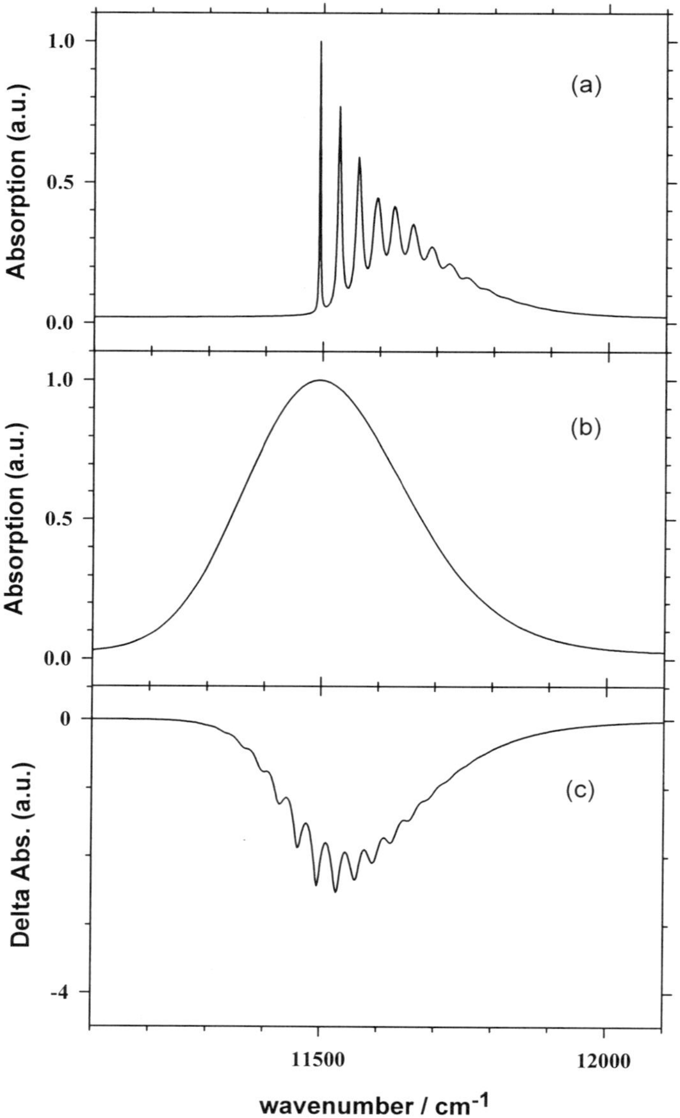

Fig. 8. Comparison of the calculated absorption and Hole-burning spectra. The panels show (a) absorption spectrum of one site of the ensemble used for P in Fig. 3, (b) the calculated spectrum of the ensemble and (c) the HB spectrum used in the panel (c) of Fig. 7. The site is chosen for the system with $\omega_{ba} = 11494.2\text{cm}^{-1}$ (870nm) .

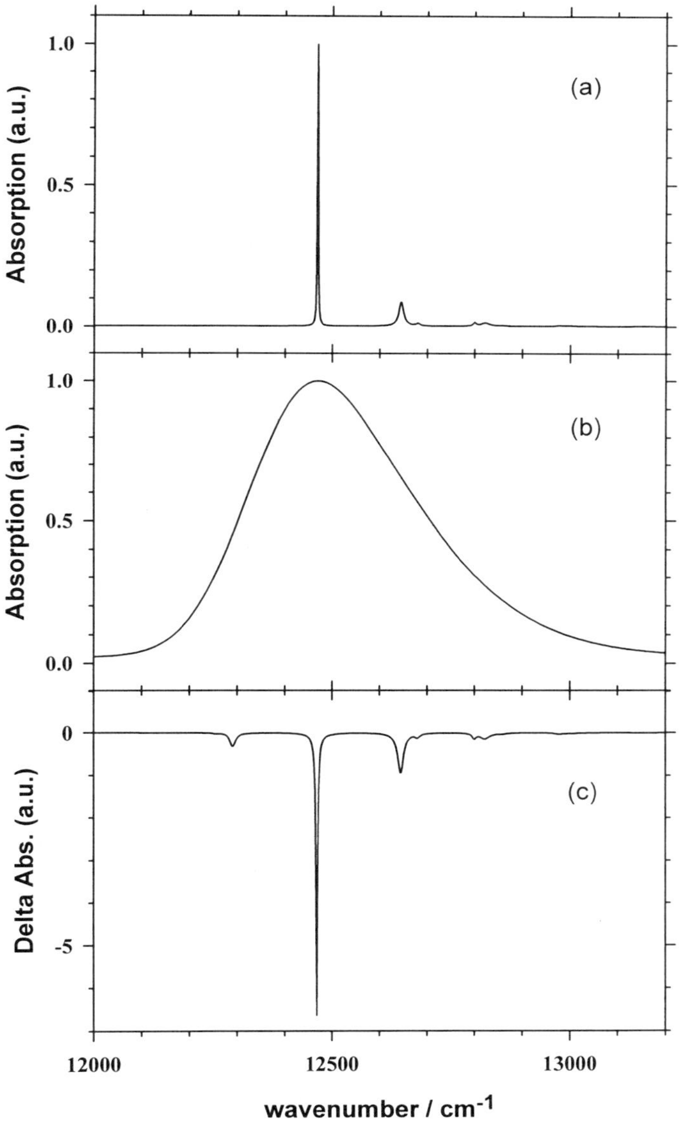

Fig. 9. Compariosn of the calculated absorption and Hole-burning spectra. The panels show (a) absorption spectrum of one site of the ensemble used for *B* in Fig. 3, (b) the absorption spectrum of the ensemble and (c) the HB spectrum used in the panel (c) of Fig. 7. The site in chosen for the system with $\omega_{ba} = 12468 cm^{-1}$ (802nm) .

mainly due to two reasons: one is that the vibrational relaxation becomes a dominant mechanism of the broadening and the other is because the low frequency mode has a relatively large coupling constant. It is also interesting to compare the bandwidth of each peak in the absorption spectrum of the RC and the HB spectra of the RCs. As already discussed in a previous Chapter, the bandwidth of the peak at $\omega_B = \omega_P$ in the HB spectrum of the RCs is two times broader than that seen in the absorption spectrum of the RC. This can be easily seen by recalling Eqs. (7-5) and (7-15). The other peaks in the HB spectrum also shows broader bandwidths compared with those seen in the absorption spectrum of the RC.

This situation can been much clearly seen in Fig. 9. Figure 9 shows a similar comparison for the Q_y band of B. A particular RC is chosen for a system with a 0-0 transition frequency of 12468cm^{-1} (802nm), which is also same as the averaged transition frequency $\overline{\omega}_{ba}$ used in the panel (b) of Fig. 9. In this case, without the low frequency mode that leads to an overlap between the peaks, the absorption spectrum and the HB spectrum of the RCs do not show broad bands as seen in Fig. 8.

9. Summary and Conclusion

The resonance Raman studies of bacterial photosynthetic RCs reveals the complexity of the problem of determining molecular parameters that are important to evaluate electron transfer constants of these systems. Since the system itself is a very complex system, in addition to heterogeneity inherent in the system, comprising of a number of the degrees of freedoms, analyses of the optical spectra are a real demanding task. Successful analyses, however, rely heavily on constructing a theoretical model that can consistently explain all the optical features found in the spectra. In previous sections, in spite of the complicated feature of the spectra observed by many workers, we concentrated on analysis of resonance Raman spectra based on our model.

It becomes unquestionably important to conduct further theoretical and experimental studies toward a complete understanding of the electron transfer mechanism of RCs. In particular, intensive analysis is needed for the assignment of vibrational modes that are crucial to the electron transfer reactions of RCs. If the high frequency modes, indeed, are relevant to electron transfer reactions, the applicability of the conventional electron transfer theory, then, should be revised carefully. Spectroscopic studies of the transient behavior of high frequency modes during the electron transfer reactions will provide us with a better understanding of the role and functions of such vibrational modes in electron transfer systems. This is currently assessed by Hochstrasser's group[81] and Zinth's group[82] by applying femtosecond time-resolved infra-red spectroscopy to electron transfer systems. Theoretical studies must be carried out to those experimental results in the future work.

Acknowledgments

We would like to thank Professor S. H. Lin for fruitful discussions and suggestions and strong encouragement to this work. This work was supported by Academia Sinica and the NSC of the Republic of China.

Appendix

Appendix I. Calculation of $\delta_{if}\vec{\varepsilon}_{\lambda\mu} \cdot \vec{\varepsilon}_{\lambda'\mu'}$

The commutation relation

$$(\vec{\varepsilon}_{\lambda'\mu'} \cdot \vec{r}_j)(\vec{\varepsilon}_{\lambda\mu} \cdot \vec{p}_j) - (\vec{\varepsilon}_{\lambda\mu} \cdot \vec{p}_j)(\vec{\varepsilon}_{\lambda'\mu'} \cdot \vec{r}_j) = i\hbar\vec{\varepsilon}_{\lambda} \cdot \vec{\varepsilon}_{\lambda'} \ . \quad \text{(A-1)}$$

can be written in the matrix form as

$$(m_j / \hbar)\sum_k \; [\, \omega_{ki} \, \{ \vec{\varepsilon}_{\lambda'\mu'} \cdot (\vec{r}_j)_{fk} \}\{ \vec{\varepsilon}_{\lambda\mu} \cdot (\vec{r}_j)_{ki} \}$$

$$- \omega_{fk} \, \{\vec{\varepsilon}_{\lambda\mu} \cdot (\vec{r}_j)_{fk} \}\{\vec{\varepsilon}_{\lambda\mu} \cdot (\vec{r}_j)_{ki}] \; = \; (\vec{\varepsilon}_{\lambda\mu} \cdot \vec{\varepsilon}_{\lambda'\mu'})\delta_{if} \, , \tag{A-2}$$

where we have used the relation:

$$\left\langle E_{n'} \middle| \vec{p}_j \middle| E_n \right\rangle = i m_j \omega_{n'n} \left\langle m \middle| \vec{r}_j \middle| n \right\rangle \; . \tag{A-3}$$

Also the commutation relation

$$(\vec{\varepsilon}_{\lambda'\mu'} \cdot \vec{r}_j)(\vec{\varepsilon}_{\lambda\mu} \cdot \vec{r}_j) - (\vec{\varepsilon}_{\lambda\mu} \cdot \vec{r}_j)(\vec{\varepsilon}_{\lambda'\mu'} \cdot \vec{r}_j) = 0 \tag{A-4}$$

gives

$$(m_j / \hbar)\sum_k \; \omega_{\lambda'} \{ \vec{\varepsilon}_{\lambda'\mu'} \cdot (\vec{r}_j)_{fk} \}\{ \vec{\varepsilon}_{\lambda\mu} \cdot (\vec{r}_j)_{ki} \}$$

$$-\omega_{\lambda'} \{ \vec{\varepsilon}_{\lambda'\mu'} \cdot (\vec{r}_j)_{fk} \}\{ \vec{\varepsilon}_{\lambda'\mu'} \cdot (\vec{r}_j)_{ki} \} = 0 \, . \tag{A-5}$$

By adding Eqs. (A-2) and (A-5), we obtain the necessary expression for $\delta_{if}\vec{\varepsilon}_{\lambda\mu} \cdot \vec{\varepsilon}_{\lambda'\mu'}$:

$$\delta_{if}\vec{\varepsilon}_{\lambda\mu} \cdot \vec{\varepsilon}_{\lambda'\mu'} = (m_j / \hbar) \sum_k \; [\, (\omega_{ki} + \omega_{\lambda'})$$

$$\times \{ \vec{\varepsilon}_{\lambda'\mu'} \cdot (\vec{r}_j)_{fk} \}\{ \vec{\varepsilon}_{\lambda\mu} \cdot (\vec{r}_j)_{ki} \}$$

$$- (\omega_{fk} + \omega_{\lambda'})\{ \vec{\varepsilon}_{\lambda\mu} \cdot (\vec{r}_j)_{fk} \}\{ \vec{\varepsilon}_{\lambda'\mu'} \cdot (\vec{r}_j)_{ki} \} \,] \; . \tag{A-6}$$

Appendix II. Time Correlation Functions

By making use of the electric dipole approximation and Eq. (A-3), then in the vicinity of resonance Eq. (2-12) can be written as

$$W_{if}^{\text{ab}} = \frac{4\pi^2}{\hbar V}\omega_{\lambda\mu}n_{\lambda\mu}\left|\langle f|\vec{M}\cdot\vec{\varepsilon}_{\lambda\mu}|i\rangle\right|^2\delta(\omega_{fi}-\omega_\lambda) \; . \tag{A-7}$$

If there is more than one initial state $|i\rangle$, that is, $|i\rangle$ can be regarded as a distribution of states, with the distribution ρ_i, the total absorption transition probability from the initially distributed states to all final states is $\sum_{ij}\rho_i W_{if}$.

Similarly, if the final state $|f\rangle$ consists of a distribution of states, with distribution ρ_f, then the total transition probability for the stimulated emission is $\sum_{if}\rho_f W_{if}$.

Therefore, the difference between probabilities is the net absorption rate

$$W = \frac{4\pi^2}{\hbar V}\omega_{\lambda\mu}n_{\lambda\mu}\sum_{if}(\rho_i-\rho_f)\left|\langle f|\hat{\vec{M}}\cdot\vec{\varepsilon}_{\lambda\mu}|i\rangle\right|^2\delta(\omega_{fi}-\omega_\lambda)$$

$$= \frac{4\pi^2}{\hbar V}\omega_{\lambda\mu}n_{\lambda\mu}(1-e^{-\beta\hbar\omega_\lambda})\sum_{if}\rho_i\left|\langle f|\hat{\vec{M}}\cdot\vec{\varepsilon}_{\lambda\mu}|i\rangle\right|^2\delta(\omega_{fi}-\omega) \; , \tag{A-8}$$

where in second equality it is assumed that molecular system is in thermal equilibrium, with $\omega_\lambda \approx \omega_{fi}$. By making use of the integral representation of the $\delta-$ function, we may then write Eq. (A-8) as

$$W = \frac{4\pi^2}{\hbar V}\omega_{\lambda\mu}n_{\lambda\mu}(1-e^{-\beta\hbar\omega_\lambda})$$

$$\times \int_{-\infty}^{\infty} d\tau e^{-i\omega_\lambda \tau} < (\hat{\vec{M}} \cdot \vec{\varepsilon}_{\lambda\mu})(\hat{\vec{M}}(\tau) \cdot \vec{\varepsilon}_{\lambda\mu}) > . \tag{A-9}$$

where $\hat{\vec{M}}(\tau) = \exp(i\hat{H}_0 \tau / \hbar) \hat{\vec{M}} \exp(-i\hat{H}_0 \tau / \hbar)$ is the dipole operator in Heisenberg picture, and $\langle \cdots \rangle$ is the dipole moment correlation function. If the incident photon intensity is $I = n_{\lambda\mu} \hbar\omega_\lambda c / V$, then the net absorption cross section is $\hbar\omega_\lambda W / I$, and the absorption coefficient is

$$\varepsilon(\omega_\lambda) = \frac{4\pi^2 N\omega_\lambda}{c\hbar}(1 - e^{-\beta\hbar\omega_\lambda})\, I(\omega_\lambda) \ . \tag{A-10}$$

where N is the number of molecules absorbing the photons and the absorption line shape $I(\omega_\lambda)$ is then defined by

$$I(\omega_\lambda) = \frac{1}{2\pi} \int_{-\infty}^{\infty} d\tau e^{-i\omega_\lambda \tau} < (\hat{\vec{M}} \cdot \vec{\varepsilon}_{\lambda\mu})(\hat{\vec{M}}(\tau) \cdot \vec{\varepsilon}_{\lambda\mu}) > \ . \tag{A-11}$$

For an isotropic medium such as liquid or gas

$$< \{ \hat{\vec{M}} \cdot \vec{\varepsilon}_{\lambda\mu} \} \{ \hat{\vec{M}}(\tau) \cdot \vec{\varepsilon}_{\lambda\mu} \} > \ = \frac{1}{3} < \hat{\vec{M}} \cdot \hat{\vec{M}}(\tau) > \ . \tag{A-12}$$

Appendix III: Recurrence Relations for $g_{\alpha'\alpha}(\tau)$

Notice that $g_{\alpha'\alpha}(\tau)$ for a multi-mode case can be written as

$$g_{\alpha'\alpha}(\tau) = \prod_j g_{\alpha'_j \alpha_j}(\tau) \ . \tag{A-13}$$

We recall Eq. (4-10);

$$g_{\alpha'_j \alpha_j}(\tau) = (\pi^2 2^{\alpha_j + \alpha'_j} \alpha'_j ! \alpha_j !)^{-1/2} (u_j^2 - w_j^2)^{1/2}$$

$$\times \exp\{ -2S_j(1-u_j^2) \} I_{\alpha'_j\alpha_j}(\tau) \ , \tag{A-14}$$

where

$$I_{\alpha'_j\alpha_j}(\tau) = \int\int dx_j dy_j H_{\alpha'_j}(u_j x_j - w_j y_j)\, H_{\alpha_j}(u_j x_j + w_j y_j)$$

$$\times \exp\{ -(x_j - s_j)^2 - y_j^2 \} \ , \tag{A-15}$$

with

$$u_j^2 = (1 + a_j \tanh \lambda_j / 2)^{-1},$$

$$w_j^2 = (1 + a_j \coth \lambda_j / 2)^{-1},$$

$$s_j = -\sqrt{2S_j}(1-u_j^2)/u_j,$$

$$a_j = \gamma'_j / \gamma''_j$$

and $S_j = \frac{1}{2}\gamma''_j d_j^2$. For simplicity in notation we hereafter omit the subscript j.

To derive the recurrence relations we may make use of the generating functions for the Hermite polynomials

$$\sum_{\alpha'} H_{\alpha'}(z_1)(t_1^{\alpha'} / \alpha'!) = \exp(-t_1^2 + 2t_1 z_1) \tag{A-16a}$$

and

$$\sum_{\alpha} H_{\alpha}(z_2)(t_2^{\alpha} / \alpha!) = \exp(-t_2^2 + 2t_2 z_2) \ . \tag{A-16b}$$

Multiplying both sides of Eqs. (A-16a) and (A-16b) together and then multiplying the resultant equation by the exponential in Eqs. (A-15), and making usr of the Gaussian integration formula, we finally obtain

$$\sum_{\alpha\alpha'} (I_{\alpha'\alpha} / \alpha'!\alpha!)\, t_1^{\alpha'} t_2^{\alpha} = \pi \exp\{ -(1-u^2-w^2)t_1^2 - (1-u^2-w^2)\, t_2^2$$

$$+ 2(u^2-w^2)t_1t_2 + 2su(t_1+t_2)\} \ . \qquad \text{(A-17)}$$

Taking derivatives of Eq.(A-17) with respect to t_1 and finding the same powers of t_1 and t_2 on both sides of the resultant equation, we easily obtain

$$I_{\alpha'\alpha} = -2(1-u^2-w^2)(\alpha'-1)I_{(\alpha'-2)\alpha}$$

$$+ 2(u^2-w^2)\ \alpha I_{(\alpha'-1)(\alpha-1)} + 2suI_{(\alpha'-1)\alpha} \qquad \text{(A-18a)}$$

and in a similar fashion we find with respect to t_2

$$I_{\alpha'\alpha} = -2(1-u^2-w^2)(\alpha-1)I_{\alpha'(\alpha-2)}$$

$$+ 2(u^2-w^2)\ \alpha' I_{(\alpha'-1)(\alpha-1)} + 2suI_{\alpha'(\alpha-1)} \ . \qquad \text{(A-18b)}$$

Substituting Eqs. (A-18a) and (A-18b) into Eq. (A-14), we, respectively, obtain the recurrence relation for $g_{\alpha'\alpha}(\tau)$

$$g_{\alpha'\alpha}(\tau) = -\ (1-u^2-w^2)(\frac{\alpha'-1}{\alpha'})^{1/2} g_{(\alpha'-2)\alpha}$$

$$+\ (u^2-w^2)(\frac{\alpha}{\alpha'})^{1/2} g_{(\alpha'-1)(\alpha-1)}$$

$$-\ 2\ (\frac{S}{\alpha'})^{1/2}(1-u^2)\ g_{(1-u^2)\alpha}\ , \qquad \text{(A-19a)}$$

and

$$g_{\alpha'\alpha}(\tau) = -(1-u^2-w^2)(\frac{\alpha-1}{\alpha})^{1/2} g_{\alpha'(\alpha-2)}$$

$$+ (u^2 - w^2)(\frac{\alpha'}{\alpha})^{1/2} g_{(\alpha'-1)(\alpha-1)}$$

$$- 2 (\frac{S}{\alpha})^{1/2} (1 - u^2)\, g_{\alpha'(\alpha-1)} \cdot \tag{A-19b}$$

In both cases, we find

$$g_{00}(\tau) = (u^2 - w^2)^{1/2} \exp\{ -2S(1-u^2) \} \cdot \tag{A-20}$$

By making use of the half-Fourier transform defined in Eq. (5-7), the recurrence relations in Eq. (A-19) may be written in the frequency domain, as

$$\tilde{g}_{\alpha'\alpha}(\omega_1) = (\frac{\alpha'-1}{\alpha'})^{1/2} \sum_k b^{2k+1} [\, \tilde{g}_{(\alpha'-2)\alpha}(\omega_1 - 2k\omega')$$

$$- \tilde{g}_{(\alpha'-2)\alpha}\{ (\omega_1 - 2(k+1)\omega' \}\,]$$

$$+ (1-b^2)(\frac{\alpha}{\alpha'})^{1/2} \sum_k b^{2k} \tilde{g}_{(\alpha'-1)(\alpha-1)}\{ \omega_1 + \omega'' - (2k+1)\omega' \}$$

$$- (\frac{S}{\alpha'})^{1/2} (1-b) \sum_k (-b)^k [\, \tilde{g}_{(\alpha'-1)\alpha}(\omega_1 - k\omega')$$

$$- \tilde{g}_{(\alpha'-1)\alpha}\{ \omega_1 - (k+1)\omega' \}\,] \tag{A-21a}$$

and

$$\tilde{g}_{\alpha'\alpha}(\omega_1) = (\frac{\alpha-1}{\alpha})^{1/2} \sum_k b^{2k+1} [\, \tilde{g}_{\alpha'(\alpha-2)}(\omega_1 + 2\omega'' - 2k\omega')$$

$$- \tilde{g}_{\alpha'(\alpha-2)}\{ \omega_1 + 2\omega'' - 2(k+1)\omega' \}\,]$$

$$+ (1-b^2)(\frac{\alpha'}{\alpha})^{1/2} \sum_k b^{2k} \tilde{g}_{(\alpha'-1)(\alpha-1)}\{ \omega_1 + \omega'' - (2k+1)\omega' \}$$

$$- (\frac{S}{\alpha})^{1/2} (1-b) \sum_k (-b)^k [\tilde{g}_{\alpha'(\alpha-1)}(\omega_1 + \omega'' - k\omega')$$

$$- \tilde{g}_{\alpha'(\alpha-1)}\{ \omega_1 + \omega'' - (k+1)\omega' \}]. \tag{A-21b}$$

For the displaced oscillator $b = 0$, $\omega' = \omega'' = \omega$, Eqs. (A-21a) and (A-21b) reduce to

$$\tilde{g}_{\alpha'\alpha}(\omega_1) = (\frac{\alpha}{\alpha'})^{1/2} \tilde{g}_{(\alpha'-1)(\alpha-1)}(\omega_1)$$

$$- (\frac{S}{\alpha'})^{1/2} \{ \tilde{g}_{(\alpha'-1)\alpha}(\omega_1) - \tilde{g}_{(\alpha'-1)\alpha}(\omega_1 - \omega) \} , \tag{A-22a}$$

and

$$\tilde{g}_{\alpha'\alpha}(\omega_1) = (\frac{\alpha'}{\alpha})^{1/2} \tilde{g}_{(\alpha'-1)(\alpha-1)}(\omega_1)$$

$$- (\frac{S}{\alpha})^{1/2} \{ \tilde{g}_{\alpha'(\alpha-1)}(\omega_1 + \omega) - \tilde{g}_{\alpha'(\alpha-1)}(\omega_1) \} , \tag{A-22b}$$

respectively. For this case, we find that $g_{00}(\omega_1)$ can be expressed as

$$g_{00}(\omega_1) = \sum_{k=0} \frac{1}{k!} \frac{iS^k e^{-S}}{\omega_1 - \omega_{ba} - k\omega + i\Gamma_{ba}} . \tag{A-23}$$

From Eqs. (A-22a) and (A-22b), for instance, the explicit expressions for $\tilde{g}_{\alpha'0}(\omega_1)$ ($\alpha' = 0, 1, 2, 3$ and 4) are, respectively, given as

$$\tilde{g}_{00}(\omega_1) = C \sum_{\ell} \binom{-1/2}{\ell} (-b^2)^{\ell} \sum_k \frac{S^k}{k!} (1-b^2)^k$$

$$\times \sum_n \binom{-k}{n} b^n \zeta_{k\ell n}(\omega_1) \ , \qquad \text{(A-24)}$$

$$\tilde{g}_{10}(\omega_1) = AC \sum_\ell \binom{-1/2}{\ell} (-b^2)^\ell \sum_k \frac{S^k}{k!} (1-b^2)^k$$

$$\times \sum_n \binom{-k-1}{n} b^n \{ \zeta_{k\ell n}(\omega_1) - \zeta_{k\ell n}(\omega_1 - \omega') \} \ , \qquad \text{(A-25)}$$

$$\tilde{g}_{20}(\omega_1) = BC \sum_\ell \binom{-1/2}{\ell} (-b^2)^\ell \sum_k \frac{S^k}{k!} (1-b^2)^k \sum_n \binom{-k-2}{n}$$

$$\times b^n \{ \zeta_{k\ell n}(\omega_1) - 2\zeta_{k\ell n}(\omega_1 - \omega') + \zeta_{k\ell n}(\omega_1 - 2\omega') \}$$

$$+ DC \sum_\ell \binom{-3/2}{\ell} (-b^2)^\ell \sum_k \frac{S^k}{k!} (1-b^2)^k$$

$$\times \sum_n \binom{-k}{n} b^n \{ \zeta_{k\ell n}(\omega_1) - \zeta_{k\ell n}(\omega_1 - 2\omega') \} \ , \qquad \text{(A-26)}$$

$$\tilde{g}_{30}(\omega_1) = EC \sum_\ell (e^{-1/2}) (-b^2)^\ell \sum_k \frac{S^k}{k!} (1-b^2)^k \sum_n \binom{-k-3}{n}$$

$$\times b^n \{ \zeta_{k\ell n}(\omega_1) - 3\zeta_{k\ell n}(\omega_1 - \omega')$$

$$+ 3\zeta_{k\ell n}(\omega_1 - 2\omega') - \zeta_{k\ell n}(\omega_1 - 3\omega') \}$$

$$+ FC \sum_\ell (e^{-3/2}) (-b^2)^\ell \sum_k \frac{S^k}{k!} (1-b^2)^k \sum_n \binom{-k-1}{n}$$

$$\times b^n \{ \zeta_{k\ell n}(\omega_1) - \zeta_{k\ell n}(\omega_1 - \omega')$$

$$- \zeta_{k\ell n}(\omega_1 - 2\omega') - \zeta_{k\ell n}(\omega_1 - 3\omega') \} \tag{A-27}$$

and

$$\tilde{g}_{40}(\omega_1) = CG\sum_{\ell} (e^{-1/2})(-b^2)^{\ell} \sum_{k} \frac{S^k}{k!}(1-b^2)^k \sum_{n} \binom{-k-4}{n}$$

$$\times b^n \{ \zeta_{k\ell n}(\omega_1) - 4\zeta_{k\ell n}(\omega_1 - \omega') + 6\zeta_{k\ell n}(\omega_1 - 2\omega')$$

$$- 4\zeta_{k\ell n}(\omega_1 - 3\omega') + \zeta_{k\ell n}(\omega_1 - 4\omega') \}$$

$$+ CH\sum_{\ell} (e^{-3/2})(-b^2)^{\ell} \sum_{k} \frac{S^k}{k!}(1-b^2)^k \sum_{n} \binom{-k-2}{n}$$

$$\times b^n \{ \zeta_{k\ell n}(\omega_1) - 2\zeta_{k\ell n}(\omega_1 - \omega')$$

$$+ 2\zeta_{k\ell n}(\omega_1 - 3\omega') - \zeta_{k\ell n}(\omega_1 - 4\omega') \}$$

$$+ CI\sum_{\ell} (e^{-1/2})(-b^2)^{\ell} \sum_{k} \frac{S^k}{k!}(1-b^2)^k \sum_{n} \binom{-k}{n}$$

$$\times b^n \{ \zeta_{k\ell n}(\omega_1) - 2\zeta_{k\ell n}(\omega_1 - 2\omega') + \zeta_{k\ell n}(\omega_1 - 4\omega') \} , \tag{A-28}$$

where

$$C \equiv (1-b^2)^{1/2} \exp\{ -S(1-b) \} ,$$

$$A \equiv -\sqrt{S}(1-b) ,$$

$$B \equiv S(1-b)^2 / \sqrt{2} ,$$

$$D \equiv b / \sqrt{2} ,$$

$$E \equiv -\{\sqrt{S}(1-b)\}^3/\sqrt{6} \quad ,$$

$$F \equiv -(3/2)^{1/2} b(1-b) \quad ,$$

$$G \equiv S^2(1-b)^4/2\sqrt{6} \quad ,$$

$$H \equiv (3/2)^{1/2} bS(1-b)^2 \quad ,$$

$$I \equiv (3/8)^{1/2} b^2$$

and

$$\zeta_{k\ell n}(\omega_1) = \frac{i}{\omega_1 - \Omega_{ba} - (2\ell + k + n)\omega' + i\Gamma_{ba}} \quad . \tag{A-29}$$

Appendix IV: Derivation of $\sigma_{ng}^{(3)}(t)_2$

Here we shall derive $\sigma_{ng}^{(3)}(t)_2$ that is appeared in Eq.(6-22).

$$\sigma_{ng}^{(3)}(t)_2 = -\frac{(-i)^3}{\hbar} \int_{t_i}^{t} d\tau_3 \int_{t_i}^{\tau_3} d\tau_2 \int_{t_i}^{\tau_2} d\tau_1$$

$$\times \sum_{k=g}^{n,m} [\, \tilde{L}_{\mathrm{I}}^{\tau 2}(\tau_2) \tilde{L}_{\mathrm{I}}^{\tau 1}(\tau_1) \hat{\sigma}_i \,]_{nk} \tilde{V}_{kg}^{\tau 3}(\tau_3), \tag{A-30}$$

where

$$[\tilde{L}_{\mathrm{I}}^{\tau 2}(\tau_2) \tilde{L}_{\mathrm{I}}^{\tau 1}(\tau_1) \hat{\sigma}_i]_{nn} = -\frac{(\hat{\sigma}_i)_{gg}}{\hbar^2} \{\, \tilde{V}_{gn}^{\tau 2}(\tau_2) \tilde{V}_{ng}^{\tau 1}(\tau_1)$$

$$+ \tilde{V}_{gn}^{\tau 1}(\tau_1) \tilde{V}_{ng}^{\tau 2}(\tau_2) \,\} \tag{A-31}$$

and

$$[\tilde{L}_{\mathrm{I}}^{\tau 2}(\tau_2) \tilde{L}_{\mathrm{I}}^{\tau 1}(\tau_1) \hat{\sigma}_i]_{nm} = -\frac{(\hat{\sigma}_i)_{gg}}{\hbar^2} \{\, \tilde{V}_{gm}^{\tau 2}(\tau_2) \tilde{V}_{ng}^{\tau 1}(\tau_1)$$

$$+ \tilde{V}_{gm}^{\tau 1}(\tau_1) \tilde{V}_{ng}^{\tau 2}(\tau_2) \,\} . \tag{A-32}$$

In the rotating wave approximation, we find

$$\sigma_{ng}^{(3)}(t)_2 = (\frac{i}{\hbar})^3 (\hat{\sigma}_i)_{gg} \{ \vec{\mu}_{ng} \cdot \vec{A}_{\mathrm{P}}^{(-)}(\omega_{\mathrm{P}}) \} \int_{t_i}^{t} d\tau_3 \int_{t_i}^{\tau_3} d\tau_2 \int_{t_i}^{\tau_2} d\tau_1$$

$$\times e^{i\tau_3(\omega'_{mg} - \omega_{\mathrm{P}})} \{ e^{i\tau_2(\omega'_{ng} - \omega_{\mathrm{P}})} e^{i\tau_1(\omega'_{gm} + \omega_{\mathrm{B}})}$$

$$+ e^{i\tau_2(\omega'_{gm} + \omega_{\mathrm{B}})} e^{i\tau_1(\omega'_{ng} - \omega_{\mathrm{P}})} \}. \tag{A-33}$$

Assuming that $\Gamma_{mg} t_i$, $\Gamma_{ng} t_i << -1$ in the integration, one can obtain

$$\sigma_{ng}^{(3)}(t)_2 = \frac{(\hat{\sigma}_i)_{gg}}{\hbar^3} | \vec{\mu}_{mg} \cdot \vec{A}_{\mathrm{B}}^{(-)}(\omega_{\mathrm{B}}) | \{ \vec{\mu}_{ng} \cdot \vec{A}_{\mathrm{P}}^{(-)}(\omega_{\mathrm{P}}) \}$$

$$\times \{ \frac{1}{\omega'_{gm} + \omega_{\mathrm{B}}} + \frac{1}{\omega'_{ng} - \omega_{\mathrm{P}}} \} \frac{e^{it(\omega'_{ng} - \omega_{\mathrm{P}})}}{(\omega'_{nm} - \omega_{\mathrm{P}} + \omega_{\mathrm{B}})(\omega'_{ng} - \omega_{\mathrm{P}})}. \tag{A-34}$$

First we consider $\sigma_{ng}^{(3)}(t)_2$ which can be simplified as

$$\sigma_{ng}^{(3)}(t)_2 = (\frac{-i}{\hbar})^3 \int_{t_i}^{t} d\tau_3 \int_{t_i}^{\tau_3} d\tau_2 \int_{t_i}^{\tau_2} d\tau_1 \sum_{\{g'\} \neq g} \sum_{k = \{n\}, \{m\}} \widetilde{V}_{ng'}^{\tau_3}(\tau_3)$$

$$\times [\widetilde{V}_{g'k}^{\tau_2}(\tau_2) \widetilde{V}_{kg}^{\tau_1}(\tau_1) (\hat{\sigma}_i)_{gg} + \widetilde{V}_{g'k}^{\tau_1}(\tau_1) \widetilde{V}_{kg}^{\tau_2}(\tau_2) (\hat{\sigma}_i)_{g'g'}]. \tag{A-35}$$

In the RWA, we find

$$\sigma_{ng}^{(3)}(t)_2 = (\frac{-i}{\hbar})^3 \int_{t_i}^{t} d\tau_3 \int_{t_i}^{\tau_3} d\tau_2 \int_{t_i}^{\tau_2} d\tau_1 \sum_{\{g'\} \neq g} \sum_{\{m\}} \widetilde{V}_{ng'}^{\tau_3}(\tau_3)$$

$$\times [\widetilde{V}_{g'm}^{\tau_2}(\tau_2) \widetilde{V}_{mg}^{\tau_1}(\tau_1) (\hat{\sigma}_i)_{gg} + \widetilde{V}_{g'm}^{\tau_1}(\tau_1) \widetilde{V}_{mg}^{\tau_2}(\tau_2) (\hat{\sigma}_i)_{g'g'}]. \tag{A-36}$$

and Eq. (A-36) can be easily integrated as

$$\sigma_{ng}^{(3)}(t)_2 = \frac{1}{\hbar^3} \{ \vec{\mu}_{ng} \cdot \vec{A}_{\mathrm{P}}^{(-)}(\omega_{\mathrm{P}}) \} \frac{e^{it(\omega'_{ng} - \omega_{\mathrm{P}})}}{\omega'_{ng} - \omega_{\mathrm{P}}}$$

$$\times \sum_{\{g'\} \neq g} \sum_{\{m\}} \{ \vec{\mu}_{g'm} \cdot \vec{A}_{\mathrm{B}}^{(-)}(\omega_{\mathrm{B}}) \} \{ \vec{\mu}_{mg} \cdot \vec{A}_{\mathrm{B}}^{(+)}(\omega_{\mathrm{B}}) \}$$

$$\times \frac{1}{\omega'_{g'g}} \{ \frac{(\hat{\sigma}_i)_{g'g'}}{\omega_{\mathrm{B}} - \omega'_{g'm}} + \frac{(\hat{\sigma}_i)_{gg}}{\omega'_{mg} - \omega_{\mathrm{B}}} \}. \tag{A-37}$$

Equation (A-37) indicates that the magnitude of this equation is less than that of Eq. (6-27) by the factor $\frac{1}{\omega_{g'g}}$ and Eq. (6-27) makes a dominant contribution to $\vec{P}^{(3)}(t)$.

Next we consider $\sigma_{ng}^{(3)}(t)_2$ which can be separated as

$$\sigma_{ng}^{(3)}(t) = \sigma_{ng}^{(3)}(t)_{12} + \sigma_{ng}^{(3)}(t)_{22}, \tag{A-38}$$

where

$$\sigma_{ng}^{(3)}(t)_{21} = -\frac{(-i)^3}{\hbar} \int_{t_i}^{t} d\tau_3 \int_{t_i}^{\tau_3} d\tau_2 \int_{t_i}^{\tau_2} d\tau_1$$

$$\times \sum_{\{n'\}} [\widetilde{L}_{\rm I}^{\tau 2}(\tau_2)\widetilde{L}_{\rm I}^{\tau 1}(\tau_1)\hat{\sigma}_i\,]_{nn'}\widetilde{V}_{n'g}^{\tau 3}(\tau_3) \tag{A-39}$$

and

$$\sigma_{ng}^{(3)}(t)_{22} = -\frac{(-i)^3}{\hbar} \int_{t_i}^{t} d\tau_3 \int_{t_i}^{\tau_3} d\tau_2 \int_{t_i}^{\tau_2} d\tau_1 \,.$$

$$\times \sum_{\{m\}} [\,\widetilde{L}_{\rm I}^{\tau 2}(\tau_2)\widetilde{L}_{\rm I}^{\tau 1}(\tau_1)\hat{\sigma}_i\,]_{nm}\widetilde{V}_{mg}^{\tau 3}(\tau_3)\,. \tag{A-40}$$

It follows that

$$\sigma_{ng}^{(3)}(t)_{21} = (\frac{-i}{\hbar})^3 \int_{t_i}^{t} d\tau_3 \int_{t_i}^{\tau_3} d\tau_2 \int_{t_i}^{\tau_2} d\tau_1 \sum_{\{n'\}}\sum_{\{g'\}} (\,\hat{\sigma}_i)_{g'g'} \tag{A-41}$$

$$\times \widetilde{V}_{n'g}^{\tau 3}(\tau_3)[\widetilde{V}_{g'n'}^{\tau 2}(\tau_2)\widetilde{V}_{ng'}^{\tau 1}(\tau_1) + \widetilde{V}_{g'n'}^{\tau 1}(\tau_1)\widetilde{V}_{ng'}^{\tau 2}(\tau_2)\,] \tag{A-42}$$

and

$$\sigma_{ng}^{(3)}(t)_{22} = (\frac{-i}{\hbar})^3 \int_{t_i}^{t} d\tau_3 \int_{t_i}^{\tau_3} d\tau_2 \int_{t_i}^{\tau_2} d\tau_1 \sum_{\{m\}}\sum_{\{g'\}} (\,\hat{\sigma}_i\,)_{g'g'}$$

$$\times \widetilde{V}_{mg}^{\tau 3}(\tau_3)[\,\widetilde{V}_{g'm}^{\tau 2}(\tau_2)\widetilde{V}_{ng'}^{\tau 1}(\tau_1) + \widetilde{V}_{g'm}^{\tau 1}(\tau_1)\widetilde{V}_{ng'}^{\tau 2}(\tau_2)\,]\,. \tag{A-43}$$

It should be noted that although $\sigma_{ng}^{(3)}(t)_{21}$ and $\sigma_{ng}^{(3)}(t)_{22}$ are not important in ordinary hole burning experiments, they may be important in transient hole burning experiments.

Notice that the contribution to $\ddot{\chi}^{(3)}(\omega_{\rm P})_1$ and $\ddot{\chi}^{(3)}(\omega_{\rm P})_2$ have opposite contributions to hole burning.

Appendix V: Inhomogeneous Broadening Mechanism

It should be noted that the absorption coefficient for the electronic transition $g \to n$ at a temperature T can be expressed as

$$\alpha(\omega,T) = \frac{4\pi^2\omega}{3a_s\hbar c} | \vec{\mu}_{e_n e_g} |^2 \sum_{u=0}^{\infty} \sum_{v=0}^{\infty} P_{e_g u}(T)$$

$$\times | \left\langle \Theta_{e_g u} \middle| \Theta_{e_n v} \right\rangle |^2 \delta(\omega_{e_n v, e_g u} - \omega) . \tag{A-44}$$

A similar expression exists for emission spectra. For simplicity, we assume that the effect of inhomogeneity is only on the electronic transition frequency $\omega_{e_n e_g}$. In this case, Eq. (A-44) becomes

$$\alpha(\omega,T) = \sum_{j=1}^{M} \alpha_j(\omega,T) , \tag{A-45}$$

where

$$\alpha_j(\omega,T) = \frac{4\pi^2\omega}{3a_s\hbar c} | \vec{\mu}_{e_n e_g} |^2 \sum_{u=0}^{\infty} \sum_{v=0}^{\infty} P_{e_g u}(T)$$

$$\times | \left\langle \Theta_{e_g u} \middle| \Theta_{e_n v} \right\rangle |^2 \delta(\omega_{e_n v, e_g u}(j) - \omega) . \tag{A-46}$$

Introducing the integral representation for the delta function, we find

$$\alpha_j(\omega,T) = \frac{4\pi^2\omega}{3a_s\hbar c} | \vec{\mu}_{e_n e_g} |^2 \int_{-\infty}^{\infty} d\tau \, \exp\{ \omega_{e_n e_g}(j) - \omega \} G(\tau,T) , \tag{A-47}$$

where $\omega_{e_n v, e_g u} = \omega_{e_n e_g} + \dfrac{E_{e_n v} - E_{e_g u}}{\hbar}$ and

$$G(\tau,T) = \sum_{u=0}^{\infty} \sum_{v=0}^{\infty} P_{e_g u}(T) \left| \left\langle \Theta_{e_g u} \middle| \Theta_{e_n v} \right\rangle \right|^2 \exp\{ \frac{i\tau}{\hbar}(E_{e_n v} - E_{e_g u}) \} . \tag{A-48}$$

We shall let $I(\omega,\tau)$

$$I(\omega,\tau) = \sum_{j=1}^{M} \exp\{ i\tau\omega_{e_n e_g}(j) - \omega \} . \tag{A-49}$$

In Eqs. (A-40) and (A-49), the summation over j refers to inhomogeneity.

For the case in which the total system consists of numerous molecules, one can introduce a distribution factor. Equation (A-49) can be rewritten as

$$I(\omega,\tau) = \int_{-\infty}^{\infty} d\vec{R}\ g(R) \exp\{ i\tau\omega_{e_n e_g}(R) - \omega \} . \tag{A-50}$$

For convenience, we shall rewrite Eq. (A-50) as

$$I(\omega,\tau) = \int_{-\infty}^{\infty} d\vec{R}\ g(R) \exp\{ i\tau\omega_{e_n e_g}(R) - \omega \} . \tag{A-51}$$

For convenience we shall rewrite Eq. (A-51) as

$$I(\omega,\tau) = \int_{-\infty}^{\infty} d\vec{R}\ g(R) \exp[-i \int_0^{\tau} ds \{ \omega - \omega_{e_n e_g}(R) \}]$$

$$= < \exp[-i \int_0^{\tau} ds \{ \omega - \omega_{e_n e_g}(R) \}] > . \tag{A-52}$$

Using the expansion, we obtain

$$I(\omega,\tau) = \int_{-\infty}^{\infty} d\vec{R}\ g(R) \exp\{ -i \sum_{m=1}^{\infty} \int_0^{\tau} ds\ c^{(m)}(s) \} , \tag{A-53}$$

where

$$c^{(1)}(s) = < L_1(s) > \tag{A-54}$$

$$c^{(2)}(s) = -i\int_0^\tau ds \; \{ < L_1(\tau)L_1(s) > - < L_1(\tau) >< L_1(s) > \}$$

$$etc.. \tag{A-55}$$

Notice that in our case

$$L_1(\tau) = \omega - \omega_{e_n e_g}(R), \tag{A-56}$$

$$< L_1(\tau) >= \int_{-\infty}^{\infty} d\vec{R}\, g(R)\{ \omega - \omega_{e_n e_g}(R) \}, \tag{A-57}$$

$$< L_1(\tau)^2 >= \int_{-\infty}^{\infty} d\vec{R}\, g(R)\{ \omega - \omega_{e_n e_g}(R) \}^2 \tag{A-58}$$

and

$$< L_1(\tau) >< L_1(s) > \;=\; < L_1(\tau) >^2. \tag{A-59}$$

To second order of L_1, we have

$$I(\omega,\tau) = \exp[\; -i\tau < L_1(\tau) > -\frac{\tau^2}{2}\{ < L_1(\tau)^2 > - < L_1(\tau) >^2 \; \}], \tag{A-60}$$

where

$$< L_1(\tau) >= \omega - \overline{\omega}_{e_n e_g}, \tag{A-61}$$

$$< L_1(\tau)^2 >= \omega^2 - 2\omega\overline{\omega}_{e_n e_g} + \overline{\omega^2}_{e_n e_g} \tag{A-62}$$

and

$$< L_1(\tau) >^2 = \omega^2 - 2\omega\overline{\omega}_{e_n e_g} + (\, \overline{\omega}_{e_n e_g} \,)^2. \tag{A-63}$$

It follows that

$$I(\omega,\tau) = \exp[\; -i\tau(\omega - \overline{\omega}_{e_n e_g}) - \frac{\tau^2}{2}\{\; \overline{\omega^2}_{e_n e_g} - (\; \overline{\omega}_{e_n e_g}\;)^2 \;\}\;]. \tag{A-64}$$

It should be noted that the Gaussian distribution is commonly employed for inhomogeneity. In this case, we write $I(\omega,\tau)$ as

$$I(\omega,\tau) = \int_{-\infty}^{\infty} d\omega_{e_n e_g}\; g(\omega_{e_n e_g}) \exp\{\; i\tau(\omega_{e_n e_g} - \omega)\;\}, \tag{A-65}$$

where

$$g(\omega_{e_n e_g}) = \sqrt{\frac{1}{\pi D^2}} \exp\{-\frac{(\omega_{e_n e_g} - \overline{\omega}_{e_n e_g})^2}{D^2}\}. \tag{A-66}$$

By comparing Eqs. (A-65) with (A-66), we find

$$D^2 = 2\{\; \overline{\omega^2}_{e_n e_g} - (\; \overline{\omega}_{e_n e_g}\;)^2 \;\} \tag{A-67}$$

and

$$g(\omega_{e_n e_g}) = \sqrt{\frac{1}{2\pi\{\; \overline{\omega^2}_{e_n e_g} - (\overline{\omega}_{e_n e_g})^2 \;\}}}$$

$$\times \exp[\; -\frac{(\omega_{e_n e_g} - \overline{\omega}_{e_n e_g})^2}{2\{\; \overline{\omega^2}_{e_n e_g} - (\overline{\omega}_{e_n e_g})^2 \;\}}\;]. \tag{A-68}$$

Notice that if we let

$$\omega_{e_n e_g}(R) = \omega^0_{e_n e_g} + \Delta\omega_{e_n e_g}(R), \tag{A-69}$$

where $\Delta\omega_{e_n e_g}(R)$ denotes the spectral shift due to inhomogeneity, then

$$\frac{1}{2} D^2 = \; < \Delta\omega^2_{e_n e_g} > - < \Delta\omega_{e_n e_g} >^2. \tag{A-70}$$

The electronic spectral shift $\Delta\omega_{e_n e_g}(R)$ has been treated by a number of workers. The spectral shift due to non-polar groups or molecules can be expressed as

$$\Delta\omega_{e_n e_g}(R) = \frac{K}{R^6}, \tag{A-71}$$

where K is a constant related to the polarizability of the inhomogeneity molecule. If the inhomogeneity are randomly distributed, then we can use

$$g(R) = \frac{4\pi R^2}{V} \tag{A-72}$$

to obtain

$$<\Delta\omega_{e_n e_g}(R)> = \frac{4\pi K}{3V}\left(\frac{1}{R_0^3} - \frac{1}{R_\infty^3}\right), \tag{A-73}$$

where $V = \frac{4\pi}{3}R_\infty^3$ and R_0 denotes the nearest distance between the relevant molecule and the surrounding molecule.

The second term in Eq. (A-73) is negligible, *i.e.*,

$$<\Delta\omega_{e_n e_g}(R)> = \frac{4\pi K}{3V}(R_0 / R_\infty)^3 \frac{K}{R_0^6} = \Delta\omega^0_{e_n e_g}(R_0 / R_\infty)^3, \tag{A-74}$$

where

$$\Delta\omega^0_{e_n e_g} = \frac{K}{R_0^6}. \tag{A-75}$$

Similarly we find

$$<\Delta\omega_{e_n e_g}(R)^2> = \frac{1}{2}(\Delta\omega^0_{e_n e_g})^2(R_0 / R_\infty)^3. \tag{A-76}$$

It follows that

$$\frac{1}{2}D^2 = (\Delta\omega^{0}_{e_n e_g})^2 (R_0 / R_\infty)^3 = \frac{V_0}{V}(\Delta\omega^{0}_{e_n e_g})^2, \quad \text{(A-77)}$$

where

$$V_0 = \frac{4\pi}{3} R_0^3. \quad \text{(A-78)}$$

Using the relation $V = V\exp(-\beta\rho)$ where ρ is the external pressure and β denotes the compressibility factor, we can see that the spectral width increases with pressure.

$$\frac{1}{2}D^2 = V_0 / V^0 \, (\Delta\omega^{0}_{e_n e_g})^2 \exp(\beta\rho). \quad \text{(A-79)}$$

Similarly Eq. (A-74) can be written as

$$< \Delta\omega_{e_n e_g}(R) > = \Delta\omega^{0}_{e_n e_g}(V_0 / V) = \Delta\omega^{0}_{e_n e_g}(V_0 / V^0)\exp(\beta\rho). \quad \text{(A-80)}$$

This indicates that the spectral maximum also shifts with pressure. Notice that

$$\frac{D^2 / 2}{< \Delta\omega_{e_n e_g}(R) >} = \Delta\omega^{0}_{e_n e_g}, \quad \text{(A-81)}$$

which is independent of pressure.

References

[1] D. C. Rees, H. Komiya, T. O. Yeates, J. P. Allen, G. Feher, Annu. Rev. Biochem. **58** (1989) 607.

[2] D. E. Budil, P. Gast, C.-H. Chang, M. Schiffer, J. R. Norris, Rev. Phys. Chem. **38** (1987) 561.

[3] J. Deisenhofer, H. Michel, Science **245** (1989) 1463.

[4] J. Deisenhofer, O. Epp, K. Miki, R. Huber, H. Michel, J. Mol. Biol. **180** (1984) 385.

[5] J. Deisenhofer, O. Epp, K. Miki, R. Huber, H. Michel, Nature **318** (1985) 618.

[6] J. P. Allen, G. Feher, T. O. Yeates, D. C. Rees, J. Deisenhofer, H. Michel, R. Huber, Proc. Natl. Acad. Sci. USA **83** (1986) 8589.

[7] C. H. Chang, D. Tiede, J. Tang, U. Smith, J. Norris, M. Shiffer, FEBS Lett. **205** (1986) 82.

[8] J. P. Allen in Light-Energy Transduction in Photosynthesis: Higher Plant and Bacterial Models (Eds: S. E. Stevens, Jr., D. A. Bryant), American Society of Plant Phosiologists, (1988) p.155.

[9] J.-L. Martin, J. Breton, A. J. Hoff, A. Migus and A. Antonetti, Proc. natl. Acad. Sci. USA **83** (1986) 957.

[10] J. Breton, J.-L. Martin, A. Migus, A. Antonetti and A. Orszag, Proc. Natl. Acad. Sci. USA **83** (1986) 5121.

[11] C.-K. Chan, T. J. DiMagno, L. X.-Q. Chen, J. R. Norris and G. R. Fleming, Proc. Natl. Acad. USA **88** (1991) 11202.

[12] N. W. Woodbury, M. Becker, D. Middendorf, W. W. Parson, Biochem. **24** (1985) 7561.

[13] R. G. Alden, M. Hayashi, J. P. Allen, N. W. Woodbury, H Murchison and S. H. Lin, Chem. Phys. Lett. **208** (1993) 350.

[14] R. G. Alden, W. D. Cheng and S. H. Lin, Chem. Phys. Lett. **194** (1992) 318.

[15] S. H. Lin, R. G. Alden, M. Hayashi, S. Suzuki and H. A. Murchison, J. Phys. Chem. **97** (1993) 12567.

[16] W. Holzapfel, U. Finkele, W. Kaiser, D. Oesterhelt, H. Scheer, H. U. Stilz and W. Zinth, Chem. Phys. Lett. **160** (1989) 1826.

[17] W. Holzapfer, U. Finkele, W. Kaiser, D. Oesterhelt, H. Scheer, H. U. Stilz and W. Zinth, Proc. Natl. Acad. Sci. USA **87** (1990) 5168.

[18] S. Schmidt, T. Arlt, P. Hamm, H. Huber, T. Nägele, J. Wachteitl, M. Meyer, H. Scheer and W. Zinth, Chem. Phys. Lett. **224** (1994) 116.

[19] G. R. Fleming, J.-L. Martin and J. Breton, Science **333** (1989) 190.

[20] C. Kirmaier, D. Gaul, R Debey, D Holten and D. D. Schenck, Science **251** (1991) 922.

[21] M. H. Vos, J.-C. Lambry, S. J. Robels, D. C. Youvan, J. Breton and J.-L. Martin, Proc. Natl. Acad. Sci. USA **89** (1992) 613.

[22] J.-L. Martin and M. H. Vos, Annu. Rev. Biophys. Biochmol. Struct. **21** (1992) 199.

[23] E. W. Knapp, S. F. Fischer, W. Zinth, M. Sander, W. Kaiser, J. Deisenhofer and H. Michel, Proc. Natl. Acad. Sci. USA **82** (1985) 8463.

[24] K. Dressler, E. Umlauf, S. Schmidt, P. Hamm, W. Zinth, S. Buchanan and H. Michel, Chem. Phys. Lett. **183** (1991) 270.

[25] C. Lauterwasser, U. Finkele, H. Scheer and W. Zinth, Chem. Phys. Lett. **183** (1991) 471.

[26] P. Hamm, K. A. Gray, D. Oesterhelt, R. Feick, H. Scheer and W. Zinth, Biochimi. Biophys. Acta **1142** (1993) 99.

[27] S. Schmidt, T. Arlt, P. Hamm, C. Lauterwasser, M. Meer, H. Scheer and W. Zinth, Biochim. Biophys. Acta **1144** (1993) 385.

[28] T. Arlt, S. Schmidt, W. Kaiser, C. Lauterwasser, M. Meyer, H. Scheer and W. Zinth, Proc. Natl. Acad. Sci. USA **90** (1993) 11757.

[29] C.-K. Chan, L. X.-Q. Chen, T. J. DiMagno, D. K. Hanson, S. L. Nance, M. Schiffer, J. R. Norris and G. R. Fleming, Chem. Phys. Lett. **176** (1991) 366.

[30] M. H. Vos, J.-C. Lambry, S. J. Robles, D. C. Youvan, J. Breton and J.-L. Martin, Proc. Natl. Acad. Sci. USA **88** (1991) 8885.

[31] D. J. Lockhart, C. Kirmaier, D. Holten and s. G. Boxer, J. Phys. Chem. **94** (1990) 6987.

[32] P. L. Dutton, K. J. Kaufmann, B. Chance, P. M. Rentzepis, FEBS Lett. **60** (1975) 275.

[33] C. Kirmaier, D. Holten, W. W. Parson, FEBS Lett. **185** (1985) 76.

[34] M. R. Wasielewski, D. M. Tiede, FEBS Lett, **204** (1986) 368.

[35] S. V. Chekalin, Ya. A. Matveetz, A. Ya. Shjuropatov, V. A. Shuvalov, A. P. Yartzev, FEBS Lett. **216** (1987) 245.

[36] M. Bixon, J. Jortner, M. E. Michel-Beyerle, A. Ogrodnik, W. Lersch, Chem. Phys. Lett. **140** (1989) 626.

[37] R. A. Marcus, Chem. Phys. Lett. **133** (1987) 471.

[38] D. J. Lockhart, R. f. Goldstein, S. G. Boxer, J. Chem. Phys. **90** (1988) 1408.

[39] T. R. Middendort, L. Mazzola, S. G. Boxer, Biophys. J. **59** (1991) 130.

[40] P. O. J. Scherer, S. F. Fischer, Chem. Phys. **131** (1989) 115.

[41] R. A. Friesner, Y. Won, Biochimica et Biophysica Acta **977** (1989) 99.

[42] C. Kirmaier, D. Holten, Proc. Natl. Acad. Sci. USA **87** (1990) 3552.

[43] M. R. Wasielewski, D. M. Tiede, FEBS Lett. **204** (1986) 368.

[44] M. Du, S. J. Rosenthal, X. Xie, T. J. DiMagno, M. Schmidt, D. K. Hanson, M. Schiffer, J. R. Norris and G. R. Fleming, Proc. Natl. Acad. Sci. USA **89** (1992) 8517.

[45] A. Warshel, Proc. Natl. Acad. Sci. USA **77** (1980) 3105.

[46] S. H. Lin, M. Hayashi, S. Suzuki, Z. Gu, W. Xiao and M. Sugawara, Chem. Phys. **197** (1995) 435.

[47] M. H. Vos, F. Rappaport, J.-C. Lambry, J. Breton and J.-L. Martin, Nature **363** (1993) 320.

[48] M. H. Vos, M. R. Jones, C. N. Hunter, J. Breton, J.-C. Lambry and J.-L. Martin, Biochemistry **33** (1994) 6750.

[49] R. A. Marcus, J. Chem. Phys. **24** (1956) 966.

[50] V. G. Levich Adv. Electrochem. Eng. **4** (1966) 249.

[51] N. R. Kestner, J. Logan, J. Jortner, J. Phys. Chem. **78** (1974) 2148.

[52] J. J. Hopefield, Biophys. J. **18** (1977) 311.

[53] Z. T. Chu, Y. Y. Lin, S. H. Lin, J. Mol. Sci. **5** (1987) 127.

[54] S. H. Lin, J. Chem. Phys. **90** (1989) 7103.

[55] X. Z. Gu, M. Hayashi, S. Suzuki and S. H. Lin, Biochim. Biophys. Acta **1229** (1995) 215.

[56] S. Suzuki, H. C. Sung, M. Hayashi and S. H. Lin, Physica A **221** (1995) 15.

[57] S. G. Boxer, T. R. Middendorf and D. J. Lockhart, FEBS **200** (1986) 237.

[58] S. G. Boxer, D. J. Lockhart and T. R. Middendorf, Chem. Phys. Lett. **123** (1986) 476.

[59] S. G. Boxer, R. A. Goldstein, D. J. Lockhart, T. R. Middendorf and L. Takiff, J. Phys. Chem. **93** (1989) 8280.

[60] S. Franzen, R. f. Goldstein and S. G. Boxer, J. Phys. Chem. **94** (1990) 5135.

[61] D. J. Lockhart, S. L. Hammes, S. Franzen and S. G. Boxer, J. Phys. Chem. **95** (1991) 2217.

[62] T. R. Middendorf, L. T. Mazzola, D. F. Gaul, C. C. Schenck and S. G. Boxer, J. Phys. Chem. **95** (1991) 10142.

[63] M. A. Steffen, K. Lao and S. G. Boxer, Science **64** (1994) 810.

[64] R. Jankowiak and G. J. Small, in Spectral Hole Burning: A Window on Excited State Electronic Structure, Heterogeneity, Electron-Phonon Coupling and Transport Dynamics of Photosynthetic Units, Academic Press Inc. Vol. II (1993) p.133-177.

[65] R. Jankowiak, J. M. Hayes and G. J. Small, Chem. Rev. **93** (1993) 1471.

[66] N. R. S. Reddy, P. A. Lyle and G. J. Small, Photosynthesis Research **31** (1992) 167.

[67] N. R. S. Reddy, R. Picorel and G. J. Small, J. Phys. Chem. **96** (1992) 6458.

[68] S. G. Johnson, D. Tang, R. Jankowiak, J. M. Hayes, G. J. Small and D. M. Tiede, J. Phys. Chem. **94** (1990) 5849.

[69] G. J. Small, J. M. Hayes and R. J. Silbey, J. Phys. Chem. **96** (1992) 7499.

[70] N. R. S. Reddy, S. V. Kolaczkowski and G. J. Small, J. Phys. Chem. **97** (1993) 6934.

[71] P. A. Lyle, S. V. Kolaczkowski and G. J. Small, J. Phys. Chem. **97** (1993) 6924.

[72] N. R. S. Reddy, S. V. Kolaczkowski and G. J. Small, Science **260** (1993) 68.

[73] N. R. S. Reddy, R. Picorel and G. J. Small, J. Phys. Chem. **96** (1992) 6448.

[74] S. G. Jhonson, D. Tang, R. Jankoiak, J. M. Hayes G. J. Small and D. M. Tiede, J. Phys. Chem. **93** (1989) 5953.

[75] A. P. Shreve, N. J. Cherepy, S. Franzen, S. G. Boxer and R. A. Mathies, Proc. Natl. Acad. Sci. USA **88** (1991) 11203.

[76] N. J. Cherepy, A. P. Shreve, L. J. Moore, S. Franzen, S. G. Boxer and R. A. Mathies, J. Phys. Chem. **98** (1994) 6024.

[77] A. D. Procyl, D. F. Bocian, Annu. Rev. Phys. Chem. **43** (1992) 465.

[78] V. Palaniappan, P. C. Martin, V. Chynwat, H. A. Frank and D. F. Bocian, J. Am. Chem. Soc. **155** (1993) 12035.

[79] V. Palaniappan, M. A. Aldema, H. A. Frank and D. F. Bocian, Biochemistry **31** (1992) 11050.

[80] J. R. Diers and D. F. Bocian, J. Am. Chem. Soc. **177** (1995) 6629.

[81] S. Matti, G. C. Walker, B. R. Cowen, R. Pippenger, C. C. moser, P. L. Dutton and R. M. Hochstrasser, Proc. Natl. Acad. Sci. USA **91** (1994) 10360.

[82] P. Hamm, M. Zurek, W. Mantele, M. Meyer, H. Scheer and W. Zinth, Proc. Acad. Natl. Sci. USA **92** (1995) 1826.

[83] B. A. Heller, D. Holten and C. Kirmaier, Science **269** (1995) 940.

[84] M. C. Zerner, G. Loew, R. Kirchner, U. Muller-Westrehoff, J. Am. Chem. Soc. **102** (1980) 589.

[85] J. Ridley, M. Zerner, Theor. Chim. Acta (Berlin) **32** (1973) 111.

[86] M. A. Thompson, M. C. Zerner, J. Phys. Chem. **94** (1990) 3820.

[87] E. W. Knapp, P. O. J. Scherer, S. F. Fischer, Biochimi. Biophys. Acta **852** (1989) 295.

[88] E. W. Knapp, S. F. Fischer, W. Zinth, M. Sander, W. Kaiser, J. Deisenhofer, H. Michel, Proc. Natl. Acad. Sci. USA **82** (1986) 8463.

[89] P. O. J. Scherer and S. F. Fischer, Chem. Phys. Lett. **136** (1987) 431.

[90] P. O. J. Scherer and S. F. Fischer, Chem. Phys. Lett. **141** (1987) 179.

[91] S. F. Fischer and P. O. J. Scherer, Chem. Phys. **115** (1987) 151.

[92] P. O. J. Scherer and S. F. Fischer, Chem. Phys. **131** (1989) 153.

[93] A. Warshel and W. W. Parson, J. Am. Chem. Soc. **109** (1987) 6143; . *ibid.* W. W. Parson and A Warshel, J. Am. Chem. Soc. 6152.

[94] W. W. Parson, Z.-T. Chu and A. Warshel, Biochim. Biophys. Acta **1017** (1990) 251.

[95] Y. Won and R. A. Friesner, J. Phys. Chem. **92** (1988) 2208.

[96] E. J. P. Lathrop and R. A. Friesner, J. Phys. Chem. **98** (1994) 3050.

[97] E. J. P. Lathrop and R. A. Friesner, J. Phys. Chem. **98** (1994) 3056.

[98] M. Marchi, J. N. Gehlen, D. Chandler and M. Newton, J. Am. Chem. Soc. **115** (1993) 4178.

[99] J. N. Gehlen, M. Marchi and D. Chandler, Science **261** (1994) 499.

[100] M. Alonso and H. Valk, *Quantum Mechanics: Principles and Applications*, (Addison-Weseley, Reading, Mass., 1973).

[101] W.L. Peticolas, Ann. Phys. Chem. **81** (1967) 233.

[102] S.H. Lin, Y. Fujimura, H.J. Neusser, and E.W. Schlag, *Multiphoton Spectroscopy of Molecules*, (Academic Press, New York, 1989).

[103] M. Weissbluth, *Photon-Atom Interactions*, Academic Press, New York, 1989.

[104] C. Eckart, Phys. Rev. **47** (1935) 552.

[105] A. Sayvetz, J. Chem. Phys. **7** (1939) 383.

[106] R. Islampour and M. Kasha, Chem. Phys. **74** (1983) 67; *ibid*, **75** (1983) 157.

[107] R. Islampour and S.H. Lin, J. Mol. Spectrosc. **147** (1991) 1.

[108] M. Born and R. Oppenheimer, Annalen der Physik, **84** (1927) 457.

[109] M. Born and K. Huang, *Dynamical Theory of Crystal Lattices*, (Clarendon Press, Oxford, 1966).

[110] A.H. Jahn and E. Teller, Proc. Roy. Soc. A**161** (1937) 220.

[111] H.C. Longuet-Higgins, Advan. Spectrosc. **2** (1961) 429.

[112] A.D. Liehr, Progr. Inorg. Chem., 3, 281 (1962); **4**, 455 (162); *ibid*, **5**, 385 (1963).

[113] M.D. Sturge, Solid State Phys. **20** (1967) 91.

[114] C.J. Ballhausen and A.E. Hansen, Ann. Rev. Phys. Chem. **23** (1972) 15.

[115] T. Azumi and Matsuzaki, Photochem. Photobiol. **25** (1977) 315.

[116] G. Fishcer, *Vibronic Coupling*, (Academic Press, New York, 1984).

[117] G. Herzberg and E. Teller, Z. Phys. Chem. Leipzig, B**21** (1933) 410.

[118] Orlandi and W. Siebrand, Chem. Phys. Lett. **15** (1972) 465 ; J. Chem. Phys. **58** (1973) 4513 .

[119] V. Lawtez, G. Orlandi and W. Siebrand, J. Chem. Phys. **56** (1972) 4058.

[120] D. Duschinsky, Acta Physicochim. **7** (1937) 551.

[121] D.P. Craig and G.J. Small, J. Chem. Phys. **50** (1969) 3827.

[122] G.J. Small, J. Chem. Phys. **54** (1971) 3300.

[123] F. Metz, M.J. Robey and E.J. Schlag, Chem. Phys. Lett. **51** (1977) 8.

[124] T.E. Sharp and H. M. Rosenstock, J. Chem. Phys. **41** (1964) 3453.

[125] E.V. Doktorv, I.A. Malkin, and V.I. Nan'ko, J. Mol. Spectrosc. **56** (1975) 1; *ibid*, **64** (1977) 302.

[126] H. Kupka and G. Olbrich, J. Chem. Phys. **80** (1984) 3163; **82** (1985) 3975.

[127] H. Kupka and P.H. Cribb, J. Chem. Phys. **85** (1986) 1303.

[128] R. Islampour, M. Hayashi and S.H. Lin, Chem. Phys. Lett. **239** (1995) 7.

[129] J.B. Page and D.L. Tonks, J. Chem. Phys. **75** (1981) 5694.

[130] G.W. Ford, M. Kac and P. Mazur, J. Math, Phys. **6** (1965) 504.

[131] R. Islampour and S.H. Lin, J. Phys. Chem. **95** (1991) 10261; J. Phys. Chem., **97**, 6793 (1993).

[132] J. Tang and A.C. Albrecht, *Raman Spectroscopy*, ed. H.A. Szymanski, 2, 33, New York, Plenum, 1970.

[133] A.C. Albrecht, J. Chem. Phys., **34**, 1976 (1961).

[134] A.C. Albrecht and M.c. Hutley, J. Chem. Phys., **55**, 4438 (1971).

[135] B. Johnson and W. L. Peticolas, Ann. Rev. Phys., **1**, 457 (1973).

[136] J.M. Friedman and R.M. Hochstrasser, Chem. Phys., **1**, 457 (1973).

[137] M. Mingardi and W. Siebrand, J. Chem. Phys., **62**, 1074 (1975).

[138] G.J. Small and E.S. Yeung, Chem. Phys., **9**, 379 (1975).

[139] Y. Fujimura and S.H. Lin, J. Chem. Phys. **70** (1979) 247; ibid, **71** (1979) 3733.

[140] R.J.H. Clark and B. Stewart, J. Am. Chem. Soc., **103** (1981) 6593.

[141] D.N. Batchelder and D. Bloor, in: *Advances in Infrared and Raman Spectroscopy*, vol. 11, ed. R.J.H. Clark and R.E. Hester, John Wiley & Sons, 1984.

[142] B.Y. Jin and R. Silbey, Chem. Phys. Lett., **197** (1992) 117.

[143] F. Markel, N.S. Ferris, I.R. Gould, and A.B. Myers, J. Am. Chem. Soc., **114** (1992) 6209.

[144] S. Hassing and O. Sonnich Mortensen, J. Chem. Phys., **73** (1980) 1078.

[145] A.B. Myers, R.A. Mathies, D. Tannor, and E.J. Heller, J. Chem. Phys., **77**, 3857 (1982).

[146] D.J. Tannor and E.J. Heller, J. Chem. Phys., **77**, 202 (1982).

[147] V. Hizhnyakov and I. Tehver, Phys. Status Solidi, **21**, 755 (1967).

[148] D.L. Tonks and J.B. Page, Chem. Phys. Lett., **79**, 247 (1981).

[149] Y.J. Yan and S. Mukamel, J. Chem. Phys., **85**, 5908 (1986).

[150] K. Kodama and A.D. Bandrauk, Chem. Phys. Lett., **80**, 248 (1981).

[151] W. Siebrand and M.Z. Zglerski, J. Phys. Chem., **88**, 4718 (1982).

[152] S. H. Lin, B. Fain, N. Hamer, Adv. Chem. Phys. **79** (1990) 133.

[153] S. H. Lin, R. G. Alden, R. Islampour, H. Ma, A. A. Villaeys, Density Matrix Method and Femtosecond Processes, (World Scientific, Singapore, 1991);

B. Fain, S. H. Lin, N. Hamer, J. Chem. Phys. **91** (1989) 4485.

[154] M. Sugawara, Y. Fujimura, C. Y. Yen, S. H. Lin, J. Photochem. Photobio. A**54** (1990) 321.

[155] S. H. Lin, R. G. Alden, C. K. Tang, Y. Fujimura and M. Sugawara in Mode Selective Chemistry, Eds. J. Jortner *et al.*, (Kluwer, Dordrecht, 1991) p.167.

[156] Z. T. Chu, Ph. D. Thesis, Arizona State University (1988); part of this work was presented at the Biophysical Society Meeting, Biophys. J. **53** (1988) 66a.

[157] R. G. Alden, Z. T. Chu, S. H. Lin, R. Islampour, J. P. Allen, J. C. Williams, Adv. Multiphoton Prc. Sectrosc. **7** (1991) 201.

[158] S. H. Lin, J. Chem. Phys. **61** (1974) 3810.

[159] H. Eyring, S. H. Lin and S. M. Lin, Basic chemical kinetics (Wiley-Interscience, New York, 1980).

[160] A. Vermeglio, J. Breton, G. Paillotin and R. Cogdell, Biochim. Biophys. Acta **510** (1978) 514.

[161] V. Rosenbach-Belkin, J. R. E. Fisher and A. Scherz, J. Am. Chem. Soc. **113** (1991) 676.